Advances in Spatial and Network Economics

Titles in the Series

Martin J. Beckmann and *Tönu Puu*
Spatial Structures

Aisling J. Reynolds-Feighan
The Effects of Deregulation on U.S. Air Networks

Åke E. Andersson, David F. Batten
Kiyoshi Kobayashi and Kazuhiro Yoshikawa (Eds.)
The Cosmo-Creative Society

Börje Johansson · Charlie Karlsson
Lars Westin (Eds.)

Patterns of a Network Economy

With 33 Figures and 44 Tables

Springer-Verlag

Berlin Heidelberg New York
London Paris Tokyo
Hong Kong Barcelona
Budapest

Professor Dr. Börje Johansson
Royal Institute of Technology, Department of Regional Planning
Östermalmsgatan 26, S-100 44 Stockholm, Sweden
and
Norwegian Institute of Technology
Department of Economics, N-7034 Trondheim, Norway

Dr. Charlie Karlsson
Jönköping University, Department of Economics
P.O. Box 1026, S-551 11 Jönköping, Sweden
and
Institute for Future Studies, P.O. Box 6799
S-113 85 Stockholm, Sweden

Dr. Lars Westin
University of Umeå, Department of Economics and CERUM
S-901 87 Umeå, Sweden

ISBN-13: 978-3-642-78900-7 e-ISBN-13: 978-3-642-78898-7
DOI: 10.1007/978-3-642-78898-7

Softcover reprint of the hardcover 1st edition 1994

42/2202-543210 - Printed on acid-free paper

Preface

In 1986 the Centre for Regional Science in Umeå, CERUM, initiated a collaborative research programme focussed on interaction networks in general and economic networks in particular. These efforts comprised both theoretical and empirical studies, with participants from a large number of Swedish, other European as well as American and Japanese universities. In 1988 this programme of network research merged with the project "European Networks" which was started and promoted by the Swedish Institute for Future Studies, IFS.

The participants in the research programme congregated in a series of workshops in Vienna and at IIASA outside Vienna, in Berlin, Boston, Kyoto, Tottori, Karlstad and Umeå. In this way the chapters in the book reflect the result of repeated interactions in social and intellectual networks extending across the globe.

The theme of the book is patterns in a network economy. It aims at a modest contribution to a renewal of economic theory by introducing the concept of economic networks and giving it a central theoretical position. The approach is designed to illuminate the interface between markets and other forms of exchange. Another ambition is to construct a tool which can be applied to the analysis of structural aspects of economic dynamics and evolution. In this context the contributions shed new light on modern theories of international trade as well as innovation processes in a global economy.

The organisers of the research effort presented in this volume are Börje Johansson, Charlie Karlsson and Lars Westin. They have drawn on series of discussions with the participants in the project on European networks, i.e., Janerik Gidlund, Jan-Evert Nilsson, Sverker Sörlin and Gunnar Törnqvist. Inspiration and intellectual support has also been offered by the contributors to the current book.

Our friends Åke E. Andersson, Kiyoshi Kobayashi, T.R. Lakshmanan and Manfred Fischer have all helped us to organise the workshops and meetings which have been instrumental in the production of the ideas presented in the book. Jennifer Wundersitz has been responsible for the editorial work and Ingrid Lindqvist has keyboarded the camera ready manuscript. We also wish to thank the staff at our own departments and at IFS for their help during the course of this work.

Stockholm in November 1993

Börje Johansson
Project leader, Royal Institute of Technology

CONTENTS

CHAPTER 1

Patterns of a Network Economy - An Introduction

Charlie Karlsson
Jönköping International Business School

Lars Westin
University of Umeå

1.1 NETWORK ECONOMICS - PAST AND PRESENT

The purpose of this book is to give a presentation of current theoretical and empirical research in the field which may be labelled "network economics". The label reflects that concepts associated with networks and networking have recently gained considerable interest in economics as well as other sciences. We may thus speak of a new field of economics in emergence, although the roots of "network economics" originate from other fields of economics as well as a broader set of disciplines. One such source of inspiration is the group of studies associated with link formation between corporations and their subsidiaries, between suppliers and their customers, and the co-operation between agents in R&D projects, e.g. Williamson (1975, 1979), Johansson and Mattsson (1987), Lundvall (1988), Håkansson (1987). Another source, dealing with more aggregated links, may be found among studies of intra-industrial trade, such as Burenstam-Linder (1961), Grubel and Loyd (1975) or Balassa and Bauwens (1988). In the first case, the demand for variety and deeper specialisation in order to obtain temporary monopolies leads to the establishment of reciprocal links between suppliers and customers. Such links represent a network capital whose territorial extension is sensitive to internodal affinities and barriers. This introduces temporal rigidities and a spatial dispersion of flows which, at more aggregated levels, may be observed as a part of intra-industrial trade.

Other inquiries into the network concept may, in this case in connection with "general system theory", be found in economic geography, especially during the sixties. The standard reference is Hagget and Chorley (1969). Worth mentioning here is that this promising development of a network approach in geography lost momentum during the early seventies. Only physical infrastructure was considered, partly as a consequence of the narrow definition of networks, and partly because of competition from quantitatively oriented research associated with production systems and input-output models in business and regional economics. Today, infrastructure networks analysis, production systems and input-output models are all regarded as instruments in the toolbox of network economics.

Parallel with those contributions, a more algorithmic oriented area of research on numerical models of infrastructure and transportation networks has also developed. Examples are found in Enke (1951), Avondo-Bodino (1962), and Harker (1985). Within mathematics, network analysis and related fields such as graph theory and combinatorics have become established as abstract and pure fields, but they have also to some extent been applied to the social sciences, e.g. in studies of transportation networks and energy systems. Pure and applied material are presented in e.g. Berge (1973), Roberts (1978, 1984), Wilson and Beineke (1979), and Chen (1990). This mathematical theory of networks has generated a large number of theorems on topics such as chains, trees, paths, cycles, connectivity, maximum flow and network design. But so far the number of effective applications to network economics of the type of problems formulated and the results obtained have been quite few.

Despite the existing contributions within various disciplines, we may today, as we initially observed above, witness a deeper interest in network theory and network methods. This is obviously driven by the growing importance of networks and network concepts in society. From an economic point of view, the following three mutually interwoven processes are in this case especially important.

- The increased intensity and complexity of human interaction.
- The decreasing importance of geographical territory as a determinant of accessibility
- The failure of markets as a means to solve inter-firm relations when customised commodities are developed in complex environments

The fundamental force behind the "new accessibility" and growth of complexity is found in the development of communication and information systems. Conditions for travelling, information handling and goods transportation are undergoing a rapid transformation. Deep relations between territorially non-adjacent actors are being developed in response to opportunities given by the new communication possibilities. The increase in long distance air traffic and telecommunications provides evidence of changing patterns at the global scale. Fax machines, optic fibre cables, and mobile phones are creating new economic structures in synergy with flexible manufacturing and transportation systems. Logistic systems and spatially dispersed production networks are furthermore slowly replacing production based on "fordism" and scale. Economic networks, i.e. reciprocal structures of co-operation and risk sharing between independent actors, are being established in order to deal with the new forms of complexity. The economy is apparently facing a period of qualitative change originating from this "fourth logistical revolution". The outcome of this structural transformation is obviously to a large extent impossible to anticipate. However, it may be possible to increase understanding and preparedness by identifying the central mechanisms governing this development. This accounts for the challenge which makes "network economics" important.

The field of "network economics" deals with this intensified interaction between communication systems, economic networks and society. Issues such as why and when economic networks are created, how they are maintained, the way in which they evolve, cooperate and compete, and the reason for their destruction form the core of the field. The identification of synergies between networks and other interactive structures is also important. Nevertheless, today the concept of a "network economy" has come to contain both new and historically marginal although growing network structures which are well inside the core of network economics. The concept also has come to contain a broad set of processes related to social interaction in general. This reflects the fact that our increasing understanding of the way information and communication technologies, increased mobility, income growth and demands for variety change the conditions for networks as well as other forms of exchange and interaction in the economy is still only in its infancy.

Our point is that network economics will contribute to our understanding of many processes in the economy, but it should not be seen as a new "general theory" of economics. It is not needed in order to understand e.g., the globalisation of traditional markets in standardised commodities. The fields of international trade, regional economics, economic geography and industrial organisation have already developed our knowledge in this and many related topics. However, those fields are dominated by theories developed for entities with territorially determined borders and spatially localised factors while industrial organisation, and international trade, too often deal with a world lacking any spatial friction. The goal of network economics is to increase our understanding of flow patterns in a modern differentiated society, where new forms of barriers, mobility, accessibility, complexity and self-organisation caused by interaction through links not always directly connected with geographical distance or territorial adjacencies dominate the development. Network properties are also essential in order to understand central parts of economic dynamics while a dynamic perspective is unavoidable in order to understand networks. In spite of this it is not especially risky to suggest that the fields mentioned above will become highly integrated with network economics and they will move closer to each other. The obvious obstacle to such integration is the fact that established social networks within each field will maintain scientific barriers as long as possible.

From an applied scientific point of view, it is also evident that the development of high speed computers and information storage systems already has changed and will continue to change the methods of science and the way we approach the analysis of economic networks. It is now possible to treat large data bases efficiently in research, a circumstance which increases the need for network models for analysing data. In addition, simulation allows dynamic systems to be investigated beyond the boundaries set by analytical mathematics. This will probably also initiate a closer interaction between pure and applied mathematics, dynamic systems analysis and network economics.

1.2 ECONOMIC NETWORKS

Nodes connected by links are the basic elements of networks. Together they constitute the structure defining a specific network. The economic theory of networks deals with links and couplings connecting economic agents at different levels. A node may thus be comprised of "agents" such as individuals, firms, organisations, cities or nations. An economic link between two nodes contains a tangible or intangible element of capital. Hence, economic links are durable; they are created by means of investment efforts and deteriorate if they are not maintained (Johansson, 1991). The durability criterion is fundamental. Most agreements between agents to interact, most explicit as well as implicit relations or contracts, provide in principle the seed of an economic link. Even behind repeated market transactions and intra-firm relations there are often some forms of economic links developed or they constituted the origin of the transaction. However, standardised occasional transactions and temporary non-lasting couplings are not durable, contain no elements of capital and should not be regarded as economic links.

Economic links should therefore be analysed as capital objects, which are basically sunk costs. Sunk costs bring rigidity and structure into the pattern of interaction between economic agents. This applies to interaction with respect to factors of production and capital equipment, as well as to exchanges of knowledge, information and energy. An active and maintained link between two nodes is not broken without a change, a form of instability such as an innovation which gives a third node superiority. To be successful, the new partner has to be superior enough to overcome the sunk cost advantages of the existing link. Hence, economic links introduce both rigidities and dependencies, which also reflect asymmetric power or symmetric reciprocity, risk transmission or risk sharing, and the distribution or concentration of options. The cost of link termination and the alternatives available to each node in such structures represent node and network vulnerabilities. This emphasises that economic networks include elements of both competition and co-operation.

The "fourth logistical revolution" has brought about both increased growth in and a dispersion of the speed of economic processes. In fast environments, buffers and inventories in the network reduce its speed and provide time for adjustment, but they also introduce cost risks due to obsolescence of inventories and missed opportunities. In the case of production systems, the demand for speed and the reduction of inventories is captured by the concept "just in time": that is, production for the customer at the moment of consumption. This concept is associated with exact and fast transportation, information handling, and decision-making in networks consisting of flexible and often small entities exploiting economies of scope. When accessibility loses its strong correlation with geographical distance simultaneously in many networks, the important factor is instead to be connected to a network. In such ultimate networks accessibility is a zero-one variable (Karlqvist, 1990). Globalisation – or initially, due to territorially determined cultural affinities, a global regionalisation of relations – is an outcome of such changes. Owing to this geographical adjacency is no longer the dominant principle of human organisation but one principle of many. Not only reciprocal networks but also traditional markets and hierarchical organisations are exploiting this change challenging traditional territorial borders. The dualistic picture of this development implies increased importance for local environments as attractors of an increasing number of footloose factors of production. Today nodes are competing to be integrated in such mobile networks.

Apparently, economic networks are multi-dimensional and may be classified according to a large set of criteria. In the following chapters taxonomies of special types of networks are elaborated. The list below contains some core variables in such a classification.

- The functional content and purpose(s) of the network
- The level of aggregation
- The localisation and extension in space
- The topology and connectivity of internal and external relations
- The degree of formalisation and stability of individual relations and assignments
- The division of power in the network
- The durability and time scale of changes in relation to the environment
- The dynamics between actors, in relation to external actors, and in relation to the function of the network

From the discussion it is evident that "economic networks" are far from clear-cut phenomena. One interpretation is associated with the niche in-between markets and corporations which is motivated by transaction costs and risk sharing in complex environments; another is the reorganisation of production from a factory system based on scale and hierarchical relations to a system based on logistics. Other networks represent combinations of these two interpretations. In addition to this an economic network may generally be regarded as a club with the aim of handling externalities, solving problems of communication and flexibility, and distributing benefits and risks. The scale of the network and the number of participants in the club, is then determined by the costs of information handling and decision making as well as the purpose of the network and its dynamics in relation to the environment.

1.3 PATTERNS OF INTERACTION AND HIERARCHIES OF NETWORKS

Beside the interpretations of economic networks above, the word "network" is also often used in connection with economic system interdependencies in general; that is, relations between actors in traditional "atomised" markets and even in hierarchical structures such as classical central place oriented city systems. Obviously, this is in contradiction with what we earlier identified as economic networks. The question is: can network economics also contribute to our understanding of pattern formation in these cases and in other interacting structures?

First, one may observe that from a topological point of view the network concept covers a broad set of structures – including hierarchies. However, in many disciplines, such as political science, sociology and linguistics, hierarchical structures are often seen as the opposite to networks. The network concept is then reserved to the case of symmetric dependence between nodes while hierarchies are associated with asymmetric dependence, control and power, giving orders downward and reporting results upwards in a structure. Hence, a broader topological definition of economic networks would in this respect allow for a framework within which structures in-between those two extremes, and transitions between states, may be analysed. The degrees of hierarchy and asymmetry of power would, for example, become functions of the purpose of the network.

Independent of such conceptual ambiguities, is it obvious that some generalised frameworks and results from studies of economic networks (e.g. at the micro level) may also be applicable to other levels of interaction and to less clear-cut or even hierarchical structures. In economic networks, the nature and the volume of the flows as well as those with whom a node has developed relations, are crucial for the growth of the node and the network. Certain spatial environments foster and stimulate the establishment and continuous renewal of a network and its projects. Networks differ in the density of their external links, with consequences in terms of their capability to realise renewal and mobilise resources. Hence, the external network structure has impacts on the prosperity and evolution of the internal structure. "Networking" is thus a part of the investment in the future of a network. Generalisations which make network evolution a function of link structure, link quality, and symmetries to interaction in its broadest sense, result in "general theories of networks". Naturally, such theories have strong connections to "general system theory". One may question whether the two really represent independent fields.

However, direct reasoning by analogy where micro related network approaches are applied to patterns of aggregated flows or to the development of network structures may not generally be possible. Instead, a distinction between the macro structure of accessibility, the actual flows which are an outcome of the interaction on this structure and the behaviour of individual units in the flows is needed for any analysis of interactive structures. An example is the interaction between economic networks and infrastructure networks in transportation systems. In this case a distinction between the physical infrastructure given by links and nodes, the flows on this network and the decisions by companies and households – some of which are organised in networks – is necessary. Hence, the use of a network and the network itself, its creation and development should be regarded as processes at different levels, interacting with each other and with other economic and social processes.

Network economics has to pay attention to this multiplicity of processes with varying time scales in interacting systems. In the case of transport systems, the physical infrastructure often adjusts more slowly than the activities using the infrastructure. From an analytical perspective it is thus advantageous to introduce a hierarchy of levels among networks and other processes and treat slow processes as structural determinants of faster ones (Westin, 1990). Gradual change at one level takes place under a structural stability determined by slow structures at higher levels, while radical change occurs during structural instability caused by changes which are often gradual at those levels. From the point of view of network evolution, one may in this respect emphasise the constructive role structural instability plays in the emergence of new structures and organisational forms. Earlier we also observed that a potential for creativity is intrinsic in instabilities and that innovations may make a link or a certain type of network obsolete. This is not to say that collective action in favour of change should only be directed towards the creation of instabilities. As is suggested by Teubal and Zuscovitch in their contribution to this volume, collective action in support of network creation is also important when externalities and transaction costs are presented in risky environments. In the international trade policy debate a similar discussion is going on in relation to "strategic trade". One argument against collective intervention is in this case that problems of project evaluation, effect measurement and the internalisation of externalities imply an obvious risk that only new arguments in favour of political protection will be developed.

Such a cautious attitude in relation to interventions is also supported by the findings by Bak and Chen (1991), which illustrate how our possibilities to explain and predict the behaviour in highly interactive systems are constrained. By simulation they show that such systems have a tendency to self-organise towards a critical, almost chaotic state where minor events initiate chain reactions leading to unexpected evolution. Under such circumstances it becomes impossible to predict the evolution of individual entities in the system although the process generating the flows has reached a stable structure, is well understood, and is even a deterministic process.

A related policy oriented question is how the costs and benefits in a network develop and are redistributed between actors over time. In an evolving network, individual and collective vulnerability changes and places pressure upon existing relations and structures for the redistribution of wealth. New individual and collective systems of insurance therefore have to be developed as substitutes for obsolete means of risk sharing, barriers and buffers. This touches upon fundamental conceptual and theoretical problems dealing with decision making, influence, democracy, and power in networks. Questions that to some extent are dealt with in the chapters on vulnerability in this book.

In the following, studies of networks, pattern formation and network evolution associated with networks of trade, production, corporations, innovation, knowledge and technology are presented. Those networks represent both tangible and intangible structures at different levels of aggregation. The contributions are grouped into three parts. The first part of the book, entitled "Economic Networks and the Network Economy", contains chapters which give general insights into the process of network formation and network evolution at different levels in society. The second part, "Trade and Production Networks" is devoted to combinations of theoretical and empirical observations in relation to international trade, interregional flows and inter-industrial linkages at the branch level. Finally, in the third part of the book, "Innovation and Technology Networks", empirical studies and models which are oriented towards the dynamic effects of knowledge and innovation networks are presented.

1.4 PART ONE: ECONOMIC NETWORKS AND THE NETWORK ECONOMY

The first chapter in part one is written by *Teubal* and *Zuscovitch*. Their chapter contains a fundamental discussion of the motives behind the formation of economic networks and their dynamics. It is suggested that technological learning and learning through the development of discrimination capabilities among nodes in a network create a complementary relation between the collective learning obtained through interaction between producers/users and the evolution of their networks. In particular, network formation and collective learning become important in connection with technologies characterised by system complexity where various sub-systems introduce hierarchies of constraints.

Product development in such complex environments makes it necessary to release several constraints simultaneously. Hence, the innovative phase brings about a need for strong interaction and co-ordination in order to facilitate collective learning. We may also generally witness how both suppliers and users demand variety and have a joint interest in the appropriate development of differentiated products. Today, this may be achieved efficiently with the help of new information technologies which allow for the exploitation of economies of scope.

In this context, Teubal and Zuscovitch introduce *sustainable differentiation* as an important goal for innovative corporations as well as other actors in the economy. Sustainability is strongly connected to the formation and evolution of networks. Networks are based on reciprocity among members, which favours specialisation and differentiation. The interpretation of reciprocity as a risk reducing mechanism is fundamental to the chapter and provides a basis for networks as a specific form of industrial and social organisation.

An important policy conclusion in the chapter is that in the case of radically new capital goods, networks will not form naturally. A central network agency which secures a transmission of individual experiences to collective learning is needed in such cases in order to guarantee a socially optimal level of learning and the establishment of the network. Such network agencies may be hybrid organisations with both public and private interests. This policy conclusion about the need for network agencies may be relevant when it comes to networks associated with collective goods as well.

In the second chapter, *Maillat et al.* take a combined space and time perspective on networks and network evolution, while focusing on the interaction between *the milieu* and innovative networks. The structure of a network, its morphology and organisation are here closely linked to the nature of the process of innovation. Existing, stable and long term relations are the sources of innovative networks, since those are based on actors making investments in a relational capital which provides the possibilities to develop creative high risk projects. The innovative network evolves with the development of the project and the constraints faced in the project. Actors are, as was also suggested by Teubal and Zuscovitch, forced to develop the network in order to relax constraints and to keep the project sustainable.

In the analysis by Maillat et al., inventors are not, as in the Schumpeterian tradition, isolated entities but heavily influenced by and dependent upon the dynamics of the regional milieu. Beside the earlier history of the actors and their relations, the milieu and the regional context play especially important roles in the "idea-forming" stage. However, the innovative character of the project will finally lead to differentiation and independence from the milieu. In the second part of the chapter a tentative typology of innovative networks, ranging from organised hierarchical networks with a leader firm to compact unstabilized networks, is introduced and analysed. Hence, networks are not solely organisations based on symmetric and equal relations. But the more innovative the process becomes, the less precise and hierarchical is the network. In this classification, the relations between actors, domination and power as well as the diffusion of know-how and financial resources become central in the description and characterisation of networks.

In the third contribution to Part One *Vedin* continues to emphasise the importance of *space and time* in network evolution. In his chapter, Vedin presents case studies of global corporations, for which "going global" has offered possibilities to explore economies of scale and scope but where communications have also caused severe problems in the global strategy. Hence, problems which arise when innovation and development are accomplished in a spatially non-adjacent milieu form the focus of the chapter. In contrast to the previous chapter by Maillat et al., the milieu itself is now viewed as networks.

Two of the central networks in the global corporation are discussed by Vedin. The analysis of the interplay between the logistic system linking consumers with producers (which in many respects must be local) on the one hand, and the global network inside the company on the other deals with issues also raised by Teubal and Zuscovitch as well as Capello and Pompili. The producer-customer relationship is developed towards "just in time" deliveries of customised commodities in fragmented markets. In this process, an early integration of customers in the development, or even in the research stage, represents the extreme case of a vertical network.

In the global corporation, the internal transfer of knowledge, competence and components becomes crucial in obtaining corporate competitive advantages. Hence, communications have to be developed in order to fulfil the need for interaction and to gain time. Vedin concludes that the key to benefits from globalisation lies in differentiation and economies of scope. As an extreme case of differentiation within a supplier network, Vedin discusses the hollow company; i.e., the industry without production. Flexibility and appropriate timing are also the means of competition here.

The fourth chapter in Part One by *Goldberg*, takes as its origin the increasing transnational flows of information and financial transactions which have resulted from the globalisation of networks, markets and corporations. "Going global" promises benefits from efficiency but also raises the question of *vulnerability* due to connected–ness, and the loss of variety and flexibility at national, regional or local levels. With examples from global deregulated financial and property markets, Goldberg argues that global vulnerability has increased. The search for efficiency through economies of scale has led to a narrowing of the boundaries of stability and instability at the local arena. The decentralisation of decision making and the application of "lowest level responses" can lead to increased stability, but since decentralised decision makers act according to the same decision-making rules, instabilities arise instead.

Goldberg does not evaluate the effects that global networks based on economies of scope have on vulnerability. Such systems, characterised by connectedness and diversity as means to obtain sustainability, may seem to be less vulnerable, or – as is suggested in the sequel by Braunerhjelm – to contain other forms of vulnerability, compared with systems based on connectedness and scale alone. Instead, Goldberg suggests that international regulations, parallel systems and combinations of local and centralised sources of supply are insurance against the instabilities connected and specialised networks introduce.

The sixth chapter by *Batten*, discusses historical parallels in the emergence of *city networks* in Europe and Japan. The global interaction created by corporations and other actors has led to a strengthening of non-adjacent city networks at different levels. These networks challenge the pattern of interaction based on spatial adjacencies, as explained by central place theory. As Batten observes, non-adjacent city networks, e.g. the Venetian trade system, appeared early in the European city system alongside more local networks based on geographical contiguity. Both the European and Japanese examples of city growth, as they are delineated by Batten, point to the need for a deeper analysis of social systems as complex evolutionary multilevel networks.

The Japanese example is also examined in the last chapter in part one by *Lakshmanan*. In the chapter, he deals with the purposes and roles of the state in the creation of technology induced development. It is argued that the Japanese state performs in a similar way to the central network agency discussed by Teubal and Zuscovitch. It secures the transmission of individual experiences to collective learning through communication, co-ordination, conflict resolution and collective visions. Hence, the Japanese state is not acting in a market-displacing way; rather the *state-market interface* is of a network and relational nature. Such a market-conforming industrial policy, dependent on information from a broad spectrum of the society, may reduce the risk of making major errors. As Lakshmanan suggests, this network nature of the state-market interface and its dependence on a national context, has to be explored further.

1.5 PART TWO: TRADE AND PRODUCTION NETWORKS

In the second part of the book, five chapters which combine theory with empirical analyses of trade and production networks are presented. In the first chapter, the effects of vertical integration between producers and customers on the *comparative advantage of integrated sectors* are studied by *Fagerberg*. The close interaction between suppliers and customers within a country is claimed to give both the export producing and technology delivering sectors internationally competitive positions. This hypothesis has a strong resemblance to the arguments presented in Part One about why learning by doing and learning by using are important in the evolution of innovative networks.

Fagerberg tests this hypothesis on twenty-three pairs of sectors in sixteen OECD countries. The result supports the assumption in thirteen of the sixteen countries. Austria, France and the UK did not support the hypothesis. Fagerberg suggests that this is because of the colonial past of France and the UK and the close interaction between industries in Austria and Germany. The support for the hypothesis also improved between 1965 and 1987. Hence, the study indicates that although global integration has increased, the importance of supplier/customer links at the national level has also been accentuated. Generally, the study emphasises the long-standing effects that historical bounds have on trade patterns. Such rigidities are also the focus of the following chapter.

The second chapter in Part Two by *Johansson* and *Westin*, emphasises the effects of *friction, barriers and affinities* on international trade patterns. The traditional theory of international trade is deeply rooted in the analysis of homogeneous commodities in frictionless markets. Johansson and Westin analyse the Swedish export pattern with a division of export sectors into producers of homogeneous price competitive and differentiated product competitive commodities. Their empirical test of the national pattern of those aggregates shows that market performance is related to sensitivity with respect to different forms of friction, barriers and affinities. The trade pattern of differentiated commodities in sectors with increasing employment is found to be dependent on cultural affinities while the trade pattern of price competitive commodities is dependent on distance and common borders. The differences between product competitive commodities with increasing and decreasing employment underline that affinities are especially important in the early innovative stage of a product cycle. The results thus support the suggestion that affinities between producers and consumers play an important role in the establishment of innovative networks.

Barriers between nations are also crucial in the third chapter in this part by *Braunerhjelm*. In this case the impact of the composition of industries based on mass production and scale economies versus production networks based on *machine flexibility and product attributes* on national vulnerability is analysed. In particular, the localisation effects of barriers between an economically integrated region and a neighbouring country, e.g. the case of Sweden and the EG, are discussed. It is assumed that the network oriented engineering sector is internationally mobile while the basic industry, based on scale, is immobile.

In such an economy, Braunerhjelm argues, two forces will influence national vulnerability. While a large sector characterised by machine flexibility has reduced vulnerability with respect to demand volatility due to changes in product characteristics, the foot-lose nature of establishments in this sector increases vulnerability with respect to differences in attractivity and barriers between the home nation and the integrated area. Hence, the discussion by Braunerhjelm to some extent complements the chapter by Goldberg in Part One. Together their chapters indicate that it may not be possible to discuss changes in vulnerability over time without distinguishing between different aspects of vulnerability. Often it seems to be the case that substitution between various forms of aggregated vulnerability may be possible to identify while changes in an overall level of vulnerability may be more difficult to evaluate.

The fourth contribution to Part Two by *Capello* and *Pompili* continues the analysis of international trade initiated in the earlier chapters. Capello and Pompili deal with the complementarities and processes of substitution between international trade flows, transnational companies and international corporate networks. The empirical part of the chapter integrates a study of trade flows at the sectoral level between Italy and Sweden with a study of direct investments by Swedish transnational companies in Italy. This study indicates, contrary to what was assumed, that a simple relation of substitution between trade and foreign investments does not exist between Sweden and Italy. Instead, trade persists while both direct investments and the number of co-operation agreements increase.

In the theoretical part of the chapter, the international corporate network is related to other forms of international relations and different types of linkages. The internationalisation of corporate networks has been understood as an importation of external know-how to a local economy or as a natural extension of local and national networks. Capello and Pompili develop a deeper explanation of international agreements as a means to solve external corporate relations with respect to trade in knowledge. Knowledge has collective properties which favour and are favoured by symmetric agreements. Such agreements are also favoured by the uncertainty caused by cultural, institutional, and other barriers between nations, despite the growth in globalisation. This uncertainty hampers the willingness to make direct investments in unfamiliar economies. Hence, an evolutionary perspective is introduced. This provides insights into the dynamics of trade, into the industrial structure and the international corporate organisation in an environment of increasing complexity in which the relative importance of different factors of production and national barriers are changing.

The last chapter in Part Two by *Okumura* and *Yoshikawa* deals with the problem of identifying industries which may strengthen regions facing growth problems. The empirical part is devoted to the Kohoku region in Japan. Okumura and Yoshikawa emphasise the importance of horizontal linkages in the formation of regional technological complexes based on technological similarity and the exploitation of existing marketing channels. Since most measures of linkages between industries in a region are concerned with vertical links, a set of measures of horizontal links is suggested. Such measures may be used to identify industries which have special prerequisites to utilise and which may become a complement to existing industries and information channels. In the chapter such industries are also suggested for the Kohoku region.

1.6 PART THREE: KNOWLEDGE, INNOVATION AND TECHNOLOGY NETWORKS

The common factor amongst the six chapters in the third and last part of this volume is knowledge and innovation diffusion. The initial chapter by *Karlsson* gives an overview of the development of the main human system for knowledge creation and knowledge diffusion – the university system – since its birth. Special interest is devoted to the development of the business school, which is shown to be a product of the mature industrial revolution. The development of the university system is related to "the four logistical revolutions". This chapter shows how universities developed as *nodes in international knowledge and technology networks* from their very beginning and how the network organisation has been a characteristic feature of higher education and research. Karlsson also maintains the view that network technology and the art of networking have become central issues for the education at modern business schools. However, he also calls attention to the fact that the network paradigm has seldom been used in overall analyses of the development of higher education and science.

While the chapter by Karlsson gives a historical overview of universities as networks, the contribution by *Fischer*, *Maggi* and *Rammer* focuses entirely on a special aspect of knowledge diffusion: namely, the major factors and barriers that affect *media choice and contact decision behaviour* in the university setting. Universities are studied since they play an important role as the creators and disseminators of knowledge. The threefold purpose is to develop a conceptual framework, to identify the major factors and barriers influencing decisions and to identify cross-national differences in contact decision behaviour. Hence, the chapter provides insights into the way in which networks of international knowledge diffusion are created and affected by cross-national differences in decision behaviour. However, the authors also indicate the need for more theoretical work concerning contact behaviour in general, and in particular among knowledge workers such as scientists.

The chapters written by Karlsson and Fischer et al. are in some sense complementary to that by *Kobayashi* and *Andersson*. This chapter, which is a theoretical contribution, proposes a computable dynamic input-output model to examine *interactions between knowledge accumulation and economic development* with endogenous technical change. The input-output model is one method to illustrate the economic networks in society. Their model explicitly introduces a sector called "the knowledge sector" into a traditional dynamic input-output system. It is assumed that the growth of knowledge is the main source of technical change. In order to increase knowledge, society needs inputs and investments just as it does to produce material capital. The model may improve our understanding of endogenous technical change when long-run growth is primarily driven through knowledge accumulation by forward looking profit-maximising agents.

The fourth chapter by *Fischer* and *Johansson* provides an attempt to assemble and systematise empirical observations in the *diffusion of information technology applications* to various sectors and locations in three countries. In this case, information technology comprises microprocessors and their electronic equivalents in the form of integrated and interlinked circuit devices. In particular, two aspects are in focus: (i) the degree of product standardisation, and (ii) the intangible economic/communication networks and tangible infrastructure networks constituting the regional environment. The regional environment determines the likelihood that innovation networks shall grow into rich and intensively utilised structures. This regional environment can be associated with the external networks of a firm. Peripheral regions provide an environment less rich than central regions and for all three countries the authors observe that such regions provide an environment in which the adoption of applications of information technology is delayed compared to other locations.

Blaas and *Nijkamp* aim to analyse whether technological innovations form the basis of the economic development of the European Snowbelt – i.e. the area around the Alps – and whether this region offers favourable *incubator conditions for new technology* in the European network system. Innovation and new technology have been focal points of economic research in the past decade and the creation of a "new technology" niche in a region has often been regarded as a guarantee for regional vitalisation. However, the innovative potential of a region is determined by a complicated interplay between the regional production environment and its production structure. For the development of innovative new firms the presence of network structures functions as an important seedbed condition. Unfortunately, the empirical part of this chapter is hampered by a lack of sufficient comparative empirical data. Nevertheless, the authors draw the conclusions that apart from being located around the Alps, there is no cohesive network structure between the regions in the different countries involved and that the reasons for having attracted new technology sectors differ drastically among these regions. Hence, the innovative potential of the regions in the European Snowbelt seems to depend more on their geographical position in the heartland of Europe than on Neo-Schumpeterian incubation conditions.

Infrastructure investments are generally suggested in order to improve regional accessibility and growth. In the last contribution, *Zhang* extends the two sector model as it was formulated by Uzawa by introducing aggregate measures of knowledge and infrastructure into the production function. The extended model is formulated in order to obtain policy conclusions on the growth effects of public investments in infrastructure and of changes in the national saving ratio. The resulting model is characterised by increasing returns to scale. The model indicates that only when the ratio between infrastructure and the stock of capital in society is small is it unambiguously beneficial to invest in collectively financed infrastructure. The effect of the savings rate on growth is complicated and ambiguous. This result supports the lack of definite conclusions from empirical studies of the impacts of savings on growth.

REFERENCES

Avondo-Bodino, G., 1962, *Economic Applications of the Theory of Graphs*, New York.

Bak, P. and K. Chen, 1991, Self-Organized Criticality, *Scientific American*, January.

Balassa B. and L. Bauwens, 1988, The Determinants of Intra-European Trade in Manufactured Goods, *European Economic Review* 32:1421-1437.

Berge, C., 1973, *Graphs and Hypergraphs*, North-Holland, Amsterdam.

Bertalanffy, L. von, 1968, *General System Theory*, George Brazillier, New York.

Burenstam-Linder, S., 1961, *An Essay on Trade and Transformation*, Almquist & Wicksell, Uppsala.

Chen, W.-K., 1990, *Theory of Nets: Flows in Networks*, John Wiley & Sons, New York.

Enke, S., 1951, Equilibrium Among Spatially Separated Markets: Solution by Electric Analogue, *Econometrica* 19:40-47.

Grubel, H.G. and P.J. Loyd, 1975, *Intra Industry Trade: The Theory and Measurement of International Trade in Differentiated Products*, Macmillan, London.

Hagget, P. and R.J. Chorley, 1969, *Network Analysis in Geography*, Edward Arnold, London.

Harker, P.T., (eds), 1985, *Spatial Price Equilibrium: Advances in Theory, Computation and Application*, Springer-Verlag, Berlin.

Håkansson, H., 1987, *Industrial Technological Development*, Croom Helm, London.

Johansson, B., 1991, Economic Networks and Self-Organization. In Bergman, E., G. Maier, F. Tödtling, (eds.), "*Regions Reconsidered - Economic Networks, Innovation, and Local Development in Industrial Countries*", Mansell, pp. 17-34.

Johansson, J. and L.G. Mattsson, 1987, Interorganizational Relations in Industrial Systems: A Network Approach Compared with the Transaction-Cost Approach. *International Studies of Management & Organization*, vol XVII, no. 1.

Karlqvist, A., (ed.), 1990, *Nätverk* (In Swedish), Gidlunds bokförlag, Värnamo.

Lundvall, B.Å., 1988, Innovation as an Interactive Process - from User-producer Interaction to the National System of Innovation. In Dosi, G. et.al., (eds), *Technical Change and Economic Theory*, Pinter, London.

Roberts, F.S., 1978, *Graph Theory and Its Applications to Problems of* Society, SIAM, Philadelphia.

Roberts, F.S., 1984, *Applied Combinatorics*, Prentice-Hall, New Jersey.

Westin, L., 1990, Vintage Models of Spatial Structural Change. PhD Diss. Dep. of Economics, University of Umeå.

Wilson, R.J. and L.W. Beineke, 1979, *Applications of Graph Theory*, Academic Press, London.

Williamson, O.E., 1975, *Markets and Hierarchies: Analysis and Antitrust Implications*. A Study in the Economics of Internal Organization, The Free Press, New York.

Williamson, O.E.,1979, Transaction-Cost Economics: The Governance of Contractual Relations, *The Journal of Law and Economics* 22: 233-261.

PART I: ECONOMIC NETWORKS AND THE NETWORK ECONOMY

CHAPTER 2

Demand Revealing and Knowledge Differentiation Through Network Evolution

Morris Teubal

The Hebrew University,
Jerusalem

Ehud Zuscovitch

BETA/Louis Pasteur University of Strasbourg
Ben Gurion University, Beer-sheva

2.1 INTRODUCTION

Economic development involves increased production efficiency and rising consumption standards. This requires capital accumulation, which allows for a better division of labor; that is, a process of ever deepening differentiation of knowledge and know-how on both the scientific/technological and user levels. Since the industrial revolution, efficiency has entailed standardization and scale economies have only been available for highly standardized goods around the production of which it was possible to conceive and develop efficient apparatus. This meant huge plants to produce physical commodities. However, even then the financial capital that was and still is a precondition for production (the physiocrats called it "advances") could only be raised through the design of a share-holding mechanism (to assure financial divisibility). The resulting mechanism was limited liability, which is also a risk-sharing device to stimulate entrepreneurship.

If we recall these basics it is because we believe that the importance of the change that we face now requires such a restatement. Networks represent an economic organization instrument through which economic agents deal with the fundamental challenge of sustainable differentiation brought about by the Information Intensive Production System. Sustainable differentiation consists of increasing efficiency and variety at the same time. This can only be dealt with through a different form of capital concentration in which the specialization process transcends the frontiers of the firm. In such a structure knowledge and know-how seeking - the learning process - is done collectively. Indeed, knowledge deepening and diffusion is not limited *per se* by the size of operations in a given production plant. A network allows the socialization of these activities, particularly in pre-competitive R&D. In a sense it also creates a flexible social capital of scientific-technological and market relevant knowledge.

Naturally, in the economic system such long run cooperative principles must be accompanied by risk reducing mechanisms. Naturally, frequent interaction in the form of information exchange is in itself a risk sharing mechanism because each agent constantly adjusts his own learning process to the change in his competitive environment and reacts implicitly in a "portfolio" manner. But risk reduction takes another social form: the network also sets a priority mechanism to reduce risk in which each agent gives some priority to members over non-members of the network. In pure markets there is of course no priority whatsoever and in pure hierarchies priority is absolute. An intermediate arrangement of priorities is in a way an economic mechanism analogous to the limited liability mentioned earlier - this time in the form of flexible organization. The similarity is recognized once we see that it also is fundamentally a credit device and credit is capital. Economic agents do enter partnership arrangements for various reasons, such as to share costs of large R&D projects or to take advantage of each other's strategic position in a given skill, or market. They create hybrid public/private agreements with government agencies or public research institutes. However, beyond the question of the different configurations or the precise arrangement undertaken, the very participation in a priority-reciprocity mechanism is fundamentally a credit mechanism that network members can employ to further specialize and gain more efficiency. Furthermore, in order to achieve a higher degree of specialization/differentiation each member must, at least partly, rely on the other specialization process and the overall human capital accumulation process becomes more social.

So the basic condition does not change. Capital accumulation must precede the differentiation process of learning; what has fundamentally changed, due to the information technology revolution, is the relationship between *diversity and efficiency.* The necessity to generate an increasingly specialized set of goods and services pushes the industrial organization towards network structures that can make this diversity economically viable. In addition to this revolution in information technology, two other factors stimulate the general interest in the subject. The first lies in the success of the cooperative/competitive attempts of Japan and Germany. The second is simply the steep rise in the number of technological partnerships, joint ventures and various other forms of participation, especially during the 1980's.

The cooperation of firms is not new of course, but this huge acceleration in cooperative agreements is indeed theoretically very appealing. The current literature on the subject tries to characterize the types of cooperation and to sort out configurations or network architectures, fields of application and micro-economic incentives for cooperation vs. competition. While it is important to understand networking mechanics we shall rather try to analyze the complementary relations between this differentiation/learning process and network evolution. We shall focus on the major change in the economic environment and try to understand first why Information Technologies lead to network arrangements (Section 2.2). Then we shall try to analyze the learning/differentiation processes in the two basic types of networks and suggest generalizations for both Learning by Doing and Learning by Using (Section 2.3). We shall then try to propose a general view of network differentiation through user-producer interaction in order to understand its basic evolution (Section 2.4). Aspects of networks which are being analyzed in the current literature will be integrated into the chapter, while the main focus remains on the evolutionary dimension of the phenomenon.

2.2 INFORMATION-INTENSIVE PRODUCTION AND THE TENDENCY TOWARDS NETWORKING

A network organization appears from the outset as a device set to face market-failures in innovation. Arrow's paper on the allocation of resources to inventive activities (Arrow, 1962a) suggests that uncertainty, indivisibility and inappropriability are the major characteristics of the innovation process from an informational standpoint. The acceleration in the constitution of industrial networks basically responds to dynamic innovative environments with strong uncertainty; it often deals with large R&D projects where indivisibility prevails and where a very extensive division of labour needs to be achieved inter and intra firm at the same time. Finally, a network is above all a device for internalizing externalities for members. If so, we have to understand the nature of the trend that has encouraged so many cooperative arrangements in the latter part of the 1980's. As Freeman justly points out (Freeman, 1990) cooperation is not a new term in economics. Still, evidence from the early eighties indicates that an increasing part of the typical R&D budget of corporations is allocated to cooperative arrangements. Systematic studies and data bank econometrics show that a substantial increase in partnership arrangements took place a few years later and that it occured in specific patterns of international trade in given sectors and key technologies. (See Cohendet and Valls, 1988; Chesnais, 1988; and Hagedoorn & Schakenraad, 1990). The latter - based on intensive efforts by MERIT to build a data bank - concentrates on information technologies, new materials and biotechnology (as we shall see later this is by no means random).

The acceleration in partnership arrangements, we claim here, is an inevitable result of the tendency towards information intensive production systems. Indeed, while previous waves of industrial transformation mainly substituted power, whether centralized or decentralized, the current transformation is about information. It is by no means new. The first industrial surveys on the performance of NC machines (Nabseth & Ray, 1974) mentioned the capacity of the new capital goods to produce small series efficiently or, in other words, to increase variety. All subsequent generations of computerized tools from robots to computer aided design and computer aided manufacturing (CAD/CAM), through flexible manufacturing systems (FMS) and the like, are not only cost benefit effective like previous mechanization technologies in terms of labour or material costs, but they also allow for production flexibility in that switching quite rapidly from one production run to another is possible.

Flexible manufacturing - unlike standard mass-production - supports and actually stimulates a much greater product variety. In order to exploit this possibility a firm has to engage expensive research both on user characteristics and on the scientific and technological characteristics that will match finely defined needs. (See also Cohen and Klepper, 1989). Micro-markets of this informational intensity can only be characterized and developed through interaction with the relevant agents - particularly users - and, as we shall see, the interactive search and learning process is the essence of new product development. The intensive search for property combinations to provide increasing variety also calls for multiple specialization. Unlike standard mass production where R&D had to produce reliable solutions for a small number of alternatives, here an increasing number of specialties must be permanently cross-fertilized to generate the solutions we need. Within the modern firm, job content, wages and more generally incentive schemes have to be redefined to stimulate the employees' creativity and innovativeness since the employer is now interested not only in the role defined by a diploma or standard skill, but rather with the full scope of the talents available. Furthermore this cannot be achieved under hierarchical principles and a team-matrix organization is needed. As soon as this occurs, modern firms themselves becoming networks.

This phenomenon is of course much stronger in inter-firm relations. Increasing variety and complexity of the relevant specialization cannot be developed alone. Any serious product differentiation for industry applications or even end users requires constant cooperation and frequent information flows. Rapid developments in so many industries make impossible fast access to unfamiliar technologies (Mowery, 1989) and in order to have an entry to the numerous specialties the firm might need cooperative agreements (as well as being less costly) are practically the only solution. Inter-sectoral technological partnerships also become more frequent as the use of complementary assets or skills ensures a much larger return to R&D investments.[1]

The flexibility introduced by the new information technologies is actually reinforced by some fundamental trends in science and technology. Advanced technologies in new materials (Cohendet et al., 1988) and biotechnologies (Joly and Zuscovitch, 1989) match this trend well. Both families of generic technologies are based on a better understanding and control of the microscopic properties of matter. Thus, microscopic control allows for tailor-made applications where, again, cooperative research development and design are required for tailored designs.

This general process of information growth in diversity (Willinger and Zuscovitch, 1988) entails more uncertainty since product life cycles become shorter too. Again, the only way the modern firms can possibly hope to cope with the diversity of bodies of knowledge and specialization is through partnership agreements and the constitution of networks. Only under such arrangements can the firm itself, although losing a part of its independence, specialize further and hope to compete. The most typical evolution in this respect is perhaps found in sub-contracting relations. If previously they were essentially governed by cheap labour considerations for the non strategic components of the main contractor, the picture is now changing drastically. More and more sub contractors need real technological skills to match the network advanced technology requirements.

To sum up our analysis of information intensity, we indeed suggest that the economic viability of information intensive production - or, using the term that we have previously employed, the sustainability of the differentiation process - depends on the ability of economic agents to share the costs of production and the dissemination of knowledge (economies of scope). However, economies of scope may not be sufficient to stimulate growth. Scale economies of some sort are also required. Informational scale economies can potentially be expected if each standardization in the information process (algorithmisation) releases intellectual resources to continue the learning process. (Ayres and Zuscovitch 1990).

Network relationships have also developed extensively in the last 15 years in connection with the development of new technological infrastructure, especially when cooperation among users - especially innovative users - is critical. In these cases the indivisibility in capabilities may relate not only to technological capabilities but also to knowledge about user needs required for developing these capabilities.

Overcoming these indivisibilities, therefore, may depend critically on achieving cooperation and networking - free and spontaneous exchanges of information among firms, even competitive firms, within or beyond the context of an explicitly joint R&D project. In Teubal et al., (1991) we termed this problem a "network creation" problem. As a result of the difficulties involved, these cooperative and networking efforts may necessitate a central network agency to coordinate and orchestrate the process ("the network PBX") and sometimes this agency lies in the public sector.

The Japanese VLSI project of 1976-79, initiated and orchestrated by MITI, demonstrated the organizational feasibility and potential usefulness of such networks. We will briefly delve into this project (following Sigurdson, 1986) in order to illustrate the principle in as concrete a setting as possible. The objective of the VLSI project was the creation of design and production capabilities for a new class of semiconductor memory devices - 1 megabit Drams. It was widely believed that the major competitor to Japanese computer firms - IBM - was planning to introduce these components into products to be launched during the eighties. This would have represented a major threat to the larger Japanese Electrical/Electronic companies - Hitachi, Fujitsu, Toshiba, Nec

---

[1] Both Teece (1986) and Milgrom and Roberts (1990) insist on the role of complementarities in the innovative process. Concerning the relationship between variety, technical change and competition see Metcalfe and Gibbons (1986) and also Cohen and Malerba (1989).

and Mitsubishi - especially if such components would not have been easily and freely available to them from outside sources. The initiative for the project and the realization of the need for cooperation among the major users came from the Electrotechnical Laboratory of MITI, which had both strategic knowledge about and insight into technology trends and a good knowledge of the problems and capabilities of the individual companies. The latter, at least initially, were apparently less convinced both of the overall need for and the possibility of achieving cooperation among fiercely competing firms (moreover, they apparently preferred to continue receiving government subsidies on a firm by firm basis).

Under the leadership of MITI a research agenda was set which involved a minority portion (about 15% of total expenditures) of R&D which was generic - to be performed in a joint lab set up for this purpose - and a majority portion of specific product development which would be performed by each one of the firms individually. MITI's role was crucial in setting the research agenda, including the identification of common generic problems in silicon crystal technology and in fine processing equipment, which were both of crucial importance for developing and producing the specific chips of each firm; in homing into a particular organizational form for the project - the temporary joint lab complemented by the R&D of individual firms; in promoting the division of labour and networking among the core group of users participating in the project; and (apparently) in promoting the diffusion of results to other users beyond those actually participating in the project. Once completed the project led to the dissolution of the joint lab; it also triggered additional cooperative efforts in the development of technological infrastructure both in Japan and abroad (with increasing proportions of private sector initiative, management and financing). The project is widely regarded as successful, both directly in terms of its explicit objectives and in terms of its indirect contribution in generating firm networking and in promoting additional cooperation in the future.

This example illustrates (i) the importance that networks of major innovative users may have in overcoming the indivisibilities associated with cutting-edge technological infrastructures, and (ii) the role played by Government agencies in the process and, in particular, in assuring effective cooperation and network creation. The role of Government here in promoting networking relates to strategic, long-term growth/structural change aspects of the economy[2]. It is thus not surprising that most of these efforts - starting with Japan's VLSI program and followed by other programs in Japan, the EEC and in the US - have focused on the emerging clusters of information technologies.

Having analyzed the trend towards networks formation in terms of information intensity and technological infrastructures we will now turn to the relationship between networks and the learning process.

2.3 TECHNOLOGICAL LEARNING AND USER DISCRIMINATION CAPABILITIES WITHIN NETWORKS

How does learning take place in a network? Our main task here is to explain the process by which knowledge and know-how deepening occur within a complex and evolving system. We shall try to make economically endogenous the process by which needs are revealed, developed and defined in the network. In the same way, we shall explain how technological capabilities differentiate on the "supply side" to offer new properties and characteristics which trigger even more need definition. Learning occurs anyway here through constant interaction among suppliers and users. However, it will be helpful to

---

[2]For a distinction between the macroeconomic, the Current Support and the Strategic Levels of Growth - promoting Industrial and Technological policies, see Justman and Teubal, 1990.

distinguish between learning-by-doing, which occurs within component supplier-assembler types of networks which characterize the automobile sector and the aeronautics and space industries (Network type N1), and learning-by-using - which is most intensive in user-producer networks (N2 type). The latter is critical in the emergence and progressive adaptation of a new capital good. We shall avoid any discussion of specific network architectures, this being more a problem of optimization of a given structure to manage information flows. We will deal with incentives in as much as the learning process is concerned.

2.3.1 Generalized Learning by Doing in N1 Networks

Learning-by-doing refers to efficiency gains that appear with practice in the production system. In a broader sense it refers to production-endogenous technical progress which consists of the numerous ways by which production is gradually improved through successive innovations and adjustments of the system. It is by no means trivial nor marginal in value. Although related to production volume as Arrow suggested in his seminal work (Arrow, 1962b), it fundamentally expresses human capital accumulation (Sheshinsky, 1967), as scientists, engineers and workers as well as the organization as such, improve their skills and professional/technological abilities. It is no wonder that the phenomenon was spotted in the aircraft industry. Learning-by-doing is particularly powerful in industries where product and production complexity involve a strong organizational dimension. Not only synchronization gains may be considerable in such a system when approaching the theoretical performance frontier; progress can also be simultaneously stimulated, generated, and diffused in many areas and divisions. In many respects the giant high-tech multi divisional corporation is a network by itself.[3] Although less "scientific" than the aircraft industry and more cost-effective in its mass production consideration, the automobile industry fundamentally presents the same articulated structure around which component suppliers and other sub-contractors define a clear network of cooperation.

The complex nature of such industries stimulates technological learning in some very specific ways in addition to the usual innovative activities that are performed here as in other economic sectors. Networks in the areas of space projects, aircraft and even automobile industries are of a considerable size. The assembler firm is usually a diversified industrial giant on its own merit, but together with its suppliers of sub-systems, instruments and components, it can set up a huge division of labour and specialization scheme that would not have been possible without this critical mass. Those projects that are close to the military/space complex receive heavy public finance support in various forms which enables them to maintain such indivisibilities. Although scientific personnel can be effectively employed in such systems once a critical size is reached the effect is a multiplication or potentialization of scientific and technological transfers from the general scientific system. As far as applications are concerned most of the big firms involved in large public or military projects are largely diversified in civil lines of business and a very significant flow of technology transfers occurs through spin-offs. What we should emphasize here is that size alone is not sufficient, just a pre-condition to specialization and diversity. A significant difference in the ability to obtain significant spin-offs exists according to the nature of their pre-existent diversified structure. (Shachar and Zuscovitch, 1990b).

The diversity of activities is not understood here merely as a means of exploiting commercially existing innovations, but even more as a condition to potentialize innovativeness in systems through permanent cross-fertilization among departments.

---

[3]Certainly this is also the case within conglomerates such as the Japanese Kereitsu - see Imai (1988).

This calls naturally for organizational flexibility and the ability to manage advanced forms of matrix and hybrid organization structures. Such structures are set to optimize creativity on one hand and efficiency on the other.

Learning-by-doing should be grasped in this framework in a much larger sense than in the strict sense of technological innovation. Scientific management methods and very strict quality control procedures are very powerful aspects too, given their importance for competitiveness. When the assembler needs to assure a given high level of reliability to the system as a whole he transmits the requirement to the various layers of equipment sub-contractors, each of them being further linked to equipment materials or services sub-contractors of their own. Since failure probability is a multiplicative form of the failure probability of its individual components, at every step quality requirements must be augmented to compensate for what can easily become an exponential loss of reliability for the system as a whole. This is an essential aspect of learning by doing. It has a tremendous impact on the partners' technological capabilities and economic performance. Furthermore, it has a consolidating effect on the structure of the network itself, for members know that they can rely on the quality performance of other members in the club.

Technological cooperation occurs in complex systems to create networks for another major reason. In modern production many products are multifunctional entities by themselves. Innovation in the system is not only determined by the spontaneous evolution of "external" science and technology bodies of knowledge, but as a function of the constraints and interdependencies exerted interactively by sub-systems. Thus automobile weight influences the material strength we can expect from the wheel bearings and chassis supports. Interdependencies lead very quickly to "lock-in" if we try to improve each component separately. In complex systems a great value is attached to releasing several constraints at a time. Innovating in a system related manner may make a block of simultaneous changes very profitable, while each individual modification would fail its cost benefit evaluation. (See also Baba and Imai, 1989, for a detailed account of system innovation in networks). Thus, when technology is complex, learning and innovation become highly interactive and require or involve considerable coordination.

Thus, network evolution is very much shaped by this generalized learning-by-doing process. Not only is the ex-ante technological division of labour further enhanced while advancing in the direction of the common goal project or product class; the nature of a firm's participation in the network has an additional effect on the specialization of agents. In a BETA (1988) study evaluating spin-offs for the European Space Agency, it was clearly shown that the role the firm fulfilled within the network (the official division of labour of the ESA) influences the value but even more so the distribution of the types of benefits that the firm extracts from its participation in the project. Although scale and skill considerations oriented the ESA's initial choice for the specific role assigned to the firm, it did occur that firms with a similar size and structure were assigned different roles and followed a different pattern of specialization in their learning (Shachar and Zuscovitch, 1990a).

2.3.2 Generalized Learning by Using in N2 Networks

The initial motivation for discussing learning by using can be found in innovation performance studies which have emphasized the importance of 'understanding user needs' and of 'product adaptation to the needs of users' for the commercial success of product innovations. In project Sappho, for example, the degree of understanding of the needs of users significantly differentiated commercially successful from failed innovations within a sample of 22 pairs of radical innovations in the Electronic instrument area (see SPRU, 1972). Moreover, marketing efforts were also significant and this relates both to the efforts of the innovator firm to understand existing needs and to efforts by it to educate users. There is no reference in this pioneering study to the degree to which users themselves know about their needs, either in terms of general awareness or in terms of any translation of general needs into specific product and feature terms. However, an additional finding - namely, that successful instrument innovations were generally second to the market (alternatively, that those radical instrument innovations which arrived first to the market generally failed) and that they were generally also less radical from a technological point of view - suggested what has been since regarded as a common sense based finding: that a major factor explaining commercial success in product innovations should be the extent to which users themselves have learned how to define and specify general needs into specific product classes, features and specifications.

Thus, if needs are only imperfectly and grossly specified - as represented, for example by the statement "I want a car" without any additional specification - the probability of providing a good which is adapted to them is low, since the supplier receives no specific qualitative signals as to whether manual or automatic transmissions are desired; etc. Thus, the second producer to attempt to produce a radically new product stands a better chance of succeeding commercially since he can learn about users' first reactions to the product. It is no surprise that success in Sappho's second group of innovation pairs - chemical process innovations -was rather associated with 'first to market', shorter lead times and a more radical technology.

An explicit consideration of the extent to which users have defined their needs in terms of product classes, features and specifications has been made in the Falk Innovation Project (FIP) which analyzed innovation performance in the Israeli Biomedical Electronic Instrumentation Sector during the second half of the sixties and the early seventies (Teubal, et al., 1976). In fact, a necessary condition for the commercial success of new biomedical electronic instruments was 'high market determinateness', i.e., the clear translation of user's general needs. One example is the general need of stimulating the recovery of post heart attack patients - in terms of product class, features and specifications - using a coronary care unit with heart rate, pulse and breathing rate functions and an on site data processing capability. In fact, it turns out that 'market determinateness' is an 'area variable' which is external to the innovation itself (or, as we have claimed in N1 networks, innovation has a strong systemic nature) and therefore beyond the categories of variables which could be considered in the Sappho project - given the methodology of comparing the successful component of a pair with the failed component of the same pair only. It was particularly useful for defining areas where infant Israeli biomedical firms could find a competitive advantage: in those areas where needs are well defined in terms of product classes, functions and features. In areas where market determinateness was low, successful product adaptation to user needs required a lot of marketing and interaction with users - e.g. endless clinical trials, etc. - a situation which small firms could not bear. Alternatively, we may say that in this situation (i) the risk of non-adaptation is higher; (ii) the learning that has to be induced in users is greater; or associated with this, (iii) the extent of market creation required by the innovator had to be greater. In short, since producer initiated and financed user learning is costly, small firms in the extremely conservative area of biomedical electronic instrumentation could not succeed when market determinateness is low.

The explicit consideration of user learning and the postulation of a specific mechanism for the process was presented in Teubal (1979). The process of User Need Determination - or of increasingly specified needs in terms of product classes first, then of functions (within the 'desired' product class and the features determining function performance) - was supposed to proceed along the Innovation Cycle and to contribute to its characterization. However, only under very specific assumptions could the classical result of 'increasing product standardization' be obtained. The reason pointed out by Teubal was that user need determination was accompanied by *endogenous* user differentiation, a process which inevitably led to submarket or niche-market creation. Thus product standardization could only occur within a particular submarket, but could not be considered as leading to homogeneous products. Need determination is considered to be triggered by one or more of the following factors: *user experience* - e.g. "sifting out from a given set of options.. a set of product features which are relevant for need satisfaction"; *user adaptation* -"when investing in learning how to use a particular product type creates an advantage to continue using that type relative to other types"; and *producer - initiated launch of new product* features - "a new product-price option appearing in a situation of user indifference between existing options of a set may trigger off need determination... An example from nuclear medicine is the improvements in the resolution of the Gamma Camera (an imaging diagnostic tool) as a result of which there has been a shift from competition between it and the nuclear scanner to a clear preference for the camera in the diagnosis of most organs" (Teubal, op. cit.).

This is the basis for user differentiation which may then arise because of (i) different uses, which may be revealed only after product innovation; (ii) differential user adaptation to products; (iii) differential preferences concerning the tradeoffs between price and quality/features; (iv) different attitudes towards risk. The framework presented in that paper anticipates some of the features of user learning and user differentiation which have become more common as a result of Information Technology; it also suggests a simple pattern of supply-demand interactions which will be further specified in the next section.

It was Nathan Rosenberg (1982) who coined the term "Learning by Using" with examples drawn from the aircraft industry. Again, the gradual improvement of the product was shown to critically depend, both for the aircraft assemblies and for some related capital goods, on the fact that these were complex systems with interdependent elementary components. Under these conditions only by actual use can one trace problems and improve both the design and various technical parameters. A significant further step in the same direction was made by Clark (1985) where complexity is again at the centre of the analysis and where the role of hierarchical constraints within the product as a complex system is made explicit. However, since we dealt with the innovation dynamics in complex systems in the above, we neglect this point here to focus more on the differentiation process to which we feel past research efforts have paid much less attention.

A final building block to the network learning and interaction processes described later on is due to Lundvall (1985, 1989). He postulated the functions that user-producer networks perform, in particular in the context of complex, product innovations. In his view networks are a device by which successful user-producer interactions - which are critical conditions for those kinds of product innovations - can take place. In fact, they have advantages over the two alternative systems of governance of resource allocation - hierarchies, i.e. the vertical integration of users and producers and markets, which are spot-relationships among independent economic agents. Networks are defined as a system of long-term relationships between users and producers, which assures significant information flows and mutual interaction. Lundvall's contribution lies in having proposed the concept of the 'organized market' (the network) and its place in the national economy. While indicating the importance of 'learning by interacting' and especially that occurring within user producer networks, he has neither specified nor characterized the nature of the learning process itself, nor its relationship with network evolution.

In a previous paper (Teubal et al.,1991) we have made a first attempt at linking a number of the traditions mentioned above: learning by using (in the form of User Need Determination) and user-producer networks (the so-called N2 networks). In that framework, the learning process itself is a dynamic force leading to network evolution, a fact that justifies attempts at characterizing the process as fully as possible. Moreover, a critical role in ensuring that individual experiences generate collective learning is played by a central network agency termed the network PBX. Underlying our framework is the view that (i) network creation is not easy - in connection with a radically new capital good, networks will not necessarily arise naturally (Lundvall's focus on incremental innovation has led him to assume that networks arise naturally since opportunistic behavior by producers or users will not payoff in the long run); (ii) that network evolution may be critical for market creation, i.e. the economic impact of the radical product innovation. Therefore, further advance depends crucially in a further understanding the learning process involved in networks - in particular, learning in a user-producer interaction setting. We propose doing this via the use of the Z variable - a user and producer discriminating capability, which is both cause and consequence of network evolution.

2.4 TOWARDS GENERALIZED LEARNING BY INTERACTION AND NETWORK EVOLUTION

The previous sections have considered aspects of the specialization, differentiation and learning processes occurring both on the supply side (producers and their suppliers) and on the demand side (users of a market) for a complex product. This analysis was carried out for each side of the market separately, that is ignoring interactions with agents belonging to the opposite side. The discussion on user differentiation, for example, focused on the effects of user experience and user adaptation to products; and if the effect of a producer-initiated new feature was considered, the latter was taken as completely exogenous. A most relevant aspect of learning pertains, however, to the interaction between producers on the one hand, and users on the other (Lundvall, 1988). This eventually will require simultaneous consideration of both learning by doing among producers and learning by using among users. Our objective in this section is more modest. First we introduce a variable in order to consider the interface between the two - Discrimination Capabilities. Learning by both users and producers is supposed to involve, among other things, an increase in these capabilities. Through this variable we describe aspects of the interaction process, especially between producers and users rather than producers and their component suppliers. We finally link our discussion of learning and interaction to network evolution.

2.4.1 Discrimination Capabilities and the Z-variable

The essence of the successful evolution of a market is coupling or adapting technology to the needs of an increasingly wide circle of users. This depends on the *discriminating capability* of users (Need Determinateness) and of producers which is related to their ability to search for, identify and home into a product option providing a satisfactory level of (net) functional utility. This capability is critical when uncertainty is great - both as to the functional utility of the various options and as far as the consequences of inappropriate choices are concerned. Its importance is also enhanced in an environment where variety has increased, for example, in response to the opportunities opened up by Information Technology.

Specifically we introduce what we term the Z-variable to represent *discriminating capability across a hierarchy of increasingly specific product types*. To clarify this, first consider a product class (="cars") belonging to the broad category "means of transportation". Associated with this product class we may introduce a set of nested subsets of product options ("product types") which conform to a hierarchy e.g. z11-small cars and z12 - large cars; z111 - small cars with automatic gear change, z112 - small cars with manual gear change... etc.[4] The hierarchy is composed of levels of increasing product type specificity, e.g. the i *(product class) level*; the *ij level* and so on. Going down the hierarchy would mean going from a more general to a more specific product type level and we will say that this involves an increase in Z.

The next element in our framework relates to users. We assume that (i) each user, at each moment of time, has a "preferred product type"; (ii) the degree of specificity of this product type (or alternatively, the hierarchy *level* to which it belongs) depends positively on his discriminating capabilities (Z); (iii) that success in coupling his needs to the product class (or alternatively 'goodness of fit' between both) depends on the specificity of the "*preferred product type*"... Once a product type, say z112 above, is "preferred" by a user, his search of product options is *focused* within this type; so is the actual option chosen. Whether his preferred type belongs to level ij ("small cars") or to level ijk ("small cars with manual gear changes") depends on his discriminating capabilities. An increase in such capability will enable the user to home into a more specific preferred product type and thereby to focus his search even more than before; we also assume that this will lead to a better choice, i.e. a needs-product class coupling involving greater (net of cost) functional utility. Conceivably maximum discriminating capability (Z) leads to a preferred product type which lies at the "bottom of the hierarchy".[5]

Up to now we have neither referred to the source of user discriminating capabilities nor to the process leading to their creation through time. This will inevitably introduce producers into the analysis, either as one of the many sources - together with user experience for example, see Section 2.3 - or as economic agents who are themselves interested in developing these capabilities. While user experience is fundamental since users have an advantage over producers inasmuch as they know their *general* needs, they may have to rely on producers to translate these needs into specific product functions, features and specifications. This is especially so with users who have little prior experience and no technological capabilities in the areas underlying the product classes concerned. Thus producers may give advice to users - "product type Zijk is precisely what you need" - or they may effectively interact with users in order to jointly learn about how to define general needs in concrete product type terms. In other words, producers may help users to learn how to better define their general needs. However, users may help producers to develop their discriminating capabilities more generally in the sense of gradually helping to establish a clear relationship between *user type* (including the relevant parameters which differentiate users) and *product type*.

A second crucial source of user discriminating capability is the activity of the network PBX (see Teubal et al, 1991) which, if successful, implies *collective learning*. Collective learning enables at least some learning by others, used in the sense of the actual transfer of experience from user to user. It may, however, involve more than this when the producer manages to generate a body of knowledge concerning the coupling of technology to needs across a variety of product types and user types (*broad*

---

[4]A hierarchy of increasingly specific product types has been proposed by Teubal et al. (1976) and by Teubal (1979) in relation to user preferences ("need determinateness"). We have already mentioned that K. Clark (1985) has convincingly suggested that hierarchy also exists in connection with product design. For the time being it may be useful to assume that both hierarchies are the same.

[5]This discussion of the Z-variable essentially follows Teubal et al. (1991).

discriminating capabilities). Thus active user participation in a user-producer network orchestrated by the producer may generate significant discriminating capabilities. These in turn may help the producer to improve his needs-technology coupling at a rapid rate, which means a rapid shift downwards along the product type hierarchy.

2.4.2 User-producer Interaction Configurations

The simplest model of interaction which also allows for an endogenous hierarchy of product types involves (i) a producer-determined menu of product type choices available at every level of the product type hierarchy; and (ii) user need determination or discriminating capability development - at the corresponding level of the hierarchy - followed by actual homing into a preferred product type. Once the latter has been ascertained, say Z*3 for level i of the hierarchy, producers also focus their efforts around this product type and arrive at a number of solutions which users use. However, producer learning surrounding the main activity of satisfying the previously defined needs enables producers to offer a variety of more specific product types associated with Z*3, say Z*31, Z*32, Z*33. These constitute the menu of product type choices available at the next (ij) level of the product type hierarchy. The process can then start all over again. A hypothetical illustrative example could begin by producers offering a menu of product type choices consisting of "large screens" (Z1); "medium sized screens" (Z2); and "small screens" (Z3) at the i-level of the hierarchy. Given these, user experience will first generate discriminating capabilities for this level and these will then enable users to home into Z3 which then becomes the 'preferred type' of that level, Z*3. Producers then focus their R&D on small screens and learn about a number of possible types, for example, as characterized by color: blue (Z*31); red (Z*32); or green (Z*33). These are then offered to users who then proceed as described above. Producers may have an additional role of helping users define their preferences by offering a string of improvements which will generate a clear advantage of one type - say Z*32 - over the others (see the effect of minor improvements on need determination in Teubal, 1979). In fact we may say that these improvements stimulate discriminating capabilities of both users and producers. The upshot is that producers' actions and producer learning stimulate these capabilities both directly - via minor improvements and indirectly and more fundamentally by determining the menu of product type choices over which the process should take place. This ZIG-ZAG type interaction is consistent with the classical innovation cycle result of progressively increasing product standardization. It is the result of one of the following assumptions: identical users and/or strong economies of scale.

More complex patterns of interaction may lead to an opposite result, namely, increased product variety and the emergence of submarkets. Thus, when the technology is dynamic and numerous product options appear there are numerous possibilities for *user differentiation* (Teubal et al, 1991). The effect may be the origin of what may be called a *product type sub-hierarchy for each user type* (rather than the menu of choices for the next level of a common hierarchy for all users). For example, the variety of screen sizes available triggers previously undifferentiated users to behave differently. Thus, U2 users will prefer Z2 and U3 users will prefer Z3. Producer response to these preferences may create an ij-level of the sub-hierarchy for U2 that refers to other features rather than the color feature which set the menu of choices for U3 users. Thereafter the industry path may continue in the same manner described up to now, the result being progressively increasing variety rather than increasing product standardization. Thus, while minor product improvements and occasional new product features may both accelerate and enhance the scope of user need determination and stabilize user similarities, a chain of more significant new feature offerings - which are the expected results of producer learning in dynamic technology areas - is likely to have a more radical effect. By stimulating user differentiation these changes from the supply side of the market will generate the need for developing *broad* discriminating capabilities. This inevitably enhances producer involvement in the generation and diffusion of these capabilities since any one user will at most be knowledgeable about his needs and by extension those of similar users, but will not be knowledgeable about the needs of other user types. The upshot of this discussion is that significant producer learning (or learning by doing within N1 networks) may significantly enhance the potential of the product class being considered. However, the materialization of this potential necessitates significant learning at the interface between technology and needs - what we have termed technology-needs coupling capabilities or *broadly defined* discriminating capabilities. This explains why network evolution strategies must explicitly be considered by producers.

As a matter of fact we should understand interaction to involve a much greater degree of innovativeness than that which was originally believed. At each round there can be a qualitative residual that redefines the properties contained in the menu which is then transmitted to the other side for reassessment. We should better understand how this residual is generated and in what way it helps to discriminate further on. Clark (1985) provides a cognitive explanation for the demand side: ... "Learning by customers is not simply a matter of finding out which technical features satisfy a set of objective/well known basic needs. Where products are new, learning focuses on development of the conceptual framework that customers use to evaluate competitive offerings and make purchasing decisions....customer problem solving is closely related to language and memory and their interaction. The formation of concepts in the new context involves the establishment of meaningful connections between the functional and aesthetical characteristics of the physical object and words stored in memory. Two aspects of the process are critical: grouping - in which the unfamiliar product is associated with other known product concepts to which it is similar or related - and distinguishing - i.e. identifying those dimensions of the product that differentiate it from the group in which it was placed. The conceptual development is based on experience and it occurs through a sequence of interactions between product and the user... differences between the new product and its conceptual relatives become sharper as the possibilities and potential of the new product become clearer (the first transistor was valued in the language and context of vacuum tubes since scientists at Bell Labs thought of the transistor as a substitute for the tube. The transistor's small size was its most distinctive characteristic... it was not until experience with transistors had deepened understanding of the possibilities of semiconductors that the new concepts like the integrated circuit arise, i.e. as learning occurs the new product eventually develops a meaning of its own".

A similar phenomenon occurs on the supply side. When facing a new substance, component or material the basic reflex of the producer is to check it within the existing constraints. In the aircraft industry when substituting an aluminum alloy by a composite material the designer will typically ask the new material to perform better at a given task (better weight/strength or weight/yield for example) within an existing product concept. Through experience, however, he will find out that the composite material has other properties he can use to alter overall aircraft design and thus, for example, take advantage of the directional properties of the composite or of their much lower material fatigue (Willinger and Zuscovitch, 1988). Hence one uses the ancient design as a benchmark through which we can find differences and similarities; and thereby enhance discriminating capabilities through differentiation of the relevant product types. One should not therefore understand the evolution of our Z-variable as monotonic or regular. Sustained differentiation resulting from technological or user learning in the innovation process occurs with discontinuities; threshold points; and sometimes even through changes in the design hierarchies themselves. When one combines the interaction principle that we have previously described with the revision of the menu due to the recognition pattern of the novelty, revealed by the qualitative residual, one obtains a more satisfactory story of market creation. Characteristics are not pre-defined as in the Lancasterian tradition where products just mediate properties. In our approach the size of the market is the outcome of a search process in which properties are revealed through network evolution and therefore the product and the market are crystallized together.

2.4.3 Implications for Network Evolution

Producers which manage to link users into an effective network will, by virtue of having established long term relationships with them, continuously be receiving user information which will be diffused among other users of the good, both actual and potential. This in itself may contribute to enhanced sales, including those to new users and even new user types, thus contributing to the diffusion of the product class . In a previous paper we discussed some of the requirements for network creation, in particular the nature of the transformation of the inventor or imitator of the new product class into what we termed the network PBX. *Thus, success in creating a network - via the collective learning that brings about broad discriminating capabilities - endogenizes to some extent the process of market creation/diffusion of the good.* More specifically, increased discrimination (or more generally coupling) capabilities enhance sales which in turn increases the experience base of the sector. If this leads to network evolution, the enhanced learning will further enhance these capabilities and through them, lead to even more sales, etc.

The centrality of the network form of organization is enhanced once we allow for interaction between learning by doing amongst producers, on the one hand, and the discriminating capabilities of both users and producers on the other. This is especially the case in relation to product classes with dynamic technological bases. In these situations the evolutionary and gradual enhancement of the discriminating capabilities of the system and of user need determination are no longer enough. This is because revolutionary changes in technology may alter in a radical way the nature of the way in which, and the framework under which, knowledge about discriminating capabilities accumulates. First, previous experience-based knowledge may become less relevant if new technology generates novel product types. We have seen this above; another even stronger example is that of submarket elimination (of those submarkets based on older product types or older 'dominant designs') and substitution by radically new product types - thus setting a completely different base for discriminating capabilities and user need determination to operate. We now confront a *radical change in the nature of the coupling or discriminating capabilities* required in order to exploit the market potential of the new technology. The function of the network PBX will be (i) to assemble as widely and as deeply as possible a variety of user and producer related experiences; (ii) to give these maximum diffusion subject to the constraints on information disclosure to competitors; and (iii) to gradually create a body of codified know-how concerning product type discriminating capabilities and need-technology coupling capabilities more generally. This body of know-how, by being increasingly anonymous, will help overcome the incentives problem associated with proprietary information. It thus will become a central stimulant to further endogenization of the diffusion process of new product classes.

2.5 CONCLUDING REMARKS

Networks are basically a new industrial organization artifact with which the economic system is responding to the changing frontier between diversity and efficiency. In such networks cooperative mechanisms are established, side by side with the competitive rules of behavior, to take advantage of collective differentiation and learning. These evolutionary economies of scope are a sine qua non to achieve *sustainable differentiation*. The economic rules of the game are indeed rapidly changing and we reason much less with representative agents both on the side of the firm and of the trade union for example. Since individual tastes can be taken, economically speaking, into consideration (demand side), personal creativity is also required on the supply side.

In this chapter we have thus tried to better understand the characteristics of the generalized learning process in such a framework. We have specified the economic mechanisms in action from a production network standpoint (generalized learning by doing) and described finer discriminating capabilities from a user network standpoint (generalized learning by using). A first tentative approach to integrate both sides in a learning by interaction mode was presented and its impact on network evolution attempted. Ongoing work on the relationship between learning, cost reduction and characteristics substitution/complementarity relations should enable us to make further progress towards this integration.

A last remark concerning policy implications from our perspective. Without speculating on all the other characteristics of the emerging new economic regime, it is quite clear from the outset, that the frontier between public and private changes in a most fundamental way. Only by hybrid entities where competition and cooperation co-exist will economic agents be able to generate and diffuse the growing quantity and qualities of information and knowledge. In such a context the rationale as well as the guiding principles for government intervention cannot be the same. Networks are an organizational artifact capable of dealing with rapidly changing patterns of production and consumption and as such we must indicate ways in which these organizations can be developed and their learning abilities stimulated through adequate economic incentives.

Since by construction the networks allow for greater dissemination of knowledge, we should assure that the learning activities that generate this knowledge are not undertaken on a less than socially optimal level.

ACKNOWLEDGEMENT

Preliminary discussions with M. Justman on some of the issues raised in this chapter are gratefully acknowledged as well as comments made by the participants of the Vienna workshop "Europe and Innovation Networks". We wish to thank in particular D. Batten and L. Westin for remarks on the evolutionary nature of our differentiation process.

REFERENCES

Arrow, K.J., 1962a, "Economic Welfare and the Allocation of Resources for Invention", in Nelsson. R.R., (ed.), *The Rate and Direction of Inventive Activity*, pp. 609-625, Princeton University Press for NBER, Princeton.

Arrow, K.J., 1962b, "The Economic Implications of Learning by Doing", *Review of Economic Studies* 29:155-173.

Ayres, R.U. and E. Zuscovitch, 1990, "Information and Technology: Chain Reactions and Sustainable Economic Growth", *Technovation*, 10:3, pp. 163-183.

Baba, J. and K.I. Imai, 1989, "Systemic Innovation and Cross-border Networks - Transcending Markets and Hierarchies to Create a New Techno-Economic System", *OECD Conference on Science, Technology and Economic Growth*, Paris.

BETA, 1988, "Etude sur les Effets Economiques des Dependenses Spatiales Europeennees", Final Report, Strasbourg.

Clark, K., 1985, "Interaction of Design Hierarchies and Market Concepts in Technological Evolution", *Research Policy* 14:235-51.

Chesnais, F., 1988, "Les Accords de Cooperation Technique Entre Firmes Independantes", *STI Review*, OECD.

Cohen, W.M. and Klepper. S., 1989, "The Stochastic Origins of Inventive Activity", Mimeo, Carnegie Mellon University.

Cohen, W.M. and Malerba, F., 1989, "Diversity of Knowledge, Innovative Activity and Technological Change", *OECD Conference on Science, Technology and Economic Growth*, Paris.

Cohendet, P., M.J. Ledoux and E. Zuscovitch, 1988, *New Advanced Materials*, Springer Verlag.

Cohendet, P. and J. Valls, 1988, "Europe-USA-Japan Triad on Axe Pacific?" *Report for the Fast Project EEC-DXII*.

Freeman, C., 1990, "Networks of Innovators - a Synthesis of Research Issues", *International Workshop on Networks of Innovators*, Montreal.

Hagedoorn, J. and J. Schakenraad, "Strategic Partnering and Technological Co-Operation" in Dankbaar, B., J. Groenewegen and H. Schenk, (eds.), *Perspectives in Industrial Organization*, Kluwer Academic Press.

Imai, K., 1988, "Japan's Corporate Networks", *Japanese Economic Studies*.

Imai, K. and Y. Baba, 1989, "Systemic Innovation and Cross-Border Networks", Paper presented at the *International Conference on the Contribution of Science and Technology to Economic Growth* at the OECD, Paris, June.

Joly, P.B. and E. Zuscovitch, 1989, "Are Biotechnologiès Economically Predictable?", *OECD Conference on Science, Technology and Economic Growth*, Paris.

Justman, M. and M. Teubal, 1990, "The Structuralist Perspective to Economic Growth and Development: Conceptual Foundations and Policy Implications", in Evenson, R. and G. Rains, (eds.), *Science and Technology in Development*, Westview Press.

Lundvall, B.Å., 1985, "Product Innovation and User Producer Interaction", Typescript, Aalborg.

Lundvall, B.Å., 1989, "Innovation, the Organized Market and the Productivity Slow-Down", *OECD Conference on the Contribution of Science and Technology to Economic Growth*, Paris, June.

Metcalfe, J.S. and M. Gibbons, 1986, "Technological Variety and the Process of Competition", *Economie Appliquée:493-520*.

Milgrom, P. and J. Roberts, 1990, "The Economics of Modern Manufacturing: Technology, Strategy and Organization", *American Economic Review*, 80, No. 3, June.

Mowery. D, 1989, "Collaborative Ventures Between US and Foreign Manufacturing Firms", *Research Policy*, 81, No. 1.

Nabseth, L. and R. Gray, 1974, *The Diffusion of New Industrial Processes*, Cambridge University Press.

Rosenberg, N., 1982, *Inside the Black Box*, Cambridge University Press.

Sheshinski, E., 1967, "Tests of the Learning by Doing Hypothesis", *Review of Economics and Statistics*.

Sigurdsson, J., 1986, "Industry and State Partnership in Japan: The Very Large Scale Integrated Circuits (VLSI) Project", Discussion Paper No 1658., Research Policy Institute, Lund, Sweden.

Shachar, J. and Zuscovitch, E., 1990a, "Learning Patterns Within a Technological Network: The Case of the European Space Program", in Dankbaar, B., J. Groenewegen and H. Schenk, (eds.), *Perspectives in Industrial Organization*, Kluwer Academic Press.

Shachar, J., and E. Zuscovitch, 1990b, "Spin-offs and Technology Policy in High-tech Networks", *NISTEP Conference*, IZU, Shimoda-city.

SPRU (Science Policy Research Unit), 1972, "Success and Failure in Industrial Innovation", Center for the Study of Industrial Innovation, London.

Teece, D., 1986, "Profiting from Technological Innovation: Implications for Integration, Collaboration, Licensing and Public Policy", *Research Policy* 15, No. 6.

Teubal, M., 1979, "On User needs and Need Determination" in Baker, M., *Industrial Innovation*, Macmillan.

Teubal, M., N. Arnon and M. Trajtenberg, 1976, "Performance in Innovation in the Israeli Electronics Industry: A Case Study of Biomedical Electronics Instrumentation", *Research Policy*, October.

Teubal, M., T. Yinnon and E. Zuscovitch, 1991, "Networks and Market Creation", *Research Policy* 20:381-392.

Willinger, M. and E. Zuscovitch, 1988, "Towards the Economics of Information-intensive Production Systems", in Dosi et al., (eds.), *Technical Change and Economic Theory*, Frances Pinter.

CHAPTER 3

Innovation Networks and Territorial Dynamics: A Tentative Typology

Dennis Maillat, Olivier Crevoisier and Bruno Lecoq
IRER, University of Neuchâtel

3.1 INTRODUCTION

3.1.1 General Problem

This chapter has been written within the framework of the GREMI (European Research Group on Innovative Milieux). The purpose of that Group's investigations and reflections is to develop a territorialized analysis of innovation by highlighting the role of the environment and, more specifically, that of the milieu in technological creation processes (Aydalot, 1986; Camagni, 1991; Maillat, Quévit and Senn, 1993). The object of this chapter is specifically to understand how the milieu is transformed through multiple interactions, to highlight the various networks involved in the innovation process and, more generally, to show how relations between the productive forces on the one hand and the urban and regional milieu on the other are transformed and what the result is. This general problem hinges principally on three fields of inquiry:

- *the innovation network*: how does it come into being (federative project), what is its architecture and strategy, how does it work and, finally, how does it evolve? We shall seek to show how interactions between the various players develop, to study the rules and principles that govern their relations and, finally, to understand the way in which learning processes develop within the innovation network.
- *relations between the milieu and the innovation network*: what is the function of the milieu in the organization of the innovation network and what is its impact on its development?
- the *effects of the innovation network on the milieu*: what is the impact on the local milieu of the learning processes that are developed by the innovation networks? More specifically, how do the innovation networks help to increase the milieu's creative abilities?

3.1.2 Innovation Networks

In very schematic terms, innovation processes may be understood simultaneously in terms of the following four propositions:
- innovation is a fundamentally collective process;
- innovation is a complex and interactive process (Amendola and Gaffard, 1988; Planque, 1990);
- innovation originates from the creative combination of know-how and specific skills (Perrin, 1990);
- finally, territorial organization is an essential component of techno-economic creation (Perrin, 1990).

Given the above propositions, market mechanisms and hierarchical rankings with their regular dynamics prove unsuited to giving rise to the innovation process. The last proposition presupposes an organizational type of innovation: we shall call this specific mode of organization an *innovation network.*

In recent years a large body of literature has sprung up on the concept of networks. Without wishing to repeat all of the discussion here, we shall dwell on the following four dimensions which help to pinpoint the concept:

- *An economic dimension*: a network can be perceived as "a mode of hybrid organization to be seen, from a theoretical point of view, as transcending the firm/market duality" (Lecoq, 1990). Viewed from this angle, the argument of transaction costs as defined by Williamson(1975) would justify the emergence of the specific modes of organization that constitute networks.

- *An historic dimension*: a network presupposes a system of long-term relations between various players, based on rules of trust and mutual recognition, reciprocity and priority. A network is a means of organizing transactions which develop over time. Therefore it does not stand still but evolves.

- *A cognitive dimension*: a network organization is the repository of a collective know-how greater than the sum of the individual players' know-how. The advantage of this method of organization is to allow collective learning processes to develop (Guilhon and Gianfaldoni, 1990).

- Finally, *a normative dimension*: all networks are characterized by a particular system of rules designed to define each member's obligations and constraints. These rules delimit a collective working area and ensure it is consistently managed vis-à-vis the network's environment.

Starting from these considerations, which relate to both the innovation process and the network concept, we shall define an innovation network. *It is an evolutive mode of organizing innovation processes - rather than one structured hierarchically or on market mechanisms - which permits continuous development of collective learning processes based on new, synergetic combinations of know-how contributed by the various partners.* The innovation network constitutes an appropriate organizational response to the complexity of the innovation process (Maillat, Quévit and Senn, 1993).

It defines a *collective working area* suitable for creative combinations of the types of know-how that are internal and external to the firm.

3.2 INNOVATION NETWORKS IN A REGIONAL CONTEXT

Innovation networks may be understood according to two axes:

- first of all, from a dynamic perspective by giving priority to a temporal reading of the process, ranging from the innovative idea (that is, the genesis of the project) to the marketing of the finished product, stressing that it is a non-linear, often chaotic process which presupposes continual movement to-and-fro between the various stages of the innovation process;

- then, according to the various fields - the technical and market environment, the division of labour and labour relations within the network and, finally, the interactions between the network and the milieu - which structure the innovation network. (See Figure 3.1).

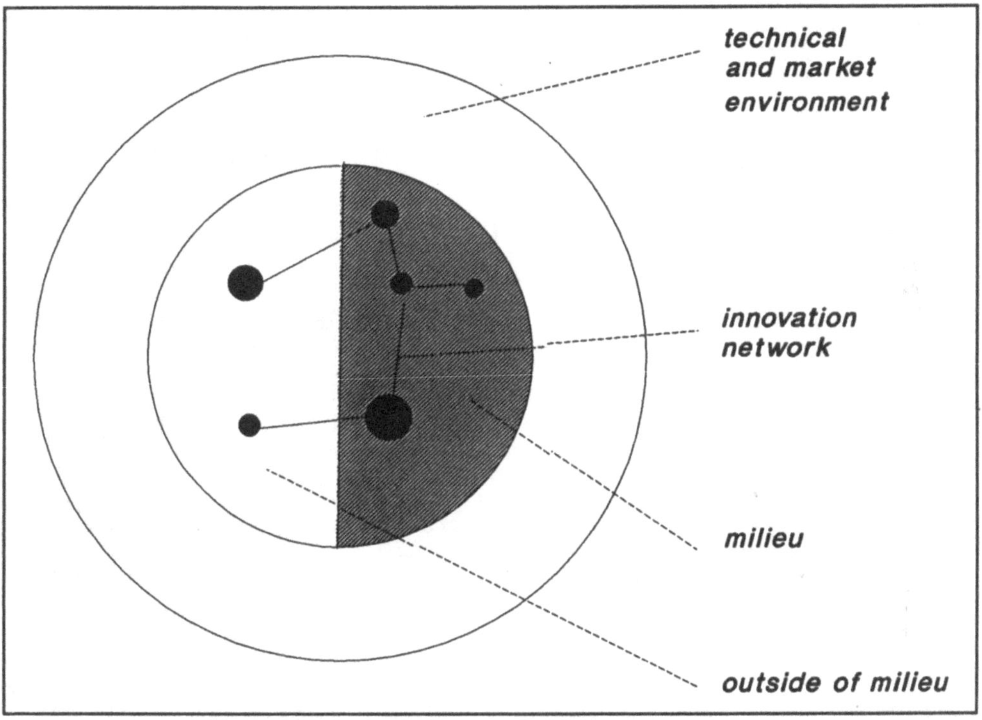

Figure 3.1: Milieu, network and environment (Source: IRER, 1993

3.2.1 Formation and Evolution of Innovation Networks Over Time

The network phenomenon and, more specifically, innovation networks must be studied from a time perspective. Such a study should reveal the network's dynamics, that is, its genesis, evolution and transformation.

The underlying idea of the notion of milieu is that innovation networks will form from existing relations. Often traditional-type sub-contracting relations, seen on a long-term basis, are a privileged medium for the emergence and formation of innovation networks. But these relations are not exclusively economic or industrial in nature: they may be professional relations which stem from belonging to the same occupation, relations established while studying or training together, or quite simply relations developed by participating in associative activities. There is thus a broad range of potential relations, ranging from the economic to the social sphere. However, the dominant characteristic of these relations that can form the medium for inter-company innovation networks is that they are lasting: the players have known each other for a long time, or have sometimes worked together, or their career paths have crossed; they respect and often like each other. Thus, a *relational capital* forms between these players; it will enable them to go farther, to expand the scope of their relations and to devise joint projects.

In this respect it is interesting to note that even if an initial relationship was a failure, this does not systematically mean that the players will break off all relations: this first experience of collective work has enabled them to get to know each other, to discover each other's abilities and deficiencies, to develop a common language and, in a word, to build the relational capital that will enable them to define other projects or subsequent forms of collaboration.

Thus, at any given point in time the innovation network is the result of a long process of learning about relationships. These observations give rise to two additional remarks:

- *A network has a history, it results from a milieu.* If we are to try to understand the genesis and development of an innovation network, we must be able to reconstitute in time the relational systems between the players. The innovation network as observed during our investigation is in fact the result of medium or even long-term relational dynamics (Maillat, Crevoisier and Lecoq, 1990).

- As soon as a network is formed, a relational capital which will allow the players to subsequently expand their collaboration and to develop new innovation projects is also formed. There is therefore a *self-maintained network dynamics* which is gradually established by updating and exploiting a relational capital.

Innovation networks generate effects as they are established and developed. In fact, hardly anything is known about the chronology of the innovative process. Although in general innovation theories take as their point of departure the decision to launch a project or even the decision to invest in a Research and Development activity, in actual fact the innovation process cannot be reduced to a sequence of precisely dated and planned stages. This sequence is random and non-linear. Consequently, it is difficult to break it down into a sequence of operations in an upstream-downstream type of logic, extending from prototype design to the development of the new product or process, to its industrialization and then marketing - stages that are perfectly joined to one another. It follows that the evolution of the innovation network is closely linked to that of the innovation project. While the network is formed and transformed, the constraints faced by the players in their innovation process are also formed and transformed.

In other words, this means that the network as we have observed it corresponds to the progress of the innovation project. The network takes different forms over the course of time, its constituent elements change and relate differently, but the network retains its identity despite these changes. *The innovation network is thus at one and the same time stable and in flux.*

If we consider the innovation process of conventional theories, we must also take other elements into account: innovation is in fact a process that combines commercial and non-commercial factors (Maillat and Perrin, 1992). In a regional context, the Gremi II survey showed that the milieu does not constitute a mere reservoir from which players derive the resources they need to innovate. *It is the dynamics of this milieu that generate innovation opportunities and stimulate certain responses from the players* (Maillat, Crevoisier and Vasserot, 1992). Seen from this angle, the firm's decision may no longer be considered as the starting point of the innovation process. The picture of an isolated inventor - a product of the Schumpeterian tradition - who inseminates the economic fabric with a heaven-sent idea is fundamentally schematic and obsolete. The innovation process is not the same in Montpellier and Bergamo, in Silicon Valley and the Jura Arc. This difference is due to the specificity of milieux, which "metabolize" technologies, markets and resources in different ways, thereby opening up new and original opportunities for a region. The stage during which the idea for a new product is born will be called the "*idea-forming stage*". This process, which is upstream of innovation proper, becomes very important as soon as the linear vision of the innovation process is abandoned. This perspective implies that, as soon as a project emerges, the innovation process groups together a large proportion of the elements needed for innovation (technology, markets, resources, etc.) at a given point in time. Of course, this combination which results from the merging of various elements must be understood as an entity, which is gradually taken out of its context and acquires a certain independence that will allow the innovation network to differentiate itself from its context. Consequently, *the innovation project is the product of a milieu*: at one and the same time it originates from a milieu, draws most of its characteristics from it, but it differentiates itself from it at a certain point in time on account of its innovative character.

3.2.2 The Different "Fields" of Innovation Networks

The object of this section is to pinpoint the various fields of reality of innovation networks. Three main fields can be distinguished: the context of innovation, formed by the technical and market environment that characterizes this innovation, the working relations within the network and, finally, the relations between this network and the milieu (see Figure 3.1).

(i) The Technical and Market Environment of Innovation

Characterizing the technical and market environment allows valuable information to be gained about innovation itself. Is it a new product, involving the implementation of new technologies and the opening-up of new markets? Or, on the contrary, are largely tried-and-tested technologies and existing markets being used? Is it actually a new technological creation or simply a technological trajectory in which firms and their networks are content to exploit and adapt existing generic technologies (Amendola and Gaffard, 1988). In particular, it is a question of assessing the impact of the innovation being studied on the product: does the innovation radically transform a product's technical and market context, or does it confine itself to only making improvements to an existing product?

Such information will in fact enable us to define more clearly the way the network is structured: in the first case, the innovation network has a creative process, even though the utilization of existing technologies participates in a combinative-type logic. Of course, the same types of know-how, the same learning methods and the same forms of network organization are not necessary.

In addition, this environment furnishes information about the relations between the network and the area outside the milieu. Indeed, we know that today markets are mostly situated outside the region, and frequently abroad. The same applies to technologies: what region can boast having a monopoly over a technology or being the only area capable of implementing it? Yet the milieu does control certain markets and technologies. They are, as it were, "endogenized", they form an *integral part* of the milieu. It is, of course, from this basis that the milieu draws support to innovate.

However, certain elements are lacking in the milieu and they have to be sought elsewhere. This "elsewhere" is not only geographical, but also refers to a little-known or unknown field of various players in the milieu, a market to be created or entered, a new technology that should be understood and tailored to one's requirements, and partners with whom one has never been in touch to be contacted.

In summary, it appears that the players are required to form networks to manage the constraints they face in their innovation process.

(ii) Relations within the Network

Working relations within the network reveal the nature of the principles on which the innovation network is based. In particular, our aim is to highlight the engines of the network, the effects of domination and power, the degree of formalization of relations (is there a contract between the protagonists, what form does it take, on what principles is it based?), the incentives that encourage each one to perform its function, the financial and material resources on which the various players have been able to rely, etc. An analysis of the relations within the network leads us to give priority to two highly complementary approaches. First, the way in which the players regulate their relations; and second, the way in which the technical consistency of the innovation process is assured.

a. Relations between Players

In our view, the relations between players must be understood as a system that manages to perform certain functions. Once again, our object here is not to lapse into functionalism: during our study we have observed successful innovation networks so we may assume that, in overall terms, the innovative process has a certain consistency vis-à-vis its environment. The success of this process is a key to understanding the ways networks are structured and organized. Indeed, the mere fact, for example, that players had performed a function and managed to be integrated into the process tells us that certain rules have been drawn up to manage the appropriation and competition problems that inevitably arise in any form of cooperation. Highlighting these rules, but only those that have played a part in the success of the process, is one of the threads that has guided the exploration of the relations within networks. A clarification is necessary at this level: these rules will be described as endogenous to distinguish them from the exogenous rules that will be examined later on. These rules are thus defined at the network level, between the players, to ensure it functions smoothly and to determine the obligations of each of the innovation partners: decision-making procedures, the recruitment of skills, cost and profit-sharing. It goes without saying that membership of a network implicitly presupposes that these endogenous, more or less formalized rules be observed. In this respect it appears that the degree of formalization of these rules is closely linked to the nature of the innovation process. It may reasonably be considered that the more the innovation process moves towards a high degree of creativeness, the harder it is to define precise and binding rules which can come into play in the management of the players' collective work in the context of their innovation process.

b. Technical Relations within the Innovation Network

The technical consistency of the innovation process may be understood in terms of the *division of labour*. Indeed, a knowledge of whether the areas of responsibility of the various innovation partners are, a priori, complementary or competitive allows their respective role to be defined more accurately and thus the technical organization of work within the network to be understood. Everyone must find his place in relation to the other partners. This procedure does not systematically run smoothly and raises difficulties that must be solved within the network. This is all the easier if each player handles a specific segment of the innovation process. In this case, the main difficulty lies in the convergence of interests. Conversely, if the various partners' areas of responsibility overlap, it is task-sharing that has to be organized. At this level, it appears that the difficulty of defining the division of labour among the partners is proportional to the complexity of the innovation process. The innovation network is therefore less structured and less organized at the beginning of the process than at the end. This organizational effect of the innovation network is very important for the transformation of the territorial production system.

A study of the division of labour allows a map of the network to be drawn and various standard configurations to be revealed: star-shaped, hierarchical, compact networks, etc. This typology reveals certain phenomena of domination and power within networks.

Indeed, a firm located in the centre of a star-shaped network is, of course, a dominant factor in the innovation process since it has an overview of the whole process. Conversely, a firm located on the periphery of the network necessarily has a more limited scope of action.

More specifically, we have to study the evolution of these rules over the course of time, our hypothesis being that the thrust of the innovation process is towards a *surplus of organization*. Having said this, it is clear that the various networks observed have very variable levels of organization at the outset. In this respect it may be said that the least structured network tends to form a protected *working area* within which the players can collaborate. In fact, seen from this angle, it is a matter of building an adequate framework, that is, of defining the overall conditions rather than determining the respective contributions of each partner. At a higher level of organization, the innovation network already presents a more clearly established, subtler division of labour, whereas the overall conditions that encompass most of the players in the first case now define only bilateral working areas: collaboration thus takes place on a two-by-two basis. Finally, a higher level of organization is characterized by a detailed specification of each partner's services, with the disappearance - or virtual disappearance - of the grey areas in the partners' services.

(iii) Relations between the Innovation Network and the Milieu

The milieu participates in the building of innovation networks and is involved in their dynamics. Reciprocally, innovation networks enrich the milieu, helping to enhance the latter's creative capacities. Three variables are thus essential to our approach: knowledge of the milieu, know-how and finally, what we call exogenous rules.

a. Knowledge of the Milieu

What do the various players know about each other? Did they know each other before, how long have they known each other, how did they meet, had they already worked together? This knowledge of the local milieu and the possibilities of mobilizing its potential play a capital role in innovation. Indeed, rules of reciprocity, trust, notions of services rendered and a sense of belonging to the same milieu afford each player opportunities in the form of mobilizable physical, financial and human resources. They come into play at every stage of the innovation process, from the idea-forming stage of the project to its industrialization. However, the cooperation of the players at the right moment is not self-evident. If the various partners manage to agree, it is because there is a convergence of interests or because belonging to the same milieu in some way obliges the partners to play the game together, knowing that later on they will have to "return the compliment". The argument whereby "you have to give in order to receive" recurs regularly during discussions. Thus, this knowledge of the field and of the potential or actual partners not only provides access to certain opportunities but also imposes obligations that are often other than financial.

b. Know-how

Know-how represents all the abilities which are used in the practical mastery of production. Different types of know-how are required in the innovation process. Where do they come from: have they been created, enhanced or, failing this, destroyed during the innovation process? Innovation does not necessarily mean know-how is created. Indeed, the mere application of a technology that is known and mastered in a field in which it has not been used before does not systematically involve new skills. Adapting a technology to a new field, on the other hand, does require learning. The creation of a new technology obviously requires a more complex learning process whose results are not easy to predict and which necessitates different organizational conditions. The same applies to markets or production techniques. By positioning themselves on existing, well-structured markets, know-how is not created and therefore does not develop. On the other hand, launching a product or process for which the market does not yet exist requires the creation of ad hoc know-how. In particular, those skills which have been found in the milieu and those which have been brought in from elsewhere should be specified. It is in fact at this level that the milieu's deficiencies or resources appear most clearly.

c. Exogenous Rules

By exogenous rules we mean rules drawn up outside the network and in this case by the milieu. We mentioned earlier that the rules used during the innovation process were drawn up as it progressed by proposing the hypothesis that they converged towards a level of organization that was greater at the end of the process than at the beginning. It should be pointed out here that these rules are learned, just as know-how is learned. Indeed, the partners may have recourse to the rules in force in their environment without transforming them, but when these rules are non-existent, too diffuse or simply unsuited to the situation, new ones have to be drawn up. These rules are generated more or less regularly and in a more or less formal manner. At this level, it is important in particular to show where these rules come from: multiple allusions to belonging to "the same trade", to "the same community" to explain how problems of work, payment, priority, checking, etc. are managed clearly highlight the importance of the rules in force in the milieu. Often, however, more precise rules governing particular subjects are drawn up within the network. Finally, a process of contractualization, with a detailed definition of the partners' services and of the procedures to be followed in the event of a problem is to be found in much more organized companies conducting much more circumscribed innovation projects.

In an innovation network, exogenous, milieu-generated rules join up with rules defined within the network itself to manage the services of the various partners. These two rule systems are closely connected; first, with the nature of the innovation process and, second, with its degree of progress in time. Indeed, it was noted at the end of the surveys that the more the innovation network is directed towards technology creation, the more the prevailing rules derive from the milieu. On the other hand, innovation networks which utilize or adapt existing technologies define a system of specific rules within the network-type organization: these rules are either produced by the network's players themselves or derive from a process of internalization of the milieu's rules. A similar process will be observed as a function of the degree of progress of the innovation project. These considerations highlight the learning processes which develop at the level of knowledge, know-how and rules, among the players involved in the innovation network. These learning processes are one of the means by which a milieu can be transformed in the course of time.

3.3 AN ATTEMPT TO DRAW UP A TYPOLOGY OF INNOVATION NETWORKS

A survey carried out in the Swiss Jura Arc has enabled us to draw up a typology of innovation networks. The examples cited in the text are taken from that survey. The object of this typology is to clarify the impact of these networks on the milieu and on the production system. Starting from the above considerations, five criteria have been selected to draw up this typology:

- *Innovation*: the object is to evaluate the impact of innovation on the product. Is innovation brought about by an addition of new elements to an existing product, by transforming a product or again by creating a product and its market? It is in fact reasonable to consider that depending on the degree of complexity of the innovation process, the creative process implies modes of organization of the specific partners.
- *The division of labour*: the object is to define the degree of specification of the services of the players involved in the network. Is the division of labour a strict one, is it modular or is it simply not yet stabilized between the various protagonists?
- *The rules*: any network implicitly presupposes a system of rules. It is necessary to assess the latter's degree of formalization and to evaluate their origin (the two criteria are closely linked). Are the rules created by a bilateral type of contract, are they defined at the network level or, finally, do they originate from a sense of belonging to a community? In the latter case they will be termed diffuse, that is, they belong to the milieu, to a working team.
- *Know-how*: innovation results from a creative combination of know-how. The aim is to evaluate the way in which learning processes develop within the network and the degree of appropriation of new forms of know-how. Do these learning processes develop individually, bilaterally or collectively?
- *The function of the milieu*: what is the milieu's contribution to the idea-forming process? Its function varies between merely providing additional skills (a supply function), an implicit contractual framework and an intermediation structure between the players (an intermediation function), ultimately an essential element in the very creation of the innovation network. An analytical table may be drawn up on the basis of these criteria (Figure 3.2).

On the basis of these criteria it is possible to determine three types of innovation networks along a continuum. Starting with the most organized network and ending with the most evolutive, we shall distinguish successively between innovation networks with a leader firm, innovation networks with a hub firm or hub group of firms and, finally, compact innovation networks.

3.3.1 Innovation Networks with a Leader Firm

Innovation networks with a leader firm are characterized by *a certain monopoly of the leader firm over the innovation project.* They determine the initiative of the project, control all the operations from design to industrialization to the marketing of the new product and control all stages of the innovation process. The use of external partners is justified mainly by the search for additional skills, the acquisition of which would take a long time and require costly investments. In this process, transactions between the prime contractor and the partners are essentially bilateral and are defined in a specification of requirements that give rise to the drafting of a more or less formal contract, depending on the partner's knowledge. Thus it is a relatively hierarchical network structure in which the leader firm manages and controls only the innovation process. To describe this innovation network one might borrow the metaphor of an orchestra in which the conductor (leader firm) writes the score alone and then individually contacts the musicians whom he asks to play the notes relating to their instrument.

By way of illustration, here is a case of collaboration to develop an expert system for a lathe centre. The description of the network comprises the following five aspects: (i) the new product and its technical context; (ii) the various partners, the division of labour and outside skills; (iii) idea-forming, the history and function of the milieu; (iv) rules governing the relations between the partners and their development; (v) the types of

INNOVATION	DIVISION OF LABOUR	RULES	KNOW-HOW	FUNCTION OF THE MILIEU	TYPE OF NETWORK
Impact of innovation on the technical and market environment	Degree of specification of partners' services	Origin and degree of formalization of rules when network is set up	Degree of appropriation of new know-how	Importance of the milieu in the idea-forming process	
Addition of elements to an existing product	Strict division of labour	Rules set up by a bilateral agreement	Individual appropriation of new know-how	The firm uses the milieu's ability to innovate	Network with leader firm
Transformation of an existing product	Modular division of labour	Customary rules in the network	Bilateral diffusion	The firm is based on the milieu's ability to innovate	Network with hub firm
Creation of a new product	Unstabilized division of labour	Diffuse rules connected with a feeling of belonging to a community	Diffusion of new forms of know-how to all the partners	The milieu creates an opportunity to innovate for the various partners	Compact network

Figure 3.2: Criteria for a typology of networks (Source: IRER, 1993)

know-how mobilized and their development; (this approach is also used in the presentation of the other examples).

(i) The New Product and its Technical Context

In this case, the innovation lies in the design of an expert system contrived to facilitate the programming of automatic screw-cutting machines. Screw-cutting, a traditional mechanical-engineering industry, consists of series production of parts by a machining technique involving the removal of chips. It is clear that one of the main problems of such screw-cutting machines is productivity. The income of the manufacturer of machined parts is directly proportional to the machines' productivity. Today screw-cutting machines are extremely complex machines fitted with numerous different tools to perform several operations simultaneously. The problem is thus to optimize the sequencing of the various operations as a function of the part to be machined, the available tools and the various machining constraints. Hence it is an optimization problem. For many manufacturers this optimization is a very complex operation. The lack of skilled labour constitutes a major obstacle to the optimal use of such machines and restricts the potential clientele.

The expert system thus enables the screw-cutter to draw the part very simply on the screen of a PC. He then defines the appropriate tools, the various operations and their synchronization, as well as the working conditions. Much simpler and faster than a CAD system, this expert system allows very expensive machines to be utilized more profitably.

(ii) The Partners, the Division of Labour and Outside Skills

The main player in this innovation is a large company, the world leader in making automatic lathes. This company did not, of course, have the skills needed to devise an expert system. It was therefore a fairly large-sized specialized foreign firm which did the computer development work. However, the nature of the expert system required the full collaboration of the various partners. This collaboration took shape with the creation of a "unit of experts" charged with providing most of the substance of the programme. In particular, this group of experts had to constitute the most exhaustive possible base of knowledge on all forms of know-how in the screw-cutting field; as for the computer experts, they mainly had to retranscribe this knowledge base into computer language and to make it compatible on a PC operation. (See Figure 3.3).

(iii) Idea-forming, the History and Function of the Milieu

The idea for this innovation stems from a need felt initially inside the company itself. The complexity of this lathe centre was causing the sub-optimal utilization of its capacity. It was therefore decided to draw up a programme to assist in the use of this machine.

The programme had to be devised externally as the company had no knowledge of expert systems, little knowledge in the field of programming at a high level of technicality and because it did not want to commit resources to this project on a long-term basis.

The search for and selection of the partner therefore took place on the basis of these constraints. This procedure took place in two stages: in the first stage, the large company tried to contact partners it had already worked with. It asked them to prepare a quotation. Depending on the constraints fixed and on a comparison of the various bids, the company found that the regional milieu (the Jura Arc) did not satisfy the constraints and that the same applied to the partners contacted elsewhere in Switzerland. In the second stage, faced with the deficiencies of the milieu, the firm had to broaden the scope of its search.

It finally chose a foreign company which was contacted through a personal acquaintance in a university. This company fully met the constraints set as it had a certain proficiency with numerically-controlled machines.

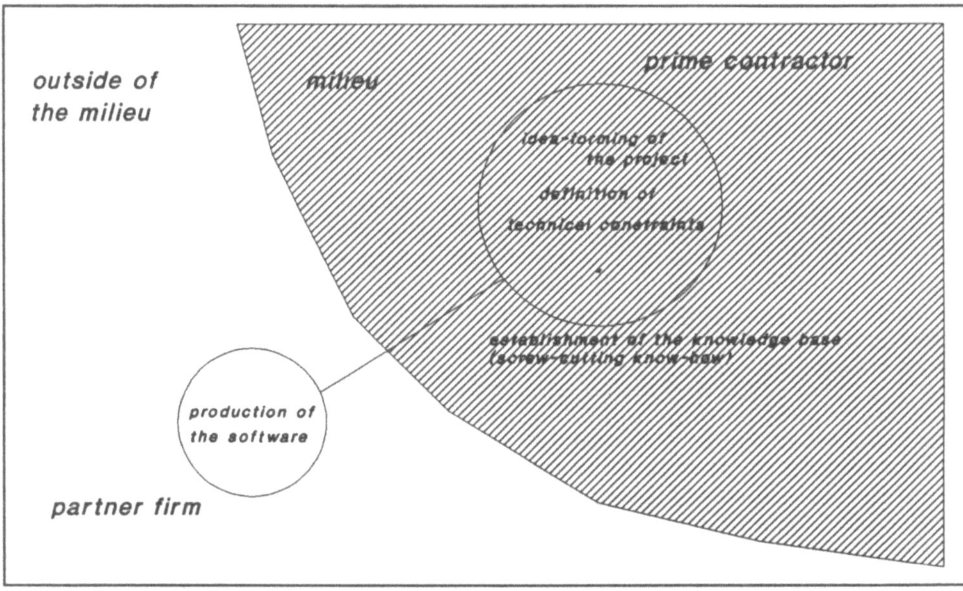

Figure 3.3: Innovation network for the development of an expert system for machine-tools (Source: IRER,1993)
* - companies visited

(iv) Rules Governing Relations between Partners and their Development

Having never collaborated before, the project managers organized several meetings to present the objectives, discuss the constraints and above all, to get to know each other before starting work. One of the partners' prime concerns was therefore the search for overall trust. Next they defined together the specification of requirements. On this basis the foreign company proposed a quotation and then a contract between the two partners was drawn up, but making sure it remained very flexible. It was in fact a framework contract intended to set out the reciprocal obligations, constraints and directions of the project, but still leaving the partner some room for manoeuvre and a certain degree of freedom in the design of the computer programme.

(v) Types of Know-how Mobilized and their Development

The types of know-how necessary for the programming of the expert system were perfectly mastered by the foreign company. So the only problem was to tailor this know-how to a specific context. The know-how connected with the machine was available in-house and did not undergo any particular developments. On the other hand, an interesting process took place with regard to the production know-how; in fact, devising an expert system requires a quantity of information and rules to be stored in a database, from which information and rules the programme will choose. In this case, some of these rules were formed by the tools used by the lathe centres. There were numerous tools and they had to go through a lengthy classification and standardization phase before they could be integrated onto a PC disk. Other rules were formed by the lathe professionals' know-how. It is easy to imagine the difficulty of codifying knowledge whose components cannot be formalized a priori. This stage took a very long time and required numerous surveys to be conducted by the group of experts about the company's workers and screw-cutting technicians. Thanks to a major standardization and formalization drive, a fair proportion of the programming operations for numerically-controlled lathes were thus automated. In so doing, not only was knowledge transferred from the milieu's operators and technicians to the machine, but productivity was increased and the work was also made more interesting.

3.3.2 Innovation Networks with a Hub Firm or Hub Group of Firms

In networks with hub firms the latter may be regarded as *the architects of the network.* Although they control the project, they study its technical feasibility in collaboration with specialist partners. The other firms act as specialist-suppliers. However, they are not controlled in every aspect of their contribution: there is a division of responsibilities rather than a division of labour. The aim is to reach agreement on the interfaces between the different modules, the inside of the module being strictly the supplier's domain. Such a network organization permits at one and the same time a division of labour and a modular design of the product. This enables the product to be developed over time without transforming its architecture, simply by incorporating new modules, including new technologies or meeting new needs. The problem with such a network is to render the various partners' know-how consistent, i.e. in this case to adjust it to the interfaces.

It is interesting in this architecture to distinguish between the centre of the network and its periphery. In the centre we have a relatively polarized structure, hinging on the hub firm or hub group of firms which manage and coordinate all the partners. At this level transactions are multilateral and based on a moral contract. On the other hand, on the periphery of the network we have bilateral relations that are often formalized. The learning processes develop according to a collective or bilateral mode at the centre of the network (expansion and deepening of know-how) whereas these processes are mainly individual on the periphery. Returning to the orchestra metaphor, we can say here that the conductor writes the score, taking account from the outset of the various musicians at his disposal. Each musician tunes up and follows the tempo of the conductor.

By way of an example, we shall describe a hub firm innovation network devoted to the design and manufacture of a machine for mounting surface components.

(i) The New Product and its Technical Context

The innovation concerns the design of an automatic and flexible machine for mounting microelectronic components on printed circuit-boards. This machine had to satisfy two essential constraints: the ability to manipulate small-sized components and to be flexible; i.e. the customer had to be able to assemble a large range of electronic circuit-boards using a large number of different components with the same process.

(ii) The Partners, Division of Labour and Outside Skills

The innovation network was organized by a small firm specializing in the design, development and manufacture of industrial robots for assembling electronic components. Armed with a good knowledge of robotics and, more specifically, of flexible automation, this firm presents itself as the main architect of this new machine. It controls its design, part of the machining and all the assembly work. However, although up to now it has possessed technical expertise in the field of assembly robots, it has not had sufficient potential to move on to the industrial stage. Becoming part of a large holding company has enabled it to find the financial resources necessary to make the transition towards the industrial stage and to do the research and development indispensable for this new machine.

The machine's design is of the "modular" type, so that it is possible to contract out its different components. So the firm at the centre of the network merely has to control the overall functioning of the machine. It defines the interfaces between the various components with its partners and leaves development and production to them. Thus the machine's frame was contracted out in the region, as were the drives, pincers and manufacturing of the electronic circuit-boards. The optical and measuring devices were purchased from recognized foreign firms specializing in these advanced fields. The use of these specialists was a guarantee of quality for potential customers.

Finally, it will be noted that this network architecture corresponds particularly well to the idea of flexible specialization (Piore and Sabel, 1984). It has the advantage of allowing the product to develop in line with the technical advances achieved for a given module without the whole product being called into question. (See Figure 3.4).

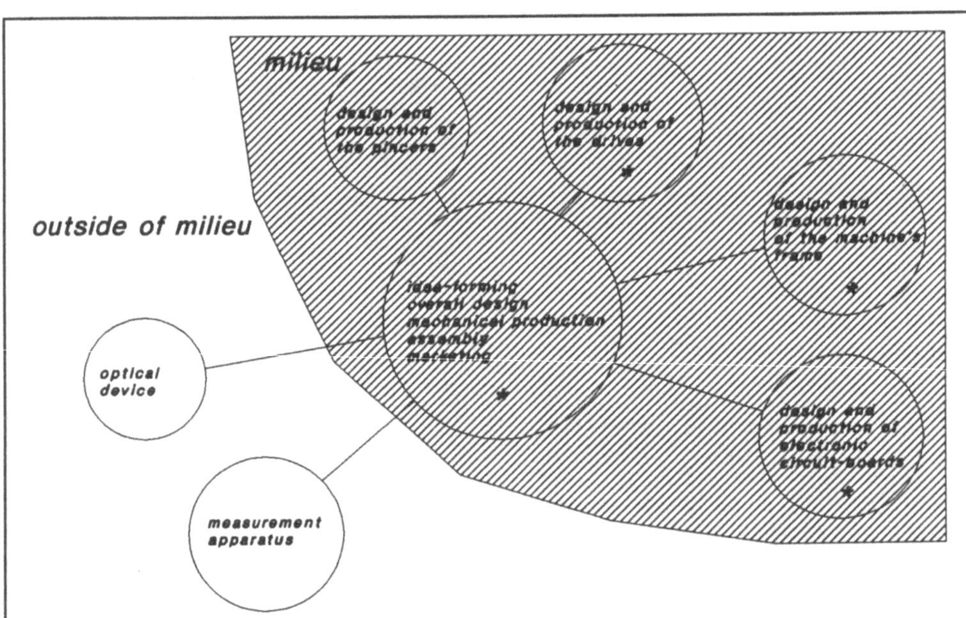

Figure 3.4: Innovations network for the design and production of a machine for mounting surface components (Source: IRER, 1993)
* - companies visited

(iii) Idea-forming, the Network History and Function of the Milieu

The idea for this machine came from a request made by two customers situated outside the region but in Switzerland. They were customers for whom this company had already made machines, but they were looking for a more flexible solution not yet available on the market. The vast majority of the sub-contractors located in the region are former partners. The other partners have all, at some time or another, shared career paths with the manager or one of the employees of the main company. As regards the manufacture of electronic circuit-boards, the firm in question is part of the same holding company. Thus there is a stable basic network of players who have already worked together and these relations were reactivated when this new machine was developed. Following this innovation, the network was strengthened and stabilized.

(iv) Rules Governing Relations between Partners and their Development

The desire to reduce all forms of contractualization of the relationship between the partners to a strict minimum is one of the special features of this organization. This was because the partners had already cooperated together on many occasions and they had known each other for a long time: a relational capital thus exists and although often-rigid contractualization unequivocally allows the respective obligations of each party to be determined, it also restricts the very development of the relationship from the outset. Often costly and taking up sometimes precious time, it thus appears to be superfluous here. Hence, there is no explicit contract between the various firms. Together the partners define an overall framework, each partner's responsibilities and obligations and a price range. The players undertake to respect these bilaterally-defined guidelines. Accordingly, the various partners subsequently work on a basis that may be described as one of agreement.

Apart from this general atmosphere of trust, two fundamental rules structure relations within the network:

- a reciprocity rule: the players undertake to exchange information on technologies and markets if it is connected more or less directly with the innovation project. This service is, of course, provided free of charge;

- a priority rule: the sub-contractors undertake not only to respect the deadlines but also to make themselves as available as possible to the prime contractor. In return, the prime contractor gives his sub-contractor regular work.

These practices make it possible to involve the various players in the project, and to create a sense of solidarity between the partners by making the relationship a more long-term one. The various partners associated in the innovation project are in some respects the core of a much more stable network whose existence easily transcends the execution of this project alone.

(v) Types of Know-how Mobilized and their Development

The know-how mobilized during this innovation process closely matches the existing division of labour. In fact, the prime contractor controls the overall design of the project, but it defines with its partners the modules whose content it does not master and for which it leaves full design and production responsibility to its partners. The learning process consists mainly of defining interfaces, and taking optimal account of the partners' skills; this interface definition leads the partners to place their know-how on a par two-by-two.

Another important development of this network's know-how was the switch to the series production of machines. This process meant that a wide range of customer requirements had to be taken into account in order to reach a sufficiently broad market; then it was necessary to define with the partners modules fulfilling the functions decided on and having specifications compatible with the overall design of the machine. Finally, each partner managed to series-produce the module of he was in charge. This standardized production led to the recruitment of unskilled personnel to do the most repetitive tasks. The production know-how was thus codified. Similarly, the overall architecture of the machine and the interfaces between the modules were codified when the work was organized between the various partners.

3.3.3 Compact Innovation Networks

This type of network is very different from the others in that there is no firm to ensure central coordination of the innovation process. There is a collective, overall conception of the innovation project. The element that brings the partners together is the notion of *project*. The contributions of the various parties cannot be evaluated a priori and therefore cannot be checked. This procedure should be seen rather as a joint experiment in which the network primarily defines *a collective working area of exchange*. Relations cannot therefore be schematized in the form of a reticular grid: the network is above all a set of partners seeking to isolate themselves so as to allow an idea or project to progress. Transactions are thus multilateral, regulated by relations of trust and reciprocity, by a joint work ethic connected with belonging to the same milieu and the learning processes develop according to a collective dynamics. To take up the metaphor used earlier, this dynamic may be likened to an orchestra that has no conductor, that is improvising, writing its score as it plays. Coordination is thus multilateral, between all the orchestra's members. The rules in force allow everyone to express optimally their capacities. The purpose of the rules is not to monitor the services.

By way of an example, we shall describe the case of a compact network for designing, developing and producing a miniature accelerometer sensor.

(i) The New Product and its Technical Context

The innovation studied through this network belongs to the field of microtechnology since it concerns the design, development and industrialization of accelerometer sensors and miniature sensor systems. Sensors are elements designed to "sense" their environment, that is, to communicate the information they capture in their milieu. To use an image, these sensors may be compared with the function of a thermometer, which is used to record the ambient temperature.

From a technical point of view, these sensors are composed of micro-mechanical parts of silicon and electronic parts. Although the technologies involved are relatively well mastered scientifically, major difficulties are encountered in the ability to combine these technologies and above all in switching from the prototype stage to an industrial stage. The industrialization stage is all the more critical since at present, although the potential market for miniature accelerometers is very large, it nevertheless still has to be created. It is important to note that these sensors are not finished products but that they will subsequently be mounted on other products. They may in fact be used for applications in medicine, in transport (underground railways, the automobile industry), in space (vibration sensors on satellites), in the recording of seismic movements, etc.. The innovation network built around this project thus faces two main constraints at one and the same time: how to create markets and tailor the product to a potential demand, and how to research the optimal conditions for the large-scale industrialization of the sensors.

(ii) The Partners, the Division of Labour and Outside Skills

At the beginning of this innovation process there was no pre-established division of labour. Each partner was able to control a large part of the project. However, the extent of the costs, the duration of the developments and the complementarity of the production equipment meant that none of the partners would have been able to execute the operation on its own. The idea is that it is not possible to consider the work sequentially. The various work sequences - prototype design, prototype development, industrialization and the marketing of the finished product have to be studied together and simultaneously. The idea is to view the product in overall terms even if the partners subsequently intervene at different stages.

As the originators of this network we have a large research centre and a medium-sized company specializing in the production of miniature electronic modules. These two partners' desire to take into account the constraints involved in the sensor industrialization process will lead to a third partner - specializing in a particular technology, the micro-machining of silicon, a basic material of electronic circuits - being integrated into the network. This company's motivation was mainly to series-produce sensors, which enabled it to use the production capacity available to it after some large investments. The relationship with the research centre in particular aimed to improve the product so as to facilitate the industrialization process. Finally, the need to research market potential at the same time led the two companies to jointly set up a small firm to handle product marketing and the customer interface. It is interesting to note in the context of this network that all the players and skills mobilized came directly from the local milieu, in which the micro-technology know-how inherited from the watchmaking tradition had been built up. (See Figure 3.5).

(iii) Idea-forming, the Network History and Function of the Milieu

Behind this project we find the desire of a research and development centre to use micro-technologies to develop sensors and sensor systems, these being perceived as a technological market niche offering considerable industrial potential. In 1982 the first research and development work was launched internally, a process that was sustained from 1984 onwards by a contract with the European Space Agency for a pilot project on satellite vibrations. This embryo network was strengthened by a project that brought together research centres and industrialists with the help of a public grant. This project brought together numerous large firms, but they withdrew later on. It appears that the role of these institutional networks was to encourage an idea and to draw up well-codified rules of the game, with more or less the same caution as a public institution. These institutional networks thus acted as a catalyst in the innovation project. The innovation network was established and developed without any a priori organization, given the particularly evolutive nature of the technologies and markets in this field. Thus, each partner must always be seen as part of an overall vision of the project, from design to marketing: "what is needed to develop and make a new product is not a sequential conception, you have to work together to reason in global terms".

(iv) Rules Governing Relations between Partners and their Development

This innovation network is supported by an extensive network of professional and personal relations. All the players know each other and have already collaborated on joint projects. The various protagonists have moved along fairly similar career paths, originally coming from the watchmaking milieu and have "maintained good neighbourly relations among people of the same trade" even if they have not seen each other for long

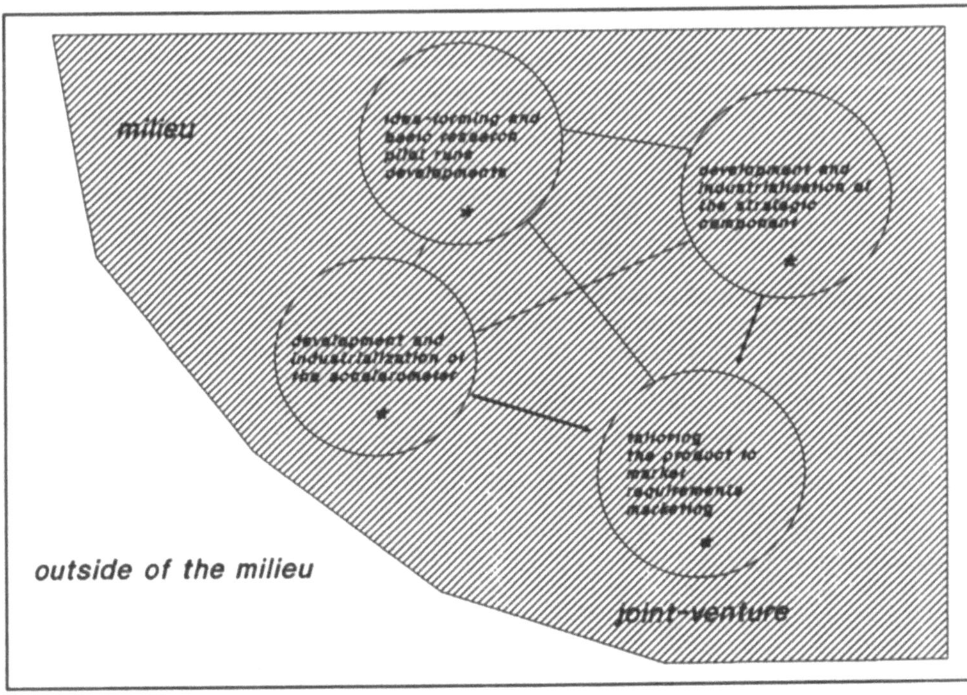

Figure 3.5: Compact innovation network for the development and production of a miniature accelerometer sensor (Source: IRER, 1993)
* - companies visited

periods. Accordingly at the base there are relations of trust or even friendship between the various partners. This general atmosphere facilitates mutual acquaintance, collaboration and the dissemination and exchange of information: "collaboration implies an exchange; to collaborate is to discover one another. Anyone who keeps secrets remains isolated". Thus there is this possibility of opening up towards each other, of disseminating one's knowledge without any risk of unilateral appropriation, as the players share one and the same work ethic, a certain moral code of business and a common desire to collaborate. Given the very nature of the innovation process, it is difficult to enclose each partner's services in a rigid contractual framework. An overall contract is drawn up among them, fixing the general lines of each partner's obligations and constraints. Of course, given the nature of collaboration, it is difficult to monitor the services of the partners. It is the feeling of belonging to a milieu, and thus observance of these implicit rules, which in a way condemns the players to "play the collaboration game without any ulterior motives"; "we receive information but we also have to provide information". Exchanges between partners are crossed and form part of a reciprocity logic. This logic of openness obliges the companies constantly to bank on innovation, as their skills are disseminated very quickly.

Given that the various partners are in competition with one another to quite an extent, as problems arise they draw up rules for distributing customers and income. Similarly, certain rules which allow partners to take an interest in their colleagues' problems have been set up. As the project is very evolutive, the rules governing working relations are often readjusted. In general, these rules aim to get the partners interested in the success of collaboration rather than to permit the monitoring of each partner's services.

(v) Types of Know-how Mobilized and their Development

The know-how is mainly in high-technology areas: research on microtechnology, the miniaturization of electronic systems, and the manufacturing of micro-components. This know-how all originated more or less directly from the research conducted in the 1960s on the electronic watch. But what is remarkable in this project is the collective creation of new forms of know-how. Indeed, development, industrialization and the creation of markets required long lead-times; skills hitherto isolated from one another were brought together and the technical and market environment was truly structured around this project. Here the learning process simultaneously concerned all compartments of production: the product, production technology, the assembly technique, tailoring to the requirements of potential customers, and the marketing network.

In summary, it is interesting to note that the various types of networks correspond to specific innovation processes. *The structure of the network, its morphology and internal organization are closely linked to the nature of the innovation process.* Whereas the network with a leader firm will develop an innovation project concerning the addition of new elements to a product already on the market, the network with a pivot firm or pivot group, in contrast, implements an innovation project in order to transform an existing product. Finally, the compact innovation network creates a new product and its market. These different modes of organization are justified by the nature and combination of the types of know-how mobilized in the context of the innovation process.

3.4 CONCLUSION

By focusing on innovation networks we have sought to understand the innovation process in its globality, transcending the sequential method which characterizes the approach of standard economic analysis. In particular, this approach has made it possible to highlight the nature of the way the partners working on the innovation project interact. The tentative typology put forward emphasizes the parallel between the degree of organization of the innovation network and the nature of the creative process. Hence, we may put forward the hypothesis that the development of the territorial production system is closely linked to the architecture of the innovation networks: whereas the network with a leader firm permits mainly individual-type learning processes and thus only a deepening of the know-how of the various partners taken individually, the other two categories of networks not only permit this deepening of know-how but also its expansion. By considering that the development and transformation of a territorial production system are closely linked to these learning dynamics, we quickly find that the different forms of innovation networks do not offer their milieu the same potential. While the very organized networks lead to an increase in terms of control and power from the industrial point of view, the more evolutive networks tend to renew the region's industrial fabric. In every case, these innovation networks lead to a surplus of organization in the region. In practical terms, this surplus covers factors such as the establishment of production networks, a certain degree of product standardization, a growth of the division of labour, a better distribution of responsibilities, rules permitting the monitoring and assessment of the partners' services, etc. This creation of organization is fundamental from the region's point of view, as it is through this that the production system is renewed and new activities are created.

Finally, whatever the structure of the network studied, it is interesting to note that the very notion of networks is justified only in the context of long-term relations. Indeed, the stability and continuity of transactions between players are prerequisites for the formation of a relational capital without which the players cannot adopt the riskier and more creative approaches that are typical of the innovation project. The notions of reciprocal trust and mutual acquaintance of the partners recur regularly in the different cases studied. Thus the network has a history and is part of a temporal dimension. This finding allows us to suggest the structuring function of the local labour market and of ascending chains of mobility in the formation of innovation networks. These networks take shape through players who know one another, as they have studied or worked together at some point in time. Career paths are thus of capital importance in the emergence of innovation networks. Thus it is the intermingled play of individuals, firms and institutions which allows the fabric of activities in a region to be renewed. Even if some firms go out of business, the individuals remain, pass on their know-how and knowledge, set up new firms and new opportunities appear. The region - and more particularly the milieu - thus emerges as the level at which the renewal of activities can be understood.

REFERENCES

Amendola, M. and J.-L. Gaffard, 1988, *La Dynamique Économique de l'Innovation*, Economica, Paris.

Aydalot, P. (ed.), 1986, *Milieux Innovateurs en Europe*, GREMI, Paris.

Camagni, R. (ed.), 1991, *Innovation Networks*, Belhaven Press, London.

Guilhon, B. and P. Gianfaldoni, 1990, "Chaînes de Compétences et Réseaux", in *Revue d'Economie Industrielle*, No. 51, pp. 97-112.

Lecoq, B., 1990, "Organisation Industrielle, Organisation Territoriale: Une Approche Intégrée Fondée sur le Concept de Réseau"", paper presented at the ASRDLF Symposium on Internationalization of the economy and development of territories, Saint-Etienne, September 1990.

Maillat, D., O. Crevoisier and B. Lecoq, 1990, Réseaux d'Innovation et Dynamique Territoriale: l'Arc Jurassien", in Dossiers de l'IRER, No. 28, Neuchâtel, GREMI III Symposium, Chaumont.

Maillat, D., O. Crevoisier and J.M. Vasserot, 1992, "Innovation et District Industriel", in Maillat, D. and J.C. Perrin, (eds.), Entreprises Innovatrices et Développement Territorial, EDES, Neuchatêl.

Maillat, D., M. Quévit and L. Senn, (eds.), 1993, Réseaux d'Innovation et Milieux Innovateurs: Un Pari pour le Développement Regional, EDES, Neuchâtel.

Maillat, D. and J.-C. Perrin, (eds.), Entreprises Innovatrices et Développement Territorial, EDES, Neuchâtel.

Perrin, J.-C., 1990, "Réseaux d'Innovation: Contribution à une Typologie", paper presented at the CREDIT-HEC Symposium Networks of Innovators, Montreal, May 1990.

Planque, B., 1990, "Les Réseaux d'Innovation. Réseaux Contractuels et Réseaux Conventionnels", a paper presented at the ASRDLF Symposium on Internationalization of the economy and development of territories, Saint-Etienne, September 1990.

Piore, M. and C. Sabel, C., 1984, *The Second Industrial Divide*, Basic Book, New York.

Williamson, O.E., 1975, *Markets and Hierarchies: Analysis and Antitrust Implications*. A Study in the Economics of Internal Organization, New York: The Free Press, New York.

CHAPTER 4

Lean Timing: Interdependencies Between Logistics, Scope, Locus and Development

Bengt-Arne Vedin

Royal Institute of Technology, Stockholm

4.1 INTRODUCTION

What does the title of this chapter signify? Logistics and locus have always been intertwined, although vertical disintegration is now placing more emphasis on suppliers. Research, development, and design have generally been regarded as separate activities, only distantly linked to the areas where production, the storage or transportation of finished goods, components, and raw materials take place. Scope, finally, represents the broadening of perspective, not just in terms of technology or markets, but in multifold ways.

This chapter intends to link these factors together, and to do so by focussing upon three popular catchphrases, which are used to describe the current industrial world, or its patterns of competition:

* competition in time
* economies of scope
* lean production

These three catchwords - that they are, rather than concepts - display a certain convergence in meaning. Competition for time is most clearly expressed in just-in-time delivery systems. The ultimate just-in-time system also encompasses the development process; it may also be described as "concurrent engineering", or more colorfully as the "rugby method", replacing "relay race" methods of work, or the matrix organization.

To achieve the necessary flexibility, less is produced in-house, more is out-sourced. And to reap the benefits from a new product before yet another, even newer product takes it place, a global approach to marketing is applied.

A broader scope implies extending to larger areas, transcending existing barriers and borders. Out-sourcing rather than vertical integration might be placed under this heading as well. And, of course, globalization is akin to transcending national borders.

Furthermore, what used to constitute well defined disciplines and fields of science has become blurred or interbred. Optronics, mechatronics, biotronics, telematics - there is a host of new words to describe how technologies are being combined in new, creative ways. Much the same is happing in science, where, in addition, new methods or phenomena, like chaos, are being applied over a broad spectrum of incongruous disciplines (in the case of chaos: meteorology, economics, fluidics, chemistry...).

The same is happening to markets. It is probably most obvious in the field of information technology, where a typewriter, a word processor, a computer terminal, and a personal computer share much the same functionality. Some companies have relied upon their basic competences to enter into a diverse set of new markets, rather than relying on old pursuits. For example, some Japanese camera manufacturers have made use of their expertise in optics, electronics, and precision mechanics to go into photocopiers, computer printers, and - in at least one case - production machinery for integrated circuitry.

4.2 INDUCING CREATIVITY - AND CAPITALISING ON IT

To get access to competences and knowledge across borders, alliances are created - even with competitors; institutes and universities get funding for projects, and collaborative research schemes are created. Suppliers, alliance partners, etc., are linked into networks, which may also encompass customers.

Customers are interested in function, but not primarily in how that function is achieved. The border between hardware and service also crumbles. The exact tailoring of the service/product package depends upon the customer, and he or she is colored by values and attitudes, which to some extent are culturally dependent. Therefore, globalisation is not the same as uniformity - to the contrary.

Lean production is a phenomenon that is described in a recent, influential book, The Machine That Changed the World[1]. It goes beyond the concept of just-in-time and describes a manufacturing system that includes a host of suppliers who produce self-contained entities, which are developed in close collaboration between the assemblying customer and the supplier. The relationship is thus a lasting one. The same holds for the output: the end customer, including the consumer, is brought into the act of developing and designing.

This is not a painless process. Through computerized tools, Boeing allowed its customers to partake in designing its new Model 777 aircraft. Boeing engineers were quite disturbed at first about having amateurs from airlines horning in.

Lean production is palpable in the sense that it uses so little space. Employees are physically close to each other, easing communication, and employee relationships are special to this production system. Drastically reduced setup times - sometimes reduced to zero - are at the core, however.

The above-mentioned "bible" of lean production takes as its prototype the automotive industry. This industry, the authors claim, is the very pacesetter of lean production, because of its economic weight, and due to the symbolic quality of its product, the car.

As a contrast to this central tenet, we would like to introduce another type of production. At the extreme, it is called the "hollow corporation", or, if we wish to attract more attention to the idea, the industry with "no production".

This is the ultimate network organization. It is represented by the computer industry, to be sure, and in that sense it might be seen as the wave of the future. But the phenomenon is by no means restricted to this particular industry. The same type of production is relied upon in the sports shoes business; in qualified machine tools; for specialised semiconductor components; and even when it comes to fashion shirts.

So if Toyota is the paragon of lean production, application specific integrated circuits, The Limited's fashion goods, Japax' highly specialised and complex machining equipment, and Nike's running shoes are all examples of organizations with little or no production performed within the company, and much networking outside it instead. Amstrad, a consumer electronics manufacturer, grew rich that way. They, like the others, concentrate on designing and marketing, but all manufacturing is

[1] Womack, James P. et al: The Machine That Changed the World. Rawson Associates, 1990, NY, NY.

subcontracted, sometimes even all assembly, although sometimes that function is retained. Much the same applies to Compaq, the PC maker.

Flexibility and scope - they are the foundations of the title of this chapter. Creativity set in motion would seem to be one concept that describes a large part of the various aspects of the three different catchphrases evoked at the start of this chapter. Competition in time to bring new creative products to bear. Broadening the scope is almost repeating Koestler's view of how the creative process works: through (in his words) bisociation, the meeting between normally disjointed developments or ideas. And the flexibility of lean production or the hollow corporation also makes it easier to adopt new, creative ideas.

4.3 THREE CASE STUDIES

(a) Turgid is global in scope though still somewhat uncomfortably heavy in their country of origin. Turgid-X is the world leader in its speciality, and this group within the huge combine is much more truly multi-national than its parent corporation. It is now active in all corners of the Triad (i.e., the U.S., Japan, and Western Europe), and the Turgid-X centers of excellence are not incorporated in the Turgid centre of corporate research.

The market for Turgid-X products is highly qualified. Customers look for qualities and features, not for price. Service is an important quality, and costs associated with the service organization translate into a rationale for having a full hand of products that can underwrite a good enough network of service centres.

Traditionally, this has been a field where contrasting competences were gathered together. With the application of computers and their associated software, the spectrum of varying professional backgrounds has necessarily become even more colorful.

Production is customized, which makes for small batches of components, instruments, systems modules etc., that can be linked to make custom design self-contained systems, in which some parts are also tailored and the software dedicated.

The market is entirely dependent upon certain government, cultural, and institutional restrictions. This means that selling points, arguments, and cost-benefit calculations are completely different between Europe, the U.S., and Japan.

In one of the four countries in which Turgid-X research and development is carried out, the company was acquired quite recently. One of the most prominent assets acquired was the entrepreneur running the company. In order to keep this asset within the organization, the corporate management went a long way to meet his requests.

This has created some tension between the different units ("favoritism"), eased by the fact that the entrepreneur's abilities are outstanding enough to meet general acclaim. He himself is aware of the potential tensions and is doing a lot to avoid monopolising research or the most interesting development projects.

Another reason for acquiring this relative newcomer was the unit's proximity to certain knowledge centers as well as to the development of important component manufacturers, producing much of the new technology - and the value added - essential to new features and qualities.

(b) Prose Inc. is also serving an institutionalised, special market. There are no governmental or other regulations here, but history, geography, and demographics make for systems that vary within the Triad. The qualified components that constitute the Prose Inc. business are made in small batches.

A large share of those components are sold internally, to the mother company or other parts of the Prose Group. Such customers are no less demanding than external customers.

Again, the combination of three Prose companies, active in the same field and now under the same umbrella, is a new phenomenon, brought about by a series of mergers, acquisitions, and buyouts, including a restructuring of the whole industry. Contrary to what was expected at the outset, no economies of production scale have been registered, nor any rationales for consolidating the production of specific component types to one or the other factory.

A problem, as well as a long-term boon, is the need for standardization and harmonization between previously diverging product designs. There is an opportunity to take the best aspects of each design although the challenge is to harmonize these features without generation shifts or other confusion being created in the market-place.

(c) While Turgid meets a market that can be said to feature an inherent multi-disciplinarity, Elne Inc. has always thrived on spanning the border betweeen two disciplines not normally combined. In recent years, the company has gone further in this direction, establishing itself through substantial acquisitions in a third field of competence, closely related to previous activities, in market terms .

This development has, furthermore, made Elne a truly global company instead of a national one selling abroad. Some of its best products were developed through international collaboration, but now such development efforts are taking place within the company in all three regions: Japan, the U.S., and Europe.

Elne's odd combination of specialties should benefit from the various strengths of the corporate and academic communities in the three parts of the Triad. Especially in Japan, where the problems of transcending disciplinary borderlines are severe.

The markets for Elne's products are quite varied. The sub-markets are distinctly different, and they are different in various European countries as well as in different parts of North America.

Contacts with academia and research institutes are important in this business. That is where several new ideas emerge. Parts of the market are found here, and academia have a profound influence on the reaction of other, perhaps more lucrative markets.

These three companies, Turgid, Prose, and Elne, all claim to struggle in their efforts to gain profits from being global. They have no clearcut answers so far, and on rainy days, they feel that the idea of leaving the national or regional base may have been a questionable one at best. But they admit to some tentative gains that they feel will point them in directions that they want to pursue.

(i) Gaining access to "multi-domestic" networks, i.e., plugging into universities, research institutes, consortia, etc, in Europe, the U.S., and Japan. This can only be done with local R&D activities offering active participation.

(ii) Sharing market and customer reactions. Even if markets are distinctly different because of regulations, product liability rules, etc, there are ways to abstract, interprete, generalise, and then adapt experiences from one market to another. The adaptation requires development, if not research. Customers can be attracted by the insights and track records obtained under other circumstances.

(iii) Getting different viewpoints. To see a market, or a development process, functioning differently from the way in which you are accustomed to, can be a painful or a liberating eye-opener (or both).

(iv) Accessing different strengths in different disciplines. Because of different histories, environments, contact networks, entrepreneurs, etc., the strengths might be complementary as well as overlapping. Even some overlap, some internal competition is seen as useful.

(v) The patchwork markets of Elne tell an interesting story of advancing stepwise, using various experiences in different market segments, with different products, etc. It is even possible to recycle failures, as it were; to advance in another market, gain new experience, then go into yet a third market, and then perhaps close the circle by returning to the first market.

(vi) The network of suppliers, providing in kind what scientists and universities provide in knowledge: new ideas as to components, software, etc. The obvious case in point is the establishment of units close to Silicon Valley to gain access to new key components early on. The multi-disciplinarity of Turgid-X or Elne seems to call for establishments in several such vicinities; and this idea is born out in reality.

(vii) Access to unique individuals. Here we are confronted with the case of Turgid-X, where it was not just a question of incorporating an acquired company within the larger structure, but a specific entrepreneur.

What, then, are some of the organizational ideas pursued by these companies? Here are some of their own answers.

Some sharing of responsibilities, i.e., some duplication of work but also some conscious complementarity. For more basic developments, the creation of centres of excellence. Sometimes this term applies to laboratories, sometimes to the production of key components or materials.

Concern for not cutting individual laboratories up into operations which are too small or fragmented. To be able to combine technologies, or to operate in very different markets, the need for multidisciplinarity must not be forgotten.

One way of doing this is to allow for "members" of one centre of excellence to be resident in some other place. Computer conferencing, electronic mail, some video conferencing, and, of course, fax and telephone, are rapidly growing in importance to enable "members" to participate in such work regularly.

Another way is to have some projects that are truly trans-national. This might be costly but it is beneficial in the sense that adaptation to local market requirements may be achieved much faster, thus gaining time on competitors. The advantage of having "many diverse viewpoints" can also be advantageous. The communication needs are met much the same way as when "outplacing" from a centre of excellence.

When alternative routes to success are feasible, allocate them to different locations. This may be particularly feasible when developing software.

When projects are entirely managed for time, i.e., when project development times must be cut short at any cost: such demands are tackled through, for example, alternate routes, duplication of efforts, and communication by electronic means, often in real time.

4.4 FROM R&D TO RDDPM

The market leaders turn out to be leaders in well focused niche markets. There they catch, on an average, a 7 per cent markup on the average price. Why? Because they feature something like five or more unique qualities. Why is the competition unable to catch up? Because when that happens, that market leader has moved on to some *new* unique features.

Thus the idea of getting the cost down along a learning curve does not work, not in its simple interpretation at any rate. The leaders would rather invent new learning curves, making the old experience at least partially obsolete. The unique features are often not just innovation but means of selling, of delivery, of servicing. The most potent current factor of competing is delivery at the appropriate time.

Here we have a direct link between all four factors in the title of this chapter. Instead of *lean timing*, we might have said appropriate timing, but it is those speedy reactions that create challenges. The customer might have been extracted as the real factor behind any reshuffling of the logistics system, of locus, development, and scope.

The last letter of the initials "RDDPM" stands for "marketing". The point is that the customer might, as we have already noted, be brought into the process even at the research or at least at the development stage. This may happen in procurement situations, or when airlines get involved in developing aeroplanes. Sometimes it happens indirectly, such as when a pertinent factor, for example price, is established first, as a goal, because that is what the market, i.e., the customer, is ready to pay. Sometimes time is also such a goal, again derived from a certain market; i.e., customer generated restrictions or indications.

Rather than performing development (or even research) in-house, the creed is to procure as much as possible from the outside. The most obvious effect is the out-sourcing of components, and then preferably of complete subsystems, with the concomitant out-sourcing of the development of such sub-systems.

Suppliers thus given more weight in the development process must be involved in this process from the outset. This implies daily participation in the development team. To secure close contact between these temporarily "outplaced" participants and their home bases which eventually will take over the responsibility for putting the results of the project into effect, a shadow group may be established that follows, repeats, and implements what is being developed for the whole system.

If the supplier is chosen for his location and for his ability, these qualities must extend also to his contributions to the development (the first "D") effort. Most often, the development team will be temporary and may be described in network terms.

When we contemplate the research ("R") stage, out-sourcing takes on other meanings. Here, the source contributes new knowledge and ideas rather than full-fledged systems or products. This is what is attempted in collaborative research efforts. Sometimes these are established between two companies, sometimes under the aegis of a university, sometimes as a permanent or a temporary "institute", sometimes as a program, catalysed or maybe even sponsored by an institution such as the European Community.

Here again, it is important to transfer new knowledge swiftly, from what is seen as the pre-competitive stage to that of competition and hopefully competitive advantage. Knowledge may be transferred in the form of documents and tools but also with people who move. Given the anchor that physical location may imply, an attractive environment and physical proximity may be of importance.

Given that the stage is pre-competitive, competitors might collaborate - and that collaboration may continue even further, as long as the rules and the dividing lines are crystal clear. Such arrangements are popular under the name of strategic alliances, and they have multiplied during the last few years. One reason is the skyrocketing of research and development costs, related to the need for a speedy payoff.

But it is not always a question of collaboration. It is no coincidence that Nippon Electric established their U.S. development laboratory next door to Bell Laboratories and the David Sarnoff Research Center.

Traditionally, it has been easy to distinguish between the development and the design stages (now for both the "D's"). The latter was defined as the routine transfer of new ideas, that had been demonstrated to work, into blueprints and instructions to make manufacturing feasible.

With the advent of computer aided design, CAD, this distinction becomes blurred, or even vanishes. The CAD systems allow for simulation. Various parts can be tested for manufacturability early on, even in complex systems. Together with this blurring of borderlines, organizational units become integrated, sometimes a painful process, and maybe at odds with the calculation methods of the economic department.

The idea of designing more easily manufactured products often leads to practical goals such as aiming for a limited number of parts. Production (our "P") people are brought into the product development process early on. The next idea, however, is the one we touched upon first - to make ideas that people want to buy, bringing even the customer into the process.

If we attempt to link this to the concept of the hollow corporation, we see the emergence of networks that may consist of formally independent units, but which work together in a long sequence of new products. The people involved tend to see the particular project and its goal, rather than their formal affiliation, as the most important parameter. Of course, when it comes to income generation, responsibilities must be well defined.

An inevitable conclusion is that the physical location of markets and customers will decide or restrict the location of different network nodes. Peoples' abilities to complement each other form other restrictions - or, conversely, forces of impetus. One might say that proximity is a variable that must be extended from the physical to the immaterial, and linguistic and cultural aspects must be brought into the picture too.

4.5 COMMUNICATION

Lightweight basic components are still transported around the world. There is no need to duplicate microprocessor or semiconductor memory plants except for expanding capacity. The same holds for sensors or actuators and, of course, pharmaceutical products.

Suppliers of more bulky items, however, are expected to deliver just-in-time. Therefore, communication is vital, both the physical delivery of goods and the involvement in the research, development, and design process via teletransmission.

Knowledge also travels the world in the form of

- people
- components
- documents-machinery
- collaborative efforts over a distance: mostly through telecommunications.

The documents are not just old fashioned books, reports, articles, and scientific papers. They might also take the form of computer conferences, video conferencing, hyper documents, etc. To describe these new ways of collaborating, of team work over great distances in space and in time, and teams of teams doing the same thing, the term groupware has been coined.

When interviewing 95 companies about their present and future needs for advanced information technology and communications services, we deliberately refrained from asking about research and development needs. The most urgent strategic concerns were, in fact, linked to the perceived fragmentation of markets. There is no longer just one mass market, but rather a host of niches. At the extreme, these fragmented markets can be seen as consisting of just one single individual. They are shifting and changing rather rapidly.

Increasingly, studies which aim to categorize such "fragments" are not just based upon written descriptions, and demographic and other statistical tables, but also on live pictures, for example, of the behavior of teenage groups, the interior decorations of the most popular new restaurants, excerpts from the most popular TV programs, etc. And the data may be used to simulate public opinion much the same way as CAD simulation works; bordering upon or developing into manipulation. Here locus comes into play in the sense that distribution systems and common local denominators such as the local economy and culture, etc., play an important role.

With such disaggregated markets, the extreme strategy would be to tailor products and services to the individual customer. That might involve the ultimate elimination of service deliveries: customers would be invited to design their products themselves.

But the development towards customer designed - as distinct from custom design! - products has already started, albeit in a small way. Using the tools of CAD and CAE (= computer aided everything), air transport companies are now designing the interiors of the planes they order all by themselves, but still, obviously, within the limitations given by the overall design of the aircraft. In Japan and France, garments are being designed by consumers themselves.

With products and services becoming more individualized for specific customers, and services and products being more intertwined, the customer is more dependent upon relations with the vendor, the vendor being selected mostly on the basis of previous experiences and reputation. Relations exist, at the end of the day, more between people than between anonymous entities.

These relationships increase the importance of communication in the development and design of the product, or the product-service package; a package possibly including education and training of the customers' personnel to make for a smooth delivery and rapid deployment of the new resources; after-installation support; and service and maintenance. The qualities of the product itself are still at the core, of course, but bad communication might destroy potential goodwill, while good communication provides information to improve the product. Those service elements involving physical delivery are obviously linked to what we have termed "locus".

One call is for new organizational structures, the other for various means of communicating - mostly through telecommunications.

The fragmentation of markets - will that entail the fragmentation of corporations as well? To some extent, as we claimed when introducing the concepts of networks and hollow corporations. But to communicate, some common frame of reference is needed; some corporate or other culture is required, both internally and in relations with customers. The concept of culture - visions, symbols, myths, precedents - is closely linked to human perception, interpretations of reality, and thus to the human side of communication.

While telecommunications can never be a substitute for those thoroughly human forms of communication, in many situations they are an adjunct to them:

*One company, in the process of merging and in the consequent sharpening and development of a new - necessarily clear, coherent, and consistent - corporate culture, claimed to have been in dire need of live video conferencing for this process, including interaction; they had sad experiences of misunderstandings and misinterpretations; units were too far apart geographically for arranging all of the visits and meetings deemed necessary

*One company, operating worldwide, depends upon video conferencing to stage global get-togethers ("beer busts"), just so the core of their culture is certain to be truly global, even if local variations do exist

*One company uses video communication for a rather peculiar task - job interviewing, especially when hiring people that are expected to interact broadly within the organisation

*One company, aware of the fact that it is living examples, actual occurrences and precedents that create culture, not orders or edicts, relies upon broadband video conferencing to communicate a selected number of precedents that are just in the process of happening, allowing for a discussion of them so that "headquarters" can see to it that the interpretation really comes out as intended globally; but staff all over the world also perceive this concern, which constitutes a "precedent" in itself; the immediacy, the "live thing" is essential

*Globalization; a need to gain time; a need for interaction - these are also parts of the pattern we see forming. Directness, eye-to-eye contact, interaction and immediacy, the need to save time come to the fore in many examples.

It is easy to point to specific industrial sectors where such advantages might be truly important. One is retailing, another is involved with certain investment decisions, construction and petroleum exploration projects.

Yet another type of communication might be called "selling and buying in spot markets". (In a way, it can be seen as the reverse of some clever retailing companies' video and computer conferencing based shedding of unattractive merchandise, "spot un-buying".) The window of opportunity only exists for some days, sometimes for just a few hours. Several people in various parts of the world must be called upon to discuss a particular offer. And it is essential to be able to read the other negotiators' faces.

The development projects of a large defense supplier require frequent exchanges of judgements between researchers and engineers in the same development project; also between the company's managers and their Defense Ministry clients; and again internal conferencing for decision making. When production is due, communication needs are heavy, too.

Their decision to install a dozen video conference rooms was, in retrospect, a very wise one. An evaluation demonstrated that the investment paid for itself in a couple months, even at a cost of between a quarter and half a million dollars per room.

Another defense contractor is responsible for custom designed (in this case - even customer designed) integrated circuits. They have regular links to several specialized producers of chips, and they have enough internal knowledge to design according to demands and specifications. Development and design include the generation of computer data for the production of diffusion masks. These data can be sent over telecommunications channels (encoded) or as floppy discs or magnetic tape to the producer of the masks. This may well be a separate entity from the establishment performing diffusion, and this again may be separate from cutting, encapsulation, and measurement, operations that may be performed offshore.

Since the chips are non-standard, they may be on the verge of that which is possible, and data obtained, sent back by satellite, are therefore inputs to the process and design parameters, calling for a new design and possibly also for adjustments in the process at the diffusion facility.

4.6 GROUPWARE

There exists a host of technological means for helping people work together. Of course, they come to nought if there is no common denominator, need, will, or cultural basis.

The point is that voice mail, telephone answering machines of increasing sophistication, cellular radios and mobile telephones, telefacsimiles, easy-to-use personal computers and laptops, video and computer conferencing, and other forms of groupware have all led to an increase in human freedom, freedom in

- space
- time
- format
- expression.

This has also meant less inhibition for the aspiring entrepreneur, who now may operate without any office at all. As we have seen, research and development teams profit from these different expressions of freedom.

There exist something like thirty different forms of groupware. Computer conferencing is the most common example. It features the possibility of asking for new entrants into a discussion or a project on a bulletin board, so projects and contributors that you were unware of may participate and help out. They may come from all parts of the world, since the communication need not be synchronous nor in the same place.

Video conferencing typically allows for different places but must occur at the same time; the same holds for ordinary telephone conferencing. Computer conferencing may take place in this way too.

Various tools for improving such conferencing have been developed, by getting to an agenda in a structured way, or by inducing brainstorming type creativity sessions. This may be done face to face, that is same place, same time, or, possibly, same time, different places.

Even the fourth mode, same place, different times may be useful, however odd it may seem. There are companies that feature a working environment that is designed so that different shifts or different teams may profit from the constant additions to that environment, in the form of additions to graphs, pictures, and items that are provided.

Let us describe these options in another way - through a scenario. The setting is a research group, delivering its report on a particular project that they have studied from outside - some time in the future.

4.6.1 A Conferencing Scenario

We have been tracking the Telia Excotel project from start to finish. It has been one of networking, collaboration, risk taking, and uncertainty reduction. Given the public and safety aspects of the undertaking, emphasis has also been on careful recording of each and every step of the project.

Our report is divided into several hyperchapters. What follows is the chapter dealing with the communications tools employed for the project. Other chapters are dedicated to describing the actual institutions involved; the background to and goals for the project; the theoretical framework for the project itself; for our research on the project; and for our theoretical description (´from uncertainty to risk to payback chart, and onwards´) of it. Hot links in hypertext are provided in the shadow manuscript, so the reader can freely jump between this chapter and the others, to see what is described in one framework against a different one as well.

Telia Service has been responsible for, and financing, Excotel. The project has involved a great many companies, institutions, and individuals, most closely Telia R&D, Ellemtel, and Erisoft. The basic link has been with the collaborative effort of the MIT, New Millenium Telecommunications, most often referred to under its nickname Abyssoft. Some of the other involved parties will be introduced when the tools for interaction are described and commented upon.

Even though it is off limits to call the Telia Decision Room the War Room, this is the obvious connotation. It is used for Telia Board and Executive Committee meetings, and has the standard set of simulation programs, decision support software, telewriting, telewalls, all sorts of conferencing facilities, etc. Since the essential task of Excotel has been geared towards the development of telecommunications network tools requiring the coordination of software, hardware, and the concomitant social processes, the War Room equipment was allowed to be amended early on to make for a description of the salient project features. These additions were kept there for the duration of the project, so that Telia executives could review progress regularly, which they had asked for, recognizing the importance of the project.

The various comments and questions of different Board and Executive Committee members were thus recorded and analyzed by standard Metatext software. Subsequent presentations were geared at responding to the concerns thus expressed, and to doing so in a presentation form, and in a language that suited the different members. Thus memos, notes, simulations, etc looked a little bit different to each member, due to individual tastes and wishes.

Independent judgement, also outside the function of the project controller, has long been regarded by the Telia top management as indispensable. Thus the system was linked to Expercom, allowing for random access to independent expertise, to get to know whether the risks suggested to be taken were worth taking, and whether or not some important points might have been overlooked. This had to be done in session, and on the spur of the moment; Board members invariably put questions of a different nature than the project management did, using the same Expercom feature but other experts.

Staffing of the team was a regular procedure; that is, everyone was hand picked for her or his individual qualities, and also, with the "producer" at the helm, with an overall concern for everyone´s ability to work on this particular team, with precisely these individuals. Some people were even lured out of retirement. Two people at the SICS institute and one at the little Metamatic consultancy were also engaged.

Initially, they were all brought together for a breaking in teamware session. This lasted exactly 16 days. The idea was to work on a similar, but "synthetic", project, not smaller, not easier, but one where the computer made for a 40 times gain in time. You might call it simulation, but problem solving, interaction, or working are just as relevant.

The point was, as always with breaking-in teamware, to learn about what personal traits might cause friction, and to reduce such future friction on the basis of this knowledge. One person could not stand the stress, and another did not fit into the team structure - neither he nor the remaining team members felt the least taken aback by this. Better to discover it beforehand!

Given the immense importance of the project, a three shift operation was foreseen. As you will learn elsewhere, it might actually be described rather as a two-and-a-half-shift, with some people tying in with the Californians, the Indians, and the Japanese at the wee hours. Anyhow, this explains the meticulous care going into the creation of the Excotel Cooproom, for a cooperative environment, which was shared by all members of the team at all times. Much like in a Japanese factory, the walls were soon full of information; but here, there were no company slogans but rather soft plans, recent problems, and breakthroughs, and of course on crystal solid displays, not on wallpaper.

When the whole team present felt like it, they might go into a collaborative session, using their entire environment. Brainstorming, in particular, was felt to gear the group up to a commonality of understanding, which was useful when they somewhat later contributed individually to either a structured groupware session, perhaps through morphological analysis, or an unstructured session, perhaps through synectics or brainstorming. Input was then mostly through pencilling notebook computers or icons. A powerful set of object icons was developed for the internal language of the group. The common color selected was a light blue, and a very particular odour was picked as the "common ground signal", *A4001+* in the Bedino notation.

Much in the same way as questions and statements from the Board and the Executive Committee were dissected, organized, and used for controlling that no important aspects were left out, the various concerns and individual traits of the team members were catered for in the information, groupware, or workflow system. The various meanings of a conversation were analyzed immediately, and a great many questions were thus necessarily resolved that otherwise would not have been. To these people, there was no magic in it; they were so used to it that they could often anticipate the questions.

The teamware was also equipped with corporate memory functions. One part is called the project information repository, PIR, allowing for all interventions to be stored, automatically classified and thus retrievable in a selective way. Some were also automatically marked so as to appear compulsorily in certain contexts, e g, Board decisions, American regulatory requirements, etc.

Another part was the Autoquery function, asking for the reasoning behind each and every step and decision. Confessedly cumbersome at first, this also became second nature to most project members. They now had a keen feeling for when to answer "intuition" as the rationale behind a certain design choice, and, according to our interviews, a firmer intuitive idea of what that intuition really might be was also forming.

Concomitant to this memory function is the backward alert; that is, an automatic signal that a problem has already been attacked. Parts of the memory went far outside the company, particularly into the MIT project, Abyssoft (see also below (***hot link 4***)). Since the rationale behind signing up for Abyssoft was precisely the opportunity to get useful inputs to Excotel, it is no wonder that we find many overlapping problems; if Telia could solve such problems, it was part of the contract that solutions be fed to the MIT consortium. All other consortia members were to do the same.

Part of the team functioned as a shadow group to the MIT project. There were some extremely interesting and challenging goals at Abyssoft, and to implement the solutions in the practice of Telia swiftly promised to be extremely profitable. Thus, the shadow team partook intimately in the development in Cambridge, Massachusetts, to be able to transfer the new findings rapidly and efficiently; some shadow team members worked within the Excotel team in Stockholm.

Telia had its own electronic mail system around the world; in fact, they had several systems. The Abyssoft project had created another such powerful network. The most obvious function of these networks is still the classical operation of sending messages, now stored not just orally or as text but rather like with the fax - since what was once fax and once computer communications have by and large merged.

But a more interesting function is what has been termed software arbitrage. The electronic mail system allows for signalling that there might be some interesting data to be acquired, some important rights to be negotiated, and so on. Speed is crucial, but so is good judgement. Therefore, electronic mail initiates the organizing of a video conference with the usual compuphernalia; i.e., computers tied to each other for the smooth demonstration of the various aspects of the proposal, project, idea, etc. Videowall, that age-old video conference catchword, is still alive and well.

The particular utility of an electronic mail or a computer conference network has not changed in another respect either: it is possible to find partners and contributors with specific knowledge and experience without knowing that they, those particular individuals, exist. The network provides its own directory; you simply have to advertise your needs and interests, and you will almost certainly get some good contacts. In this case, we have enumerated some of the Excotel Swedish contacts, but should add, for Europe, France Telecom and Association Descartes; for the U.S., Institute for the Future, Pacific Telesis, Wharton, NEC R&D, Interclass, and Bellcore; for Asia, NEC, Osaka Labs, and Kyocera. And Rao Mediware in Delhi.

Thus these various institutions and companies have agreed to share information to a certain degree; that degree is determined contractually, but they all have profound experiences of each other and know how far trust will go. There are also very few competing interests within the group, which helps.

First, they all share all of their external, open information; i.e., what may be available from open sources, even though it may be hard to come by or even to know the existence of it. For the Japanese material that NEC and Kyocera reside on in abundance, the new translation programs have proven essential.

Second, they all share internal, proprietary reports pertaining to the area of collaboration. These are exchanged through a separate set of software, allowing for brief announcements, tailored information, and select searches when a problem crops up. Without this software, going back some generations to the information lens once developed at MIT, the amount of information would be insurmountable.

Third, what has been said already about problem solving and corporate memory also applies, but in a selective way. One reason for this is information overload, the other corporate strategy. Frustrating as it may seem, there are occasions when the system only tells, say, Telia that, say, Osaka Labs has been working on precisely that problem. Then there can be an exchange as to the appropriateness of discerning that information.

To conclude, the efficient and timely execution of the Excotel project would simply not have been feasible without Teamware. It was also vital that Teamware, in turn, has the computer and telecommunications network as an underlying assumption, even if telecommunications cannot always be relied upon as the exclusive resource.

Obviously, electronic mail and computer conferencing resemble what they looked like several years ago. What has changed is the value added, the various functions for searching, sifting, adding impulses to a brainstorm, for automating information handling and suggestions directly out of the network, because of its links to the project's repository of information and its corporate memory of concerns - as well as to more general information. Color, high resolution, speed, sound, multi-media also make for more colloquial messages than just signs on a screen.

At the same time, people's inhibitions in communicating via a computer have become even more reduced. The only problem is with oral communciation, partly because the computer still has problems with natural language and particularly speech; mostly because in a work environment, a lot of whispering might be disturbing, especially in a concentrated brainstorming event.

Sometimes computer conferencing may be performed in real time. This we have not mentioned explicitly, but quite often, that type of dicussion does not require face-to-face quality but rather simulation, brainstorming, lateral thinking, or cross-impact structured inputs. There is a whole range of software for this, and most, perhaps all, of what is available has been applied within this project.

One example is something functionally resembling the team breaking-in groupware for Excotel: the CultCom method, attempting to create a commonality of framework and language between people in different settings, national, institutional, professional. It was applied early on in Abyssoft.

It is virtually self-evident, after the fact, that different communications media will display different impacts and follow different behavioral codes in different cultures. So at first there were a great many misunderstandings, to use a modest word, between people with different backgrounds. This being the case in encounters face-to-face, it is hard to understand that people did not realize that this problem would be magnified in electronically mediated metings!

Sometimes, however, face-to-face meetings seem indispensable. This is when video conferencing applies; again, some useful communication has actually been recorded off line also. There were some great communicators in the project, including a few from Japan!

Finally, we return to the same place, different time application of the Excotel project room in the vicinity of Stockholm. Which may also be used for same place, same time efforts, as indicated. It would be pointless to ask how the project might have been carried out without these technical tools. It would not have been carried out; it would not have been organized this way; and it would not have been completed in such a short time span.

4.7 EXAMPLES

Let us round off our discussion with a few examples.

*The Swedish producer of custom made chips ABB Hafo has established a subsidiary in California. The major reason is that American customers feel much more at ease when they can contact, an "American", i.e., local, company by phone or through travel. It is also not by chance that the location is that hotbed of electronics development, California, securing close contact with sources of knowledge, components, equipment, competence, and manpower.

*The successful supermarket chain WalMart employs computers, video conferences, and satellites to secure the efficient handling of goods - selling out slow-moving merchandise quickly, getting all relevant managers on the line simultaneously, and training the personnel by means of satellite.

*When the Canadian telecommunications company Northern Telecom received their first large international order from Swedish Telecom, they located a newly established development unit in Silicon Valley to work with the Swedes. This was because of the location's attractiveness as a source of knowledge and competence when it came to factors influencing new generations of electronic subscriber exchanges. The Swedes engaged in the project had to live there for something like a year at a time. Then it turned out to be a problem to implement the new production technology in factories in

Sweden, where no electronics, only electromechanical competence was available. Electronics engineers had to be recruited or transferred from other parts of the company.

*When, in the early 70's, the Swedish Power Board procured a computer control system for the Swedish power grid from the U.S. company TRW, the matrix project organization of TRW turned out to be too weak to ascertain that software and hardware worked together smoothly enough. Internal technical auditors, representatives of the Swedish customer that stayed with TRW for long stretches of time, and some adjustments of the systems specifications saved the day. Today, coordination between Sweden and teams in several locations in the U.S. is deemed easier, due to the development of computer conferencing, fax etc.

*The Swedish telecommunications company Ericsson now coordinates some 30 development centers in 18 countries. These are regularly linked together into something that functions as one local area network, i.e., as if people were communicating internally in one single building. This means that different systems components - hardware as well as software - may be developed in different parts of the world and yet brought together, tried and tested together in this convivial system (Australia, Germany, the U.S., as well as Sweden, are represented). The reason for having those many centers around the world is, of course, to be where the customers are; this decentralised approach for R&D is fairly recent for Ericsson.

*The establishing of Hewlett-Packard's computer printer factories in Asia has meant that bringing to the market new models adjusted to kanji and various other systems of printing in the region now happens five times more rapidly, resulting in twice the growth rate compared to the U.S. Hewlett-Packard is also one of the investors in factories established by the semiconductor giant Texas Instruments (TI) in Asia. TI wants to be there to be close to customers who need custom designed chips. And some of these customers, such as Taiwanese computer producer Acer, have put up a large part of the investment for TI, as has Canon in Japan. Hewlett-Packard's alliance with TI is duplicated in several parts of the world, much like Japanese car manufacturers who bring their suppliers to the U.S. and to Europe.

*We have touched upon customers, such as United Airlines, tying into Boeing's computer aided design system for the 777 aircraft. This is actually a prime example of where simulation implies transcending the borders between development and design. While development and production engineers displayed mutual distrust initially, the borders broke down so much that some of them even started sharing offices. Speaking of locus! But the system stretches even further, across the borders of the company itself, to its suppliers as well. The computers' links span the Pacific to names like Mitsubishi, Fuji, and Kawasaki Heavy Industries. In this case, members of the teams formed are from several companies and they meet about every two weeks.

* Large companies offer fairly regular types of projects and project meetings. Through the use of specific groupware, IBM and Boeing have cut lead times for typical projects and costs (mostly staff time) for meetings by an impressive 90 per cent.

*The observation from the development of Apple's recent laptop computer PowerBook reads like a footnote. Sony in Japan was contracted for the job, and with the Gulf War, management reluctantly shifted to teleconferencing. This worked so well that it will probably be a mainstay. Again, two different companies, continents, and cultures were bridged, much because Apple needed the competence for miniaturising. The year-long project was just two weeks late.

4.8 CONCLUSION

Most customers raise particular requests, one of which is speedy or at least appropriate delivery. These special requests are translated into the need for broadening the scope, in several dimensions, while time schedules get compressed.

Time compression applies to delivery, to just-in-time systems, but also to development. Special requests are met through involving the customer in the development and design phase. The demand for quality and the need for speed result in the requirement that things get done right the first time - there may be no second chance. That is one reason why production people are brought in at the development and design stages, technology allowing this to happen - through computer tools.

Thus we have, as always, two logistics systems, inbound from suppliers, and outbound to the market. They are but forces in determining where activities should take place. There is also an inbound stream of knowledge and competence, of scope, sometimes embedded in concrete components or machinery. And there is an outbound delivery of service, information, maintenance, all those additional features that make a package of qualities of what was previously just a piece of hardware.

Perhaps our hollow corporation might be said to manage the several interlinked networks that produce these flows of goods and of ideas, knowledge, and competence. One prime knowledge factor is, then, perception of market needs.

Quite a few international companies have been astonished and sometimes disappointed when they could not reap the benefits of the economy of scale when active "globally". Now several of them have discovered the beauty of the economy of scope instead: how development, and logistics, have to be tuned together in various local markets - around the globe.

REFERENCES

14th ARRB Conference 1988

Abramson, J.B. et al., 1988, *The Electronic Commonwealth*, Basic Books, New York, N.Y.

Batten, D.F and R. Thord, 1989, *Transportation for the Future*, Springer-Verlag, Berlin.

Blackburn, J.D., (ed), 1991, Time-Based Competition. Business One Irwin, Homewood, Ill.

CALS Phase I.2, Conferences 20 April, 27 April, 2 May 1989. NIST 1989.

Chesebro, J.W. and D.G. Bonsall, 1989, *Computer-Mediated Communication*, The University of Alabama Press, Tuscaloosa, Ala.

Clark, K.B. and T. Fujimoto, 1991, *Product Development Performance*, Harvard Business School Press, Boston, MA.

COM(87) 351 Final Commission of the European Communities (Drive), Brussels, 1987.

Davis, S.M., 1987, *Future Perfect*, Addison-Wesley, New York, N .Y.

Defense Logistics Agency Information Resources Management (IRM), NIST, May 90.

Drexler, K.E., 1986, *Engines of Creation*, Anchor Books, New York, N.Y.

Dutton, W.H. et al., (eds.), 1987, *Wired Cities*, G K Hall, Boston, MA.

Eliasson, G. et al., 1990, *The Knowledge Based Information Economy*, IUI, Stockholm.

Engelberger, J.F., 1989, *Robotics in Service*, Kogan Page Ltd, London.

Finkelstein, J., 1989, *Windows on a New World*, Greenwood Press, Westport Ct.

Freeman, C., 1988, *The Factory of the Future*, SPRU, May.

Giaoutzi, M. and P. Nijkamp, 1989, New Information Technology and Spatial Transport Development, Free University of Amsterdam.

Goldratt, E.M, 1990, *The Haystack Syndrome*, North River Press, Croton-on-Hudson, N.Y.

Grimson, W.E. and R.S. Patil, 1987, *AI in the 1980s - and Beyond*, The MIT Press, Cambridge, MA.

Hayes, R.H., 1990, Introducing Programmable Automation Requires Rethinking the Whole Business, Mimeo, June 4.

Hirano, H. and J.T. Black, 1989, *JIT Factory Revolution*, Productivity Press, Cambridge, MA.

Information Technology and New Growth Opportunities, OECD, Paris., 1989.

Information Technology and the Conduct of Research: The User's View. Report on Information Technology and the Conduct of Research, Institute of Medicine, Washingon D.C., 1989.

Johansen, R., 1988, *Groupware*, The Free Press, New York, N.Y.

Johansen, R. et al., 1991, *Leading Business Teams*, Addison Wesley, USA.

Johnson, H.T. and R.S. Kaplan, 1987, *Relevance Lost - The Rise and Fall of Managerial Accounting*, Harvard Business School Press, Boston, MA.

Keen, P.G.W., 1988, *Competing in Time*, Ballinger, Cambridge, MA.

Kiesler, S. and L. Sproull, 1991, *Connections*, The MIT Press, Cambridge, MA.

Leebaert, D., 1991, *Technology 2001*, The MIT Press, Cambridge, MA.

Lyon, D., 1988, *The Information Society*, Polity Press, Cambridge, MA.

March, J.G. and L. Sproull, 1989, Technology, Management, and Competitive Advantage, Mimeo, January 15.

Orishimo, I. et al., (eds), 1988, *Information Technology: Social and Spatial Perspectives*, Springer-Verlag, Berlin.

Pennings, J.M. and A. Buitendam, *New Technology as Organisational Innovation*, Ballinger, Cambridge, MA.

Phillips, F.Y., 1991, Concurrent Life Cycle Management. IC^2 Institute, The University of Texas at Austin.

Proceedings of the 4th International Conference on Automated Materials Handling edited by R.H. Hollier 20-21 Sept 1989.

Proceedings, 5th International Conference on Automated Guided Vehicle Systems.

Rheingold, H., 1991, *Virtual Reality*, Summit Books, Simon & Schuster, New York, N.Y.

Rogers, D.S., 1990, An Investigation of Information Technology Adoption: The Impact of Logistics Structure and Strategy. Dissertation, Michigan State University.

Route Guidance and In-Car Communication Systems, OECD, Paris, 1988.

Sethi, V., 1988, The Development of Measures to Assess the Extent to which an Information Technology Application Provides Competitive Advantage. Dissertation, University of Pittsburgh .

Stalk, jr, G. and T.H. Hout, 1990, *Competing Against Time*, The Free Press, New York, N.Y.

Tanja, P.T., Telematica-Ontwikkelingen en Integrale Logistieke Beheersing, *Tijdschrift voor Vervoersweenschap* 1990:2.

Technology and Global Industry, 1987, National Academy Press, Washington D.C.

Telematics in Goods Transport, ECMT-OECD, Paris, 1989.

Thord, R., (ed.), The Future of Transportation and Communication, Swedish National Road Administration, Arlöv, 1991.

Resources for Tomorrow's Transport (symposium Sept 1988), ECMT-OECD, Paris, 1989.

Triantis, A.J., 1988, Contingent Claims Valuation of Flexible Production Systems. Dissertation, Stanford University.

Tse, E. and W.E. Cralley, Management of Risk and Uncertainty in Product Development Processes. June 1989, Institute for Defense Analysis, Alexandria, Va.

Ungerer, H., Telecommunications in Europe, Commission of the European Communities, Luxembourg, 1988.

Weill, P.D., 1989, The Relationship between Investment in Information Technology and Firm Performance in the Manufacturing Sector, Dissertation, New York University.

Wilson, D.D., 1988, Assessing I.T. Performance: What the Experts Say. MIT Program: Management in the 90S, 88-050, Cambridge, MA.

Womack, J.P. et al., 1990, *The Machine That Changed the World,* Rawson Associates, New York, N.Y.

CHAPTER 5

Trading Local Autonomy for Economic Efficiency or the Faustian Bargain of Choice at the Close of the Twentieth Century: The Case of Globalized Financial and Property Markets

Michael A. Goldberg
The University of British Columbia, Vancouver

5.1 INTRODUCTION

The growing trend toward integration and complexity in modern societies has led to unprecedented increases in global incomes, outputs, and interactions. While the benefits of this integration and complexity are generally lauded as "progress" (and therefore, by assumption, "good"), there are also significant dangers and costs lurking amid this complexity and interconnectedness.

The contributions in this volume and their associated research agenda have been designed to identify some of the pluses and minuses attendant to increasing the interconnectedness of modern societies. One of the most compelling difficulties posed by heightened articulation among nations and regions is the loss of flexibility and variety resulting from the specialization that has accompanied the development of global markets and integration. Specialization has led, in a largely predictable but unacknowledged manner, to the increasing vulnerability of regions and nations, and thus of the global system itself to major shocks and disruptions.

Thus, I want to begin by outlining an earlier argument (Goldberg, 1975) that addresses trade-offs in efficiency and flexibility which was expanded considerably subsequently (Goldberg, 1989). With this argument delineated and illustrated, the chapter proceeds to examine two markets where growing global integration has been generally applauded and held up as exemplary of the benefits to be derived from "going global". These examples provide the substantive base to which the earlier argument can be applied to illustrate the dangers of global efficiency and connectedness. The chapter closes with some suggestions as to how we can reintroduce variety, flexibility and buffering to the global system to provide greater safety and stability in the longer term.

5.2 THE ARGUMENT IN BROAD OUTLINE: THE INEFFICIENCY OF BEING EFFICIENT

5.2.1 The Key Ingredients

Several notions are central to the discussion which follows. The first of these focuses on the importance of looking at wholes instead of parts. This has been called "systems thinking", referring to the focus on whole systems rather than on component parts. By understanding systems we can start to see why there have been so many unexpected consequences attendant to component-oriented approaches.

Closely related to systems perceptions are questions of stability. Stability, in turn, has been linked with diversity (the number of different parts that comprise a system) and complexity (the number of alternative ways in which the parts are connected). Stability/diversity relationships are also a fundamental element in the discussion. The ideas come initially from the biological and physical sciences, but also crop up in political science, the management of investment portfolios, and in strategic planning in virtually all areas of human activity.

Uncertainty is the last major building block explored. The future, being unknown and unknowable in advance, is uncertain. As a result, decisions with outcomes in the future are made in an environment of uncertainty. We can never know beforehand, with certainty, the effects of our decisions. In spite of this uncertainty, we have continued to make large-scale, far-reaching and long-run decisions *as if* their outcomes could be either known in advance or assumed away (which is to say that we do know in advance there will be no important side-effects). This mistaken presumption is closely related to many of the unintended effects that have accompanied our past decisions, virtually all of which were made under the presumption of knowledge instead of ignorance. I propose that we acknowledge the irreducible uncertainty that surrounds us and seek actions that are less likely to return to haunt us later.

5.2.2 Problems with Past Solutions

Four elements typify recent decision-making relating to complex multifaceted problems:

1. Problems are isolated from the larger systems of which they are functioning parts.
2. The objective or problem is narrowly defined.
3. The chosen solution is usually the simplest and most easily implemented.
4. The decisions have increasingly produced a number of undesirable and unexpected consequences.

Transportation planning in the 1950s and early 1960s provides a prototypical example of the past approach to decision-making. The transportation system was initially isolated from the land use, social, economic, political and environmental systems that jointly comprise an urban area. Transportation problems were then usually defined in terms of one particular mode of travel (automobiles), one network (freeway systems), one link in the network (a congested bridge or section of road) and one variable (capacity). Questions of land development, air pollution and social hardships resulting from obtaining rights-of-way through low-income neighbourhoods were almost always ignored. So were questions concerning the quality, as opposed to the mere quantity, of transportation services.

It should be noted before proceeding that an equally narrow definition of excess demand on the specified link would have led to a much more environmentally friendly solution, namely the reduction of demand through raising prices, the encouragement of alternative modes of travel, and the integration of transport and land use planning. However, a public clamouring for more private travel at lower prices was not willing to entertain such an alternative definition, nor were the engineers and planners whose professional lives and training centred on supply expansion, not demand moderation. Times have changed drastically in the ensuing decades and the public and the engineering and planning professions are ready to address these issues and accept the fact that limits to growth may indeed exist and lie in the not too distant future.

It seemed completely rational in this framework to opt for the simplest and most direct solution: that is, to increase capacity by building more freeways, bridges or parking areas. This created a number of side-effects. Foremost among these was the traffic induced by land development in areas with newly acquired accessibility (Putman, 1974). The form and scale of our cities were radically altered. This was particularly true in North America, where cities, especially those in the sun-belt areas, were still evolving. In the face of freeway development, cities spread out from existing urban cores into low-density, largely single-family, residential suburbs.

One of the unanticipated consequences of such a dispersed urban form has been that we find ourselves now locked into an extremely inflexible mode of transportation -the freeway and the private automobile - which is highly susceptible to disruption from changes in demography, fuel availability and preferences in public lifestyle. Alternative modes of travel and patterns of land development have been largely preempted. In an attempt to counter the freeway boom of the post-war era, a transit boom was set in motion in the mid-1960s. Unfortunately, this transit thrust was characterized by many features of freeways and thus did not represent any basic departure in thinking, only in mode of transport.

The San Francisco Bay Area Rapid Transit System (BART) began the transit boom in 1962. Huge, new, rail-rapid-transit systems followed in more than a dozen other American cities. The BART system is a worthy example because of its failure to coordinate with other transit and transportation modes, and its lack of integration into the land-use and urban-development planning process. Although started in 1962, the seventy-five miles of track which make up BART did not open until 1975: it was nearly six years late and almost $1 billion over budget. The BART system is a classic example of how high technology and large-scale efforts can go awry. Stories about BART's technical problems were frequently reported by the San Francisco newspapers during BART's early years. BART was expensive ($1.7 billion), inflexible (fixed rails and fixed modules) and initially a financial disaster (Webber, 1976).

The urban transportation problem requires diversity in definition and cannot be solved simply by switching from one single-purpose mode of transportation to another to solve one narrowly defined problem or another. The "*engineering*" mind that gave us freeways in the 1950s and 1960s was also the progenitor of BART (and of numerous other rail rapid transit systems built in the 1970s and 1980s). BART is functionally and conceptually a freeway system on rails, monofunctional (single purpose) in its design, with the same kind of stability properties and shortcomings.

Finely tuned temperate-zone agricultural systems also illustrate the pitfalls of narrowly defined problems and solutions. Borgstrom and others have pointed out (Borgstrom, 1973) that agricultural miracles have not been achieved without some sacrifice in the quality of crops. The traditional approach to agricultural problems is analogous to that taken by freeway planners. The problem has been narrowed by focusing on the yield per acre. High-yield systems are as delicately balanced as upside-down pyramids. Climate, for instance, is subject to change without notice. Methods of maximizing agricultural yield which depend for their success upon the stability of climate are vulnerable to expected climatic changes. A massive crop failure that occurred in the Ukraine in 1972 points out just how disruptive, and likely inevitable, these disturbances can be (National Academy of Sciences, 1974). Growing concern over global warming raises similar issues on a much larger scale with much larger consequences looming (Ehrlich and Ehrlich, 1991, and Bormann and Kellert, 1991).

The trend toward narrow, single-purpose thinking is partially traceable to the growing emphasis on analysis as an end in itself. Precision and rigor now dwarf holism and generality as accepted methods of good research. Because only small and well-bounded problems can be rigorously analyzed, the tendency of science and analysis to reduce problems to their component parts has greatly contributed to and fostered narrow decision-making. The late author and critic of contemporary science, Arthur Koestler (1978), pointed to the orthodoxy surrounding research in the life sciences, in particular the narrowness of the assumptions and methods employed in such research. Similarly, philosopher of science Thomas Kuhn (1970) suggested that this is a general phenomenon in science, noting that "Perhaps the most striking feature of the normal research problems...is how little they aim to produce major novelties, conceptual or phenomenal."

In view of past excesses of analytic rigor and narrowness, I will argue for a shift toward greater holism and systemic thinking. There is, I believe, a good case to be made for being roughly right about an issue or problem of some substance and importance, rather than being precisely right about trivia. I do not wish, however, to ignore the importance of analysis and of the need for more powerful analytical tools: such tools are essential for constructing any meaningful systemic and synthetic view. However, balance and perspective must replace narrow orthodoxy. The choice is not *between* analysis or synthesis but rather how to draw effectively on both methods of inquiry.

5.2.3 Stability and Diversity In the Spotlight

In simplest terms, I argue here with Aesop against putting all of one's eggs in the same, or even similar baskets. We will draw on concepts of stability and diversity alluded to earlier.

Beginning with stability, there are two aspects worth noting. First, there is the speed with which a system returns to its usual state after being disturbed. One pendulum, given a push, for example, might return more rapidly than others to its resting point or to its normal rhythm - what might be known as an "equilibrium point" in the first case, or a "stable cycle" in the second. A second aspect of stability relates to the size of the disturbance that a system can withstand and still return to its recognizable form. We can imagine, for instance, that some pendulums could withstand quite violent pushes while others would fall off their stands or break apart and cease to be pendulums in the face of such pushes. The first case we can refer to as the speed of recovery of a system, the second case as a system's flexibility (analogous to ideas of Bateson, 1972, and Holling, 1984). Other things being equal, a system with a faster speed of recovery is more stable than one with a slower speed of recovery. Similarly, the system which can absorb the larger shock, again other things being equal, would be considered more stable.

A trade-off exists between these two types of stability. When comparing the stability of systems in which flexibility and speed of recovery can both vary we need to devise a method of determining the *relative* importance of each of these elements. A mechanism must be found that can combine them for comparison. Modern society has implicitly tended to give most weight to speed of recovery at the expense of flexibility. I argue here for a more balanced weighting scheme. However, whatever scheme we choose will ultimately depend upon our own values and biases, and therefore on *judgement*.

A prairie grassland nicely illustrates the idea of stability. Grasslands are complex and evolve from simpler systems like marshes or semi-arid deserts. As the grassland develops, more species make their home in the system, and the biomass increases. The resulting grassland contains a variety of plant and animal species. It is extremely flexible. It can withstand disasters such as prairie fires, floods, droughts or insect infestations, and yet return each time to its recognizable grassland state. This return is slow, but insistent.

Compare this slow process with modern agricultural methods for the cultivation of grains. These monocultures (one plant species per field) are in fact arrested grasslands. Their development is stopped at the point of maximum growth. This is achieved by reducing competition through widespread use of herbicides, pesticides, and chemical fertilizers. The artificial elimination of competition promotes the survival of the monoculture, and chemical fertilizers increase the growing capacity of the soils. The capacity limits and competition which characterize naturally occurring grasslands are thus superseded in the search for higher yields per acre. In the past, these yields have been obtained by converting fossil-fuel energy (oil) into wheat and other foods, through mechanical planting, harvesting and spraying (Odum, 1971).

In this vein, biologists Paul and Anne Ehrlich and John Holdren note: "Advances in agricultural technology in the last hundred years have not resolved the ecological dilemma of agriculture; they have aggravated it. The dilemma can be summarized in this way: agriculture tries to manage ecosystems in such a way as to maximize productivity, nature "manages" ecosystems in such a way as to maximize stability, and the two goals are incompatible. In short, productivity is achieved at the expense of stability." (Ehrlich, Ehrlich and Holdren, 1973, pp. 163-164).

5.2.4 Learning from the Examples: Toward Some Generalizations

We can draw a number of conclusions from these examples. First, the decision-making process led to large-scale and rapidly implemented decisions. The systems upon which such decisions were imposed reacted with unexpected consequences. These unexpected and often undesirable consequences resulted largely from the simplification of the system that is caused by large, fast, simple and direct decision-making processes.

Second, the outcome of such decisions is not always predictable: we do not have the necessary information about the system and its varied interactions. Such information, I suspect, will never be available in many cases, and there will continue to be surprises if we insist on making decisions in this manner. We can never anticipate all of the possible outcomes of our interventions in the complex human and environmental systems in which we live. We are often unable even to piece together what has happened after the fact. Uncertainty is a central part of human existence. We can never eliminate it. At best we can make flexible decisions that give scope for absorbing surprises rather than being devastated by them.

5.2.5 Conclusion: The Faustian Bargain - Trading Output for Flexibility

Looking back over the argument to this point, it appears that we have made a kind of Faustian bargain: we have traded off flexibility for greater output and for higher material standards of living. We have replaced persistent, though lower yielding naturally evolved systems with more fragile technological systems which require narrowly constrained and large human intervention in order to consistently achieve their impressive yields. At the extreme of this trade-off, we are perched on the head of a pin, with instant returns to an equilibrium, but with an infinitesimally small region of stability from which we can return.

Any disturbance could destroy the frail balance which exists between precisely determined and artificially induced mixes of inputs. In the extreme example above, the various input combinations all lie within the range of tolerance of the system, that is, they all lie within the *boundaries of stability*. These boundaries of stability stake out the conditions under which the system remains recognizable as being pretty much what it was originally. If the system is disturbed, but the disturbance remains within the system's boundaries of stability, then the system, at each level of speed of recovery, will return to its equilibrium mix near the centre of the bounded region of stability. If the system is subjected to a shock which is sufficiently large to force the system outside its boundaries (perhaps an enormous flood, or a massive overdose or underdose of fertilizer) then the system as we knew it will not be restored: it will not naturally return to its former characteristics.

As we move from the grassland to the wheat field, we trade off speed of recovery for vulnerability, so that the size of the shock from which the system can return becomes vanishingly small. Ironically, by focusing on the high yield equilibrium point and devising policies to get to it quickly, we have endangered the system by causing the boundaries to close in around us. When at the equilibrium, we find ourselves, paradoxically, closer to destroying the system than we were before we imposed our equilibrium-seeking decision on the system.

For convenience, the speed-of-return equilibrium-point approach will be called the equilibrium-centred approach; and the alternative that is concerned with distance from the boundary will be called the boundary-oriented approach. Large-scale optimizing frameworks are of the equilibrium-centred variety, whereas biological control and market-oriented economic policies and other more decentralized management strategies are consistent with a boundary-oriented view. Here I am thinking about the classic competitive market typified by a large number of small buyers and sellers, operating with heterogeneous expectations formed from essentially certain information, dealing in a homogeneous product/service so that no buyer or seller can affect market prices or quantities. Clearly, this simplified market setting differs from actual markets, particularly those in financial services and products where there are large buyers and sellers and where uncertainty and herding instincts often lead market participants to act in concert. This action results in speculative bubbles bursting and instability rather than the stability that competitive markets are expected to achieve through large numbers of small and instantaneous competitive adjustments.

Accordingly, it would seem that even where most of the preconditions for perfectly competitive markets are met (large numbers of buyers and sellers, etc.), as long as the possibility exists for market participants to form homogeneous expectations about market impacting events, then these competitive markets can be highly destabilizing, particularly when they are closely integrated as is the case with financial markets today. Creating more heterogeneous expectations through more diversity in information sources and decision rules can help to avoid these undesirable outcomes of competitive markets. Another approach would be to develop some supranational bodies to overcome the weaknesses of fragmentation and homogenization of decision rules much as has been done by the Bank for International Settlements (BIS). It has created international capital adequacy standards for banks to avoid large-scale systemic instabilities brought on by the nearly identical gearing and profit-maximizing behaviours of international banks. These issues are essentially variants of the classic fallacy of composition confronting markets in a dynamic setting which economists (and philosophers) have tackled for decades. Their generality and ties to significant bodies of research in other fields makes them particularly important and susceptible to treatment when properly defined (Goldberg, 1992).

The exigencies surrounding national financial markets and their global integration provide a strong case for the sort of reorientation argued for here. Specifically, I would hope that market decision-makers can increasingly monitor their progress not in relation to an equilibrium point (eg., expected yields) or even with respect to a rigidly defined decision rule (eg., program trading), but rather in relation to the boundaries that define the system of concern (eg., the likelihood that their decisions can lead to catastrophic losses). The important thing is to remain within the boundaries, and to maximize our distance from them. An equilibrium-centred focus has the opposite effect because the tactics that get us to the equilibrium point quickly are negatively related to the boundaries of the system. Tactics that simplify the system invariably sap it of flexibility and narrow its boundaries. Tactics that acknowledge boundaries cannot be of this simplifying kind.

A boundary-oriented view of complex systems will likely lead us away from simple monofunctional decision rules. It will promote the making of smaller and less quickly implemented decisions. By doing things slower and at a smaller scale, we enable diversity/complexity relationships to be maintained, and we also allow for uncertainty. We have much more latitude to adjust and readjust to small and slow perturbations than we do to larger and more rapidly imposed interventions. Finally, from a boundary-oriented approach, we tend to take the lowest-level effective response available. Higher-level responses are more disturbing to the system and can lead to simplification and instability.

5.3 GLOBAL MARKETS AND THE REALITIES OF THE 1990s: OVERVIEW AND EXAMPLES OF SOURCES OF CONCERN

5.3.1 Global Integration of Markets: Financial Markets as a Prototype

Recent changes in financial markets provide examples to illustrate the points being made above. The growing (and now almost total) integration of the world's previously separated national financial markets has led to much larger (and cheaper) global financial transactions. It has also increased the likelihood that financial stresses and disorders in one system will now be transmitted to many others. The October 1987 global stock market crash is a compelling case in point where a market decline in New York on Monday, October 19 spread across the globe and amplified itself with each daily pass around the world. In a deregulated and highly integrated global financial marketplace, previous buffers that could have damped these signals as they moved from one market to the next, were not present and amplification, instead of dampening occurred.

Given the extraordinary connections among the world's financial markets and the exceptional volumes of data, information, and travel among these markets, future global disturbances along the lines of the 1987 stock crash are likely to occur again. Thus, a review of the evolution of financial markets from national to global is in order and is instructive because many of the forces that have globalized financial markets are at work in other markets and sectors. Therefore, while the story below is specific to the emergence of global financial markets, there are important parallels in other areas as well, so that the following review has general value in considering questions of global connectedness and vulnerability. Financial markets, moreover, nicely illustrate the arguments being made here that efficiency and vulnerability go hand in hand. It should be noted in passing however, that efficiency and vulnerability need not be at odds with each other if efficiency were to be more broadly defined to include the ability to respond to changing circumstances as a design criterion for an efficient system. For example, recent advances in manufacturing have seen production operations become more flexible and in the process more efficient in coping with rapidly changing product lines and input mixes. This broader kind of definition of efficiency is one that is entirely consistent with the boundary-oriented strategies being argued for here, although it is really at odds with the more common notion of economic efficiency where large-scale specialized manufacturing processes are seen as being more efficient because they can produce more output at lower unit costs for a given product and set of inputs. Their very scale and specialization however, make them vulnerable to changing market conditions, tastes, technologies and input and output prices, therefore clearly less efficient in fact in an uncertain and changing world.

a) A Brief Review of Major Forces Leading to Globalization in Financial Markets

The globalization of financial markets has paralleled that in the real estate sector of the economy. Each has tended to be both the cause and effect of the other. We can thus use financial markets as a prototype, being the most highly integrated markets globally, and also exhibiting all of the features that accompany and/or cause globalization of other product and service markets.

One of the earliest globalizing forces for financial markets traces to the 1950s when the Soviet Union held U.S. dollar balances outside the U.S. to avoid possible seizure of the assets (Levi, 1990). This practice spawned the Euromarket, which is very much the heart of global banking. From about $110 billion in 1970, to $450 billion in 1975, the Eurocurrency market is now estimated to be in excess of $2.5 trillion (Economic Council of Canada, 1989). More broadly, international trade surpluses have been another contributing factor in the growth of international lending. "Petro-dollars" are the most frequently cited example along with recent Japanese trade surpluses.

Another oft-cited cause of the globalization of financial markets is the breakdown of the Bretton Woods agreement with the move to floating exchange rates after 1971. Floating rates increased foreign exchange risk leading to a bevy of financial instrument hedges (eg. currency futures and options). By 1990 daily foreign exchange transactions were estimated at U.S.$640 billion (Hong Kong Trader, 1990).

Globalization would have been impossible though without the use of advanced technology in the areas of telecommunications and computers in financial services (Hamilton, 1986). The first such advance in 1970 was the creation of the Clearing House Interbank Payments System (CHIPS). In 1987, 25 million fully automated transactions were handled for 138 member banks. Average daily volumes were $300 billion (Compton, 1987). In 1972 the Society for Worldwide Interbank Financial Telecommunications (SWIFT) was formed for international payments instructions. Starting in May 1977, it carried some 800 banking transactions daily rising to nearly one million transactions in the late 1980s (Economic Council of Canada, 1989). Progress in telecommunications has also enabled stock, option and commodity exchanges to link up further facilitating global trading. In 1984, the Singapore International Monetary Exchange (SIMEX) and the Chicago Mercantile Exchange (CME) linked to offer financial futures and options so that market players could take a position on one exchange and liquidate it on the other, creating a twenty-four hour market (see The Banker, 1984).

Technology created the possibility of a truly global financial marketplace leading to a more competitive industry by permitting financial institutions to achieve economies of scale (The Economist, 1990). Technology has fostered new financial instruments and, arguably (Wachtel, 1986), led to greater market efficiency via instantaneous global dissemination of information (Bryant, 1987).

Tax and regulatory regimes also encouraged financial market participants to move funds into, or out of, certain countries. Tax havens like the Cayman and Channel Islands are examples. Similarly, Singapore emerged as an international financial centre (IFC) in the early 1970s primarily through the tax concessions it offered financial institutions active in the "offshore" Asian dollar market (Skully and Viksnins, 1987). "Offshore" banking centres have subsequently been established in Tokyo, Vancouver and Vanuatu, providing tax free zones for offshore banking transactions that do not involve domestic residents. In 1987, these centres held $973 billion (21%) of the $4.73 trillion of foreign liabilities.

Deregulation and the liberalization of financial markets in the 1970s and 1980s is a final factor in the globalization of finance. Deregulation has integrated domestic financial systems into a single truly global financial system, allowing no country to regulate its financial sector in isolation from developments in the international financial community. Britain's "Big Bang" of 1986 and the improved access of foreigners to Japan's capital markets are testimonies to this fact. So are developments in Thailand, Malaysia, South Korea and Taiwan allowing (eventually) domestic firms and investors to borrow and invest abroad freely, and foreigners to have access to their stock and bond markets.

Taking all of the foregoing factors into account leads to the conclusion that globalization will most likely intensify, not diminish, as borrowers seek the best sources of funding, and investors and lenders strive for portfolio diversification, enhanced yields and new investment opportunities. A self reinforcing (positive feedback) system seems to have been set in motion which will be difficult, if not impossible, to reverse or undo significantly. This adds significantly to the past instabilities built into domestic economic systems. Moreover, the mechanism identified for propagating and transmitting these crises in historical settings (Kindleberger, 1989, and Hubbard, 1991) needs to be modified significantly to account for the essentially instantaneous transmission that exists today and the virtual absence of national regulatory barriers (buffers) to slow or halt the transmission between countries.

b) Benefits and Costs Deriving from the Globalization of Financial Markets

The benefits or gains that result from the globalization of financial markets are the standard sorts of efficiency gains cited in the economic literature that derive from making markets more economically efficient. In the present case, lower borrowing and lending costs result from bringing together borrowers and lenders on a global scale and from the scale economies inherent in such global activity. Other gains derive from the explosion in international trade that has been made possible, and which has in turn fostered the globalization of financial markets. Booming world trade has in turn led to the fuller exploitation of local comparative advantages with resulting increases in income and employment world-wide. Another related set of gains comes from increases in global investment associated simultaneously with more efficient financial markets and with growth in trade. Further, new telecommunications technology and falling regulatory barriers lead to a business environment which creates new investment vehicles and financial instruments, which lower the costs of borrowing, increase returns and opportunities for investment and develop new client countries where these instruments can be marketed. All in all, this is a considerable set of *short term economic* achievements.

Such gains, despite economic orthodoxy and articles of faith to the contrary are not without costs. The costs however are often longer term and non-economic. The greater connectedness associated with globalization also adds greatly to local and global vulnerability resulting from the global spread of previously localized shocks. The October 1987 global stock market crash is a compelling case in point. There is also a loss of local regulatory control further eroding the safety of financial institutions. This increases the risk that such institutions can and will fail, which when combined with global connectedness, also increases the likelihood that such local failures will spread globally. The U.S. Savings and Loan debacle is a case in point where local financial failures spread to the whole U.S. leading to greater global borrowing needs and fears of global liquidity crises that were inconceivable previously.

All of these are economic issues. There are also concomitant social issues that result from having local economies plugged into global ones making previously buffered local economic systems subject to global instabilities. The loss of local autonomy and the social costs related to destroying previously stable local social and economic systems are high and mounting. They flow directly from the loss of local buffers associated with globalization.

The case of urban property markets illustrates this point and serves to reinforce the argument being made here. Specifically, when formerly largely independent markets become integrated into larger and global urban systems, costs in terms of longer term stability and the loss of autonomy can become large.

5.3.2 Global Integration of Urban Property Markets: Local Enclaves are Brought into the Global System

The globalization of financial markets described above, has been a key contributor to the emergence of a truly international property market. The consequences of such a global property market are equally significant to those stemming from the global integration of financial markets. This follows because property markets were long thought to be local and largely independent of the ebbs and flows of global economic forces. This is most definitely not the case today, so that we see at present a global network of cities tied together not only by the firms and public institutions that span them, but also by the real estate investment capital flows among them which directly influence their scale and style of development (Goldberg, Helsley and Levi, 1989). Once again, an overview of the global property market will be instructive and relevant for the broader concerns of vulnerability and interconnectedness which are the foci of this volume.

a) Some Impressionistic Indicators of The Global Integration of Real Estate Markets

Diverse indicators point to the development of a global real estate market. Business and professional media have been inundated with features about foreign investment, especially from Japan, as well as with articles on foreign markets, especially in Asia (The Economist, 1989; and Urban Land, 1990a,b,c). Globalizing financial markets has also directly affected real estate markets via growing international trading of real estate assets and real estate-backed financial instruments (Hendershott and Van Order, 1989).

Analogous impressions can be gleaned from reviewing the Asian media (Asiaweek, 1991; and Far Eastern Economic Review, 1989). Given the growth in global integrated activities (eg. multinational firms, international finance) it should be obvious that these firms all need urban space. As global markets develop, more and more market information flows around the world relating to all markets, not just the product and service markets in which firms are directly involved. This global information explosion combines with the need to extend and diversify investment, to yield a truly global property market.

b) Some Harder American Evidence

In the U.S., Europeans are the largest investors followed by the Asians, particularly the Japanese (Bacow, 1987 and 1988; Leventhal, 1989, 1990 and 1991; Shopping Centers Today, 1991). The Japanese have been garnering most of the headlines (as they have been doing elsewhere) with their U.S. holdings expected to exceed U.S.$60 billion by the end of 1990. In specific U.S. cities like San Francisco, ethnic Chinese and not Japanese are a major focus (Yannie, 1988). Americans too are taking a broader view with more being written about offshore property markets, particularly those in Asia, reflecting once again the international nature of information on property markets, a necessary condition for a global market.

c) Summing Up and Interpreting the Trend toward a Global Property Market

The most powerful trend emerging from the above is that local urban property markets cannot escape the global integration process and the trend should continue and increase in future. More urban property development and investment will have an international component so that local land use and development policies will need to evolve in this broader international perspective. This will likely apply not just to the largest cities but also to much smaller ones (eg. Victoria, B.C). It will be particularly apparent across the Pacific given past and growing family and business links among major cities on both sides of the ocean. Growing affluence can be expected to increase investment in major Pacific property markets by ethnic Chinese, Koreans, and of course by the Japanese.

Once again, there is clear evidence that the world is indeed becoming more closely integrated. Local property markets are being drawn into the global maelstrom. As a result, the macro policy environment facing urban policymakers includes not just traditional regional and national economic, political and social forces, but increasingly international pressures as well. With the growing globalization of financial and real property markets, urban policymakers can no longer ignore global flows of capital and information, and global economic and political factors when making policy decisions. The growing connectedness of local economies to global forces increases the complexity of local policymaking and exposes local areas to the vagaries of global markets in new ways that make it evermore difficult for local communities to chart their own future without considering carefully these global connections.

For example, the current savings and loan debacle in the U.S. has had major deleterious consequences on property markets all over the U.S., not just in those areas which suffered initially. This systemic downturn in U.S. property values is part and parcel of the recession in which the country is currently mired. However, that recession has clearly been worsened by the vicious cycle set in motion by the failure of virtually the entire savings and loan system and the enormous weakening of the U.S. banking system to the point where mortgage credit was unavailable to fund real estate purchases. This put downward pressure on real estate prices which led to additional failures and weaknesses in the banking and saving and loan industries which further depressed property prices, and so on. This scenario is fairly typical of property downturns. The present circumstance is much more severe though, which can be explained in large part by the globalization of both the financial and property sectors. As the U.S. downturn proceeded, foreign buyers withdrew to the point where Japanese net investment in property has actually declined slightly this year compared with the net annual growth of as much as nearly U.S.$50 billion in 1989. The withdrawal of a major foreign investor weakened such sectors as office buildings and hotels further adding to the downward spiral. Additionally, with U.S. financial markets firmly plugged into their global context, inflows of capital also declined as the U.S. recession advanced, again heaping further downward pressure on an already depressed economy. Prior to globalization the upward pressures during the boom and the subsequent downward pressures during the bust would have been more moderate being the result of purely domestic demand and supply conditions. Under the globalization scenario, these upward and downward pressures were exacerbated, destabilizing both the property and the financial markets to a greater degree than would have otherwise been the case. In short, growth in the up cycle was purchased at the expense of decline in the down cycle: the more efficient global property and financial markets led to more vulnerable and volatile local and national markets in the U.S.

By looking at globalization from this more local perspective, it is readily apparent from visiting the core of major cities anywhere in the world and seeing the symbols of multinational enterprise and international financial firms that we can get a strong indication of the local consequences of the globalization of economic activity (Daniels, 1986; O'Connor and Edgington, 1987). Cities can certainly benefit from their heightened ability to attract outside investment and population; however, just as these waves of capital and people wash onto the local economy, they can wash out again, in the process increasing the cyclicality of already cyclical local economies. Thus, global financial capital flows cannot and will not readily be separated from other international investment flows even in areas long thought to be insulated from areas of global pressure, such as urban property markets and the related urbanization and urban development processes had been thought to be (Daly, 1982; and McGee, 1984).

5.4 LESSONS, EXTENSIONS AND SOLUTIONS: Vulnerability, Areas of Growing Risk, and Approaches to Restoring Resilience in Modern Society

5.4.1 The Lessons to be Learned from the Global Integration of Markets and Politics

The integration of the world's financial and property markets which we reviewed above is a metaphor for the integration rapidly occurring in other product and service markets and by extension in the political arena. As a result of this growing connectedness the world would seem to be a much more vulnerable place despite the rapid increases in world income and trade. It would be desirable in many ways if there existed a reasonably unambiguous definition of vulnerability from which we might derive a vulnerability measurement standard or index. Such is not the case however, and moreover is not likely to be the case. Vulnerability is inherently a multidimensional and complex concept dealing with both the stability and survival properties of global socioeconomic and ecological systems (basically the dynamics of these systems) as well as with the levels of output and throughput of numerous economic, social, political and ecological subsystems (essentially the comparative statics of these subsystems). Accordingly, finding an index weighting scheme that successfully combines these highly varied sets of properties into one meaningful and usable index is highly unlikely, even conceptually.

As a result, we are left instead with a general impression that there have been major, and most often unconscious, trade-offs between gains in outputs/throughputs and the sustainability and stability of these gains. The primary purpose of this chapter is not so much to try to quantify this sense of heightened vulnerability (although that would be a most challenging and interesting subsequent step), but rather to raise this trade-off to consciousness to point out that we have been achieving higher levels of output or throughput at the expense of the resilience of our socioeconomic and ecological systems. Thus, vulnerability is really an additional dimension that needs to be considered in the future as we make investment and managerial decisions. Optimistically, this chapter can advance present practice by putting more narrowly defined efficiency decisions concerning output and throughput into a broader context, which brings vulnerability explicitly to the fore as another decision element which must be balanced against gains in the narrower economic efficiency arena.

With the ability to transmit instantaneously enormous quantities of information and the parallel and equally instantaneous ability to have counterflows of enormous quantities of financial capital comes the kind of vulnerability outlined above. The resulting ebb and flow of huge waves of financial capital can swamp local economies and severely disrupt national ones as well. Volatile interest and exchange rates are part and parcel of this world of hyperconnectedness. These rates in turn spread their impacts into all of the goods and services markets with increases yet again in volatility and vulnerability. The casualties are particularly heavy in the less developed world where large debts encumber growth and prosperity. In developed countries lower paid workers with a fixed income and older workers are the groups most vulnerable to the costs of increased volatility and rapid change. Global economic efficiency has indeed been a Faustian bargain. In the end, there were no free lunches after all.

5.4.2 Extensions: Some Additional Sources of Growing Vulnerability in Modern Society

The kinds of vulnerabilities identified above are certainly not confined to global disruptions that may result from the international spread of financial and economic signals and decisions. They are disturbingly general and ubiquitous and perhaps they are even more prevalent at the local, regional and national levels where our quest for efficiency and the low cost provision of public and private goods and services has made us particularly open to disruption and instability.

Specifically, let us consider the broad area of local, regional and national infrastructure. Hard services are particularly vulnerable. Consider our water supplies - for example - which are acutely susceptible to sabotage and natural disasters which could poison, block, or largely destroy key reservoirs and aqueducts. Similarly, power supplies are readily interrupted by minor natural events like storms. Major natural phenomena such as earthquakes and floods, and of course wilful sabotage can clearly damage or destroy pipelines, power lines, dams, and powerplants and eliminate the energy upon which modern society is so dependent.

However, vulnerability extends beyond these illustrations to virtually all of the hard services upon which we have come to depend such as our complex and highly interconnected telecommunications systems where teleports, cables and switching stations are all exposed to disturbance from atmospheric events like sunspots, natural radiation, earthquakes, and floods, let alone sabotage and acts of war. Even in more mundane networks such as urban road and rail networks we see almost daily disruptions caused from minor sources like congestion, accidents and derailings, to more debilitating events like storms that preclude using airports, and air traffic controllers who work to rule.

Despite modern medicine and its bevy of high technology diagnostic and treatment regimes, is has not reduced our vulnerability to epidemics and the spread of disease. Beyond AIDS and "legionnaires disease," extensive air travel enables strains of disease to be spread quickly well beyond the geographic scale of past eras. Additionally, the development of strains of bacteria that are resistant to many forms of drug treatment (a direct result of past medical successes), further exposes us to health risks.

Finally, our advanced capitalist societies have evolved highly efficient distribution systems to bring us goods and services cheaply and efficiently. A usually hidden cost though is increased risk to poisoning as demonstrated in recent cases of botulism in tainted tuna, and the tampering with sealed containers as in the "tylenol crisis" of the 1980s.

The point should be clear from the foregoing: all of these heightened vulnerabilities faced by modern society result from the large scale and integrated nature of these systems and from the efficiency criteria used to develop and maintain the systems resulting in limited production nodes and monofunctional distribution channels that are prone to disruption because of their scale and lack of redundancy and resilience. These systems in search of efficiency lack the redundancy, diversity, and flexibility of earlier and nominally less complex systems. However, the earlier systems were much more decentralized and less prone to disruption precisely because they tended to be simpler and more independent of each other so that disruptions were more localized. It is doubtless the case, that earlier highly localized and locally dependent systems were prone to disruption *at the local scale*. The integration of local systems into national and global ones has enabled previously local systems to endure shocks that earlier might have been disastrous by importing needed resources from the external national and global systems into which they have become integrated. However, such local gain has often been obtained at the cost of the sorts of global vulnerabilities cited here arising directly out of the increasingly tight global integration of local and national systems. It should also be noted that the formerly localized economies were characterized by lower outputs and incomes, although once again, present day higher incomes and outputs did not come free as the foregoing discussion about added vulnerability stresses.

5.4.3 Restoring Resilience, Flexibility and Longer Term Stability to the Global System: Toward Solutions

a) Rebuilding International Financial Buffers

Regaining lost diversity and resilience will not be easy given the institutions and societal values that have evolved based upon more economic output and narrowly defined efficiency. However there are several steps that can be taken to move the process forward. First, a new set of buffers needs to be devised to insulate local and national economies from the spread of the large economic shocks associated with global economic integration. Some economists are calling for a return to a system of exchange rates that are fixed, not rigidly as before, but rather within bands (boundaries) so that exchange rates can vary but not as wildly as they have in the past decade. While no one is arguing for a new set of trade barriers to provide buffering, there are growing demands for a broadly based review of the useful roles which a modern national and international regulatory system might play in stabilizing global financial markets by preventing large scale collapses and their global spread.

The capital adequacy standards of the Bank for International Settlements (BIS) provide a concrete example of reregulation and the willingness of competing nations to cooperate when faced with the possibility of disaster. The specific disaster that the BIS standards were designed to avert centred on, but was not limited to, the completely unregulated euro-dollar market largely managed in London. It was estimated that at its height in the mid-1980s this market was leveraged as much as 60 times on the capital base of the participating banks. This is clearly a house of very shaky cards, which if it failed, would have certainly spelled disaster and collapse for the global financial system, which was already weakened by the follies of third world debt, the emerging U.S. savings and loan crisis, and the October 1987 stock market crash. Under the new BIS rules which took effect in 1990, the banks of signatory countries (basically all of the developed nations which account for the overwhelming majority of global financial transactions in any event) had to increase their base capital, adjust their balance sheets by assigning a risk rating to previously unrated risks such as letters of credit and other guarantees, and be more generous in their loan loss provisions for potential problem loans (Levi, 1990).

Demands are similarly being made to regulate environmental matters more stringently both nationally and internationally so that major environmental disasters like Chernobyl and the Exxon Valdez can be averted and contained. Global reregulation is also required to protect the public from analogous unsafe and potentially unsafe products that can be disseminated globally in very short order today such as oil and chemical spills and atmospheric leakages.

Finally, there needs to be a major thrust to put an end to the ideology that has spread the gospel of completely efficient and perfect markets and of the parallel evils of government intervention and regulation. Markets do work well, *in general*. The economic revolution sweeping Eastern Europe and other planned economies bears witness to this truth. However, markets do not work perfectly. They impose costs that are not assignable to those who create them. They are also frequently unstable as a result of a variety of fallacies of composition which plague all markets where there are large numbers of independent actors. Unfortunately, these actors can often act upon the same decision rule (eg. sovereign nations do not default on their debts so there is no risk in lending to them). When apparently independent economic units act in this way the net effect in the aggregate is often at odds with the anticipated outcome. The classic example in economics is the case of farmers, who all plant wheat after seeing high wheat prices in a given year, which leads to a bumper crop at harvest and depressed prices.

A variant of this problem is the "prisoners dilemma" (Campbell and Sowden, 1985) where it is in the interests of each market participant to seek the lowest regulatory standards for their operations to reduce costs. However, when all banks respond accordingly, the system is put at significant risk, as are the competing banks. Once again, if one competitor can operate in this manner while all his other competitors are unable to, the least regulated bank will have a significant advantage. However, the world is not that simple. Countries have competed to provide low cost regulation while banks sought out these regulatory havens. Both the host countries and banks have consequently been caught in a classic prisoners dilemma.

The only way out of these dilemmas (Quine, 1976) is to have a supranational body intervene to break the deadlock. The BIS standards illustrate the point. So do the growing networks of international regulation of security markets and the sharing of information and regulatory standards by national and regional securities regulators.

Taking all of the foregoing together leads one to conclude that new approaches are called for. Specifically, individuals and societies must adopt more balanced and workable views. The new approach must be pragmatic rather than ideological, it must seek a balance between market efficiency and progressive regulation aimed at increasing market efficiency in those circumstances where markets are inefficient. The government must also intervene directly where private markets do not function well in equitably distributing income. The twin ideologies of unfettered markets and completely controlled markets must give way to a more sustainable view built on intelligent effective regulation and appropriate government intervention when markets fail.

b) Providing Buffering and Protection at More Local Levels

Several strategies can be put forward to try to build in greater flexibility and resilience at the local level where hard services are most densely provided. For example, in areas with rail rapid transit, portions of redundant trackage must be provided to allow inoperable vehicles to be taken out of service without closing down the entire system. Providing parallel bus and rail rapid transit service also adds diversity, choice and options if one mode fails as the other can absorb the resulting added passenger load. Going on one step further, greater attempts should be made to mix land uses and densities to encourage multimodal travel since many residents will be able to walk to work and the higher density will make various forms of public mass transit viable for those who work at a distance.

Turning to hard site services like phone, water and power, competing utility companies should be allowed for and encouraged so that there is choice in the case of the disruption of one system. The added redundancy can also foster competition in rates. The present long distance phone competition in the U.S. between A.T.T. and U.S. SPRINT is an example which could be encouraged locally. Analogously, governments should encourage local community heating networks - as has been proposed over the past decade - to save energy and to make use of waste materials locally. This would include better insulation and the use of solar collectors to make each building less dependent on the energy network, making the system as a whole more resilient in the process since its individual building elements would be less dependent on a large-scale centralized network.

Similar designs should be sought vis-a-vis local water systems as back ups to avoid system vulnerability, perhaps using rainwater runoff and sewage - which might be collected, processed and stored locally, once again reducing the dependence upon a large-scale vulnerable centralized system.

Finally, to provide some heightened protection for large gas, power and water lines that pass through seismically active, storm prone, or low-lying areas these lines should be built to particularly high engineering standards to minimize risk. In some cases the facility should be duplicated to provide an alternative course through a nearby but less hazardous area.

The foregoing discussion relates to hard physical infrastructure services. However, analogous policies can be developed on the economic front to try to build in more robustness and resilience in local policymaking and fiscal stability. For example, national and regional governments should encourage diverse and largely independent local policy experiments that deal with similar issues such as building and design guideiines. In this way a diversity of guidelines will find their way into local practice avoiding the imposition of the same standards across one or more regions. Thus there will be diversity and a flawed guideline will be localized and not imposed on a large region where its costs are greater and it is much more difficult to change.

On the fiscal front, senior governments should not allow local deficits, and capital borrowing programs should only be allowed if approved by public referenda. If local units are fiscally sound, then it follows that there will be no call on senior levels of government to provide a strong fiscal base for the whole public spending process. Similarly, local governments should not be allowed to speculate in land directly or indirectly through tax financing (putting in services in the expectation that the private sector will subsequently purchase the service land and the local government will get back its servicing investment). Local governments are vulnerable enough to the vagaries of local economies; they should not exacerbate this vulnerability by acting imprudently through taking on excessive debt and speculating. This is why I would further encourage local governments to provide as many hard services as possible locally, where they can control costs and provide locally appropriate services (in keeping though with some broader regional servicing standards). Such local, rather than regional, service provision will build additional redundancy into the system as well as diversity that reflects local tastes and abilities to pay.

5.5 CONCLUDING COMMENTS

The long and short of these examples is that we should try to avoid large-scale simple systems that are vulnerable to broken links in their long complex chains. Instead, we should try when possible to build in greater redundancy so that these high volume systems are less prone to disruption than they are now. Reduced vulnerability does not come free. These smaller scale more localized systems are likely to be more expensive, although if we view this added cost as a form of insurance premium it can be seen as being totally rational in narrow economic terms to act this way. Obviously, if the costs are excessive then the gains from reduced vulnerability may not outweigh the added costs. The point of the foregoing discussion is not to reach a definitive answer on the question, but rather to raise the issue and note that it may well not make economic sense. But it is worth exploring in any event as an alternative infrastructure development strategy to that which we use currently and which has resulted in often efficient, but often equally vulnerable large scale systems.

Smaller scale and slower implementation of solutions has another benefit: it allows for learning. Small scale experimental or prototype solutions enable decisionmakers to assess the consequences of their decisions and to make adjustments. Large-scale discrete decisions are much more difficult to monitor and to adjust because of their scale and the speed with which they are typically imposed on the socioeconomic systems with which they deal.

In looking back over this chapter, it becomes obvious that there are really no simple answers to the questions being posed, nor are there even any simple and yet precise definitions of such core concepts as vulnerability, flexibility and stability. Moreover, it is not the purpose of this chapter to provide either simple answers or precise definitions. Rather, it has been our intent here to raise these issues in an intuitive fashion to have vulnerability and flexibility considered explicitly as important elements for decisions. Indeed, significant losses of flexibility and increases in vulnerability have likely resulted from past successful decisions to increase outputs, incomes and narrowly defined economic efficiency.

Accordingly, we are suggesting here that vulnerability become an explicit decision-making criterion and that we devote serious efforts to defining it operationally and to providing vulnerability with a non-trivial weight in future decision processes.

REFERENCES

Asiaweek , 1991,"Property: Buying into the Neighbours", May 17, pp. 56-60.

Bacow, L.S., 1987, "Understanding Foreign Investment in U.S. Real Estate", Working Paper No. 12, M.I.T. Center for Real Estate Development, Cambridge, Mass.

Bacow, L.S., 1988, "The Internationalization of the U.S. Real Estate Industry", Working Paper No. 16, M.I.T. Center for Real Estate Development, Cambridge, Mass.

The Banker, 1984, "Singapore relies on mutual offset", May.

Bateson, G., 1972, *Steps To An Ecology of Mind*, Ballantine Books, New York.

Borgstrom, G., 1973, *The Food and People Dilemma*, The Duxbury Press, Duxbury, Mass.

Bormann, F. H. and S. R. Kellert (eds.), 1991, *Ecology, Economics, Ethics: The Broken Circle*, Yale University Press, New Haven, Conn.

Bryant, R.C., 1987, *International Financial Intermediation*, The Brookings Institution, Washington, D.C.

Campbell, R. and L. Sowden (eds.), 1985, *Paradoxes of Rationality and Cooperation: Prisoner's Dilemma and Newcomb's Problem*, University of British Columbia Press, Vancouver, B.C.

Compton, E., 1987, *The New World of Commercial Banking*, Lexington Books, Lexington, Mass.

Daly, M.T., 1982, *Sydney Boom, Sydney Bust*, Allen and Unwin, Sydney, Australia.

Daniels, P.W., 1986, "Foreign Banks and Metropolitan Development: A Comparison of London and New York", *Journal of Economic and Social Geography*, 77:4, pp. 269-287.

Economic Council of Canada, 1989, *A New Frontier: Globalization and Canada's Financial Markets*, Economic Council of Canada, Ottawa, Ontario.

The Economist, 1989, "Japanese Property: A Glittering Sprawl", pp. 25-28 and "Japan's Big Property Grab", October 3, pp. 85-86.

The Economist, 1990, "A Survey of International Banking: A Question of Definition", April.

Ehrlich, P., A.H. Ehrlich and J.P. Holdren, 1973, *Human Ecology*, W.H. Freeman, San Francisco, pp. 163-64.

Ehrlich, P. and A.H. Ehrlich, 1991, *Healing the Planet*, Addison-Wesley, Reading, Mass.

Far Eastern Economic Review, 1989, "Focus: Property", 143:11, 16 March, pp. 41-65

Goldberg, M. A., 1975, "On the Inefficiency of being Efficient", *Environment and Planning A*, 7:6, pp. 921-939.

Goldberg, M.A., 1989, *On Systemic Balance: Stability and Diversity in Socioecological Systems*, Praeger Publishers, New York, N.Y.

Goldberg, M.A., 1992, "Issues Facing The Housing and Urban Development Systems in Canada in a Globalized Economy: Canadian Housing and Urban Development

Policy as Examples of The Recurring Problem of Governing the Local Tents in the Global Village", A paper presented at the 5th International Research Conference on Housing, July 7-10, Montreal, Canada .
Goldberg, M.A., R. Helsley and M. Levi, 1989, "The Location of International Financial Activity: An Interregional Analysis", *Regional Studies*, 23:1, pp 1-10.
Hamilton, A., 1986, *The Financial Revolution*, The Free Press, New York, N.Y.
Hendershott, P.D. and R. Van Order, 1989, "Integration of Mortgage and Capital Markets and the Accumulation of Residential Capital", *Regional Science and Urban Economics*, 19:189-210.
Holling, C.S., 1984, "Resilience of Ecosystems: Local Surprise and Global Change", mimeographed, International Institute for Applied Systems Analysis, Laxenburg, Austria.
Hong Kong Trader, 1990, "Territory's Foreign Exchange Market Ranks High in Survey", December, p. 7.
Hubbard, R. G. (ed.), 1991, *Financial Markets and Financial Crises*, University of Chicago Press and the National Bureau of Economic Research, Chicago, Ill.
Kindleberger, C.P., 1989, *Manias, Panics, and Crashes: A History of Financial Crises*, Revised edition, Basic Books, New York, N.Y.
Koestler, A., 1978, *Janus: A Summing Up*, Vintage Books, New York.
Kuhn, T.S., 1970, *The Structure of Scientific Revolutions*, Second edition, University of Chicago Press, Chicago.
Leventhal, K. and Co., 1989, "1988 Japanese Investment in United States Real Estate", Kenneth Leventhal and Company, Los Angeles, Calif.
Leventhal, K. and Co., 1990, "Japanese Investment in U. S. Real Estate Declined Nearly 11% in 1989, Mainly Because of Reduced Office Investment", *Real Estate Newsline*, 7:3 and 4, March/April.
Leventhal, K. and Co., 1991, "Japanese Investment in U.S. Real Estate Declined 12% in 1990, but Exceeded $13 Billion", and "In 1990, for Second Consecutive Year, U.S. Hotels and Resorts Captured Largest Share of Japanese Investment", *Real Estate Newsline*, 8:4, April, pp. 1, 4-7.
Levi, M. D., 1990, *International Finance*, Second Edition, McGraw-Hill Book Co., New York, N.Y.
McGee, T.G.,1984, "Circuits and Networks of Capital: The Internationalization of the World Economy and National Urbanization", Mimeographed, The Institute of Asian Research, University of British Columbia, Vancouver, B. C.
National Academy of Sciences, 1974, *News Report* 24:10, p. 4.
O'Connor, K. and D. Edgington, 1987, "The Location of Services Involved with International Trade", *Environment and Planning A*, 19, pp. 687-700.
Odum, H. T., 1971, *Environment, Power and Society*, John Wiley and Sons, New York.
Putman, S., 1974, "Preliminary Results from an Integrated Transportation and Land Use Models Package", *Transportation* 3:2, pp. 192-224.
Quine, E. V., 1976, *The Ways of Paradox and Other Essays*, Harvard University Press, Cambridge, Mass.
Shopping Centers Today, 1991, "The World", November, pp. 67-78.
Skully, M.T., and G.J. Viksnins, 1987, *Financing East Asia's Success: Comparative Financial Development in Eight Asian Countries*, St. Martin's Press, New York, N.Y.
Urban Land,1990a, "Real Estate Capital Flows", 49:1, January, pp. 32-33.
Urban Land, 1990b, "First International Property Market Held in France", 49:7, July, pp. 21-22.
Urban Land, 1990c, "Outlook for European Investment in U.S. Real Estate", vol. 49, No. 9, September, pp. 28-29.

Wachtel, H., 1986, *The Money Mandarins: The Making of a Supranational Economic Order*, Pantheon Books, New York, N.Y.

Webber, M.M., 1976, "The BART Experience - What Have We Learned?", *Public Interest* 45:79-108.

Yannie, V.J., 1988, "The Effects of Hong Kong's Transfer to China in 1997 on San Francisco Real Estate", A Graduate Student Essay prepared for Julian J. Studley as part of the requirements for a Master's Degree in Political and Economic Assessment of Pacific Basin Economies, Dominican College, San Rafael, California, August 16.

CHAPTER 6

The Evolutionary Network Economy: Historical Parallels from Europe and Japan

David F. Batten

Royal Institute of Technology, Stockholm

6.1 AN INTRODUCTION TO THE NETWORK ECONOMY

It has long been evident that two different representations of space are possible. In the continuous model, location is defined as a point in a spatial coordinate system. By way of contrast, the discrete model approximates space using an index of location. Each index stands for one region within a set of interacting (often contiguous) regions. Regional input-output analysis is a typical case in point.

Although both the continuous and the discrete models do provide helpful insights, in reality most interactive socio-economic systems do not correspond to either of these two representations. As a third alternative, it is instructive to model a spatial economy as a system of networks consisting of (1) the economic activities servicing or located within a set of specific places, called nodes, (2) the existing links which connect these nodes, and (3) the flows occurring on these links. The desire for this network perspective arose firstly from a perceived need to integrate transportation analyses with those of economic systems. It turns out that there is an even more compelling rationale for a network approach, namely that many interactive socio-economic systems are actually structured in network formations which are more-or-less observable.

It is also clear that some networks are quite local in character, being confined to a few relatively small zones in geographical space. Others are much more extensive, linking key nodes in our urban, national and international economies. At each of these different levels of aggregation, the pertinent network is nothing more than a set of nodes and the links which connect these nodes. The intensity of networking relates to the density of the flows on these links. There is also the question of the quality of networking, as well as the important relationships which exist between different networks at different levels. Some nodes belonging to a more extensive network actually contain a less extensive network consisting of nodes and links at a lower system level.

The above characterization provides a useful prescription for subsequent model construction. It stresses that today's complex network systems are interdependent and hierarchical in character and do not function simply according to traditional principles of geographical contiguity. The ideas contained in this chapter are founded on the assumption that a network approach to economic analysis is a fruitful and realistic one, albeit complex. This complexity stems partly from our limited knowledge of the myriads of contacts and latent exchange processes taking place across various networks, and partly from the strongly interdependent and hierarchical character of networks at different levels of economic activity. Many of the cosmopolitan centres that serve as key nodes in the international economic system nowadays have much stronger links to one another than to their surrounding regions (see, e.g. Batten and Törnqvist, 1990).

Thus we are stimulated to consider a whole range of interactive elements within and between various nodes of our economic environment. Some pioneering attempts have been made to grapple with the basic elements of the network economy (see, e.g. Jacobs, 1984). Multilevel networks have also been addressed elsewhere (see, e.g. Johansson, 1989; Batten and Törnqvist, 1990). Although these earlier works help to provide a foundation for future analytical research, it is nevertheless important to convince ourselves that the network economy is an important and observable one. To a greater or lesser extent, we need to be able to identify the key elements involved - namely the various nodes, links and flows which have governed economic development over many centuries.

The purpose of this chapter is simply to compare a few of the key elements and stages in the evolution of the European and Japanese network economies, as a justification for ongoing analytical attempts to describe the evolutionary possibilities in today's global network economy. Thus history is our primary source for the specific characterizations which follow. The German historian, Christian Meier, recently suggested that Europeans must gain the capacity to view European history from the outside: "to take over the view from the non-Europeans". This is a step in that direction.

6.2 THE SEEDS OF EUROPE'S NETWORK ECONOMY

Although the very first discernible seeds of a network economy were probably sown as far back as the 9th century in T'ang China, writings describing that era are very sparse. It is more enlightening for our present purpose to examine Europe in the 11th century. According to economic historians such as Pirenne (1925, 1936) and Braudel (1982), the revival of cities in Medieval Europe coincided with the expansion of trade. At that time, a rather striking contrast existed between much of Western Europe - where land stood for everything and commerce virtually nothing - and Venice, a city bereft of arable land living solely on its gains from trade. The city's location on the sandy islets of a lagoon on which nothing would grow, together with the contagious nature of commerce, virtually guaranteed that the practice of commercial exchange would grow rapidly. Carrying the practice of trade with them, the Venetians gradually acclimatized it to other parts of Italy. The merchants of Venice prompted imitators. Long-distance traders multiplied rapidly and established their own guilds which provided the engines for urban growth.

Many obstacles confronted the movement of people and goods during this formative period. Traffic was constrained both by the deteriorating state of the otherwise admirable Roman network of roads and by fiscal exploitation. The dangers and deficiencies of land transport made waterways the premier trade routes. Coastal shipping was gradually augmented by inland river trade. The emerging industrial centre of Flanders, the ports of Genoa, Venice and Bruges, the fairs in Champagne and other towns, the Italian colonies in the Levant, and the Hanseatic towns can be said to collectively constitute the principal nodes of Medieval Europe's earliest network economy. (See Figure 6.1).

Scandinavian trade kept Flanders in close contact with the North Sea and the Baltic countries. Bridges, dikes, quays and canals constituted key infrastructure elements of this period. There is no evidence of any desire to favour national trade by protecting it from foreign competition. National economics was not meaningful at all. The internationalization which characterized the Medieval civilization well into the 13th century was one which encouraged free trade to a much greater extent than do most trading nations of today. It was conducted via sea and land networks linking key nodes in the form of fairs, markets, ports and town centres. This was a network economy in a rather pure form.

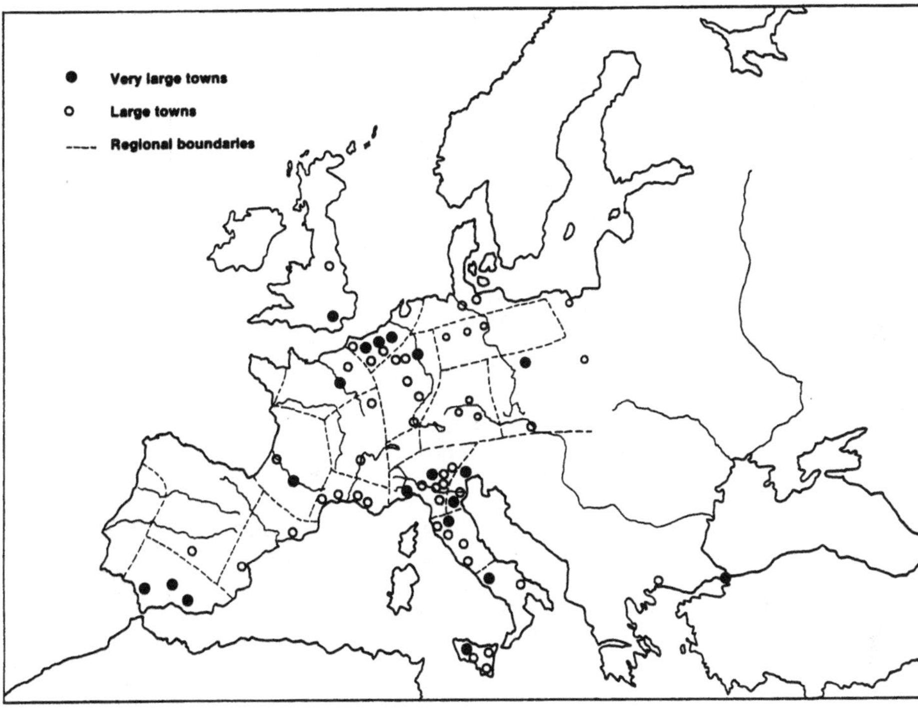

Figure 6.1 Europe in 1350 with key nodes (cities and towns)

One should not lose sight of the fact that the Medieval economy was also a multilevel network economy, supporting several networks of different character. For example, the network of fairs was quite distinct from the network of smaller markets which emerged throughout the whole of Europe by the beginning of the 9th century. These local markets merely serviced the daily needs of the district population (cf. the gravity model). They were held weekly, attracted a limited circle, and concentrated on small retail activity. By way of contrast, the fairs were periodic (e.g. annual or semi-annual) meeting places for professional merchants, were major centres of wholesale exchange, and set out to attract the greatest possible number of people and goods. Their radius of attraction was by no means huge, although the Champagne fairs did attract merchants from the whole of Europe. In theory each fair was open to all trade, just as each seaport was open to all shipping. Thus towns serving as ports or fairs may be viewed as belonging to a higher level network than towns restricting themselves to local market activity.

The Medieval city grew as a result of trade, but trade itself grew out of the growing willingness of merchants to move over long distances. This was made possible by safer and more efficient means of transportation. Thus infrastructure (in its various primitive forms) played a crucial formative role in shaping Europe's earliest network economy, just as it continues to influence the development trajectory of today's network economy.

What is perhaps most fascinating about this development in Europe is that it was basically repeated five centuries later on the other side of the world.

6.3 THE NETWORK ECONOMY OF TOKUGAWA JAPAN

Some historians have argued that the emergence of various networks to facilitate trade and migration in different parts of the world was far from coincidental. City development in Japan prior to and during the Tokugawa period seems to have been influenced partly by the urbanization trends in China many centuries earlier, and partly by the arrival of the first European traders in 1542. Although this remarkable period of economic and spatial integration came many centuries after the revitalization of Europe, Japan's patterns of development bore a striking resemblance to those urban developments which took place in Medieval Europe.

Kamakura and Kyoto were the two principal cities of Japan from the end of the 12th century to the middle of the 14th century. During the Kamakura period, certain ports in the Kinki region developed quickly into large cities. Some of these - notably Yodo, Otsu and Hyogo - had previously served as transshipment points for the transport of taxes and shoen (estate) goods, but they suddenly acquired greater gateway significance.

Accompanying the growth of Kinki ports came the scheduling of periodic markets in most areas of Japan. They were located at administrative centres and along water and land routes. Estate goods were converted into cash at these markets, and then large shipments of goods were transported to intermediate port cities and sold there at much higher prices. New commercial areas developed not only in ports and estates (shoen) but also at post stations on the main roads and near the gates of famous temples and shrines. In the 14th century, rising agricultural production coincided with the emergence of free fairs and local markets and the formation of commercial spheres linking cities and also periodic village markets. Thus the foundations of a new network economy of Japan were laid in a manner strikingly similar to the transition within Europe. Once again, the network structure was multilevel. Merchants in Japan also sought the protection of guilds, with guild leaders emerging as powerful players in the game of urban development.

But Japan was essentially an unstable feudal society during most of the 14th and 15th centuries. It was only with the advent of guilds during the late 15th and 16th centuries that their cities were able to achieve a form of self government comparable to the "free" cities of feudal Europe. Kakamura was destroyed in 1334, and Kyoto almost destroyed at the end of the 15th century, but many smaller ports expanded and new transportation routes spread to the nation's main artery, the Inland Sea. Additional links encouraged more overland traffic between the Japan Sea and Kyoto. The structure of the network economy was becoming much more visible.

By the end of the 16th century, Japan had entered possibly the most remarkable period of urban development in its history. But the preconditions for this sudden city growth were again concealed. It was the era of castle towns, triggered by the siting of periodic markets at castle locations. The first Europeans to reach Japan in 1542 found her in the midst of internal conflict. But the feudal lords were nevertheless keen to trade with the Europeans. Furthermore, domestic commerce began to grow with the removal of old feudal barriers by Nobunaga (Takekoshi, 1930; Sheldon, 1958). Travelling merchants were no longer taxed. Nobunaga was a pioneer of Japanese road construction, also abolishing the monopolistic economic system whereby merchants were the only legal long-distance traders.

The three great cities of Kyoto, Osaka and Edo (later Tokyo) were known collectively as the "Three Metropolises" (santo) or "three ports" (sangatsu). The multitude of remaining urban centres or provincial towns - which amounted to several hundreds - included castle towns, minor ports, religious centres, market towns and highway post towns (Moriya, 1990). This two-tier system of settlements, together with the many highways and byways that linked them together, constituted an elaborate communication system that produced and disseminated new knowledge. It is small wonder, then, that so many Japanese people became literate during the Tokugawa period, developing a keen interest in the state of the world around them.

The primacy of western Japan was quite evident at that time. Kyoto monopolized political, economic and cultural authority. It also served as the greatest industrial city during the early modern period (mainly because of its textile industry). As the key centre of Japanese culture, it was the hub of Japan's scholarship, writings, and religious thought. Today there are still over 2000 Buddhist temples and Shinto shrines in the city. Thus Kyoto was an imposing knowledge centre in terms of industrial production, culture and religion.

By way of contrast, the city of Osaka grew into the key centre for economic information. It initially revived as a place of religion, but the real urban development came under Toyotomi Hideyoshi. He built the great Osaka Castle, which capitalized on the city's well-favoured geographic location at the head of Osaka Bay. Osaka was able to control shipping on the Inland Sea. It also straddled the mouth of the Yodo River, connecting it directly to Kyoto. When Osaka Castle was torn down and rebuilt by the Tokugawa bakufu in the early seventeenth century, much of its former area was converted into a residential area for the urban commoners. There were many construction projects which transformed the city into a great metropolis. Urban commoners, mainly merchants, were then willing to relocate to the city from Kyoto and nearby towns along the shores of Osaka Bay. This resulted in a transfer of merchant capital into the city. (See Figure 6.2).

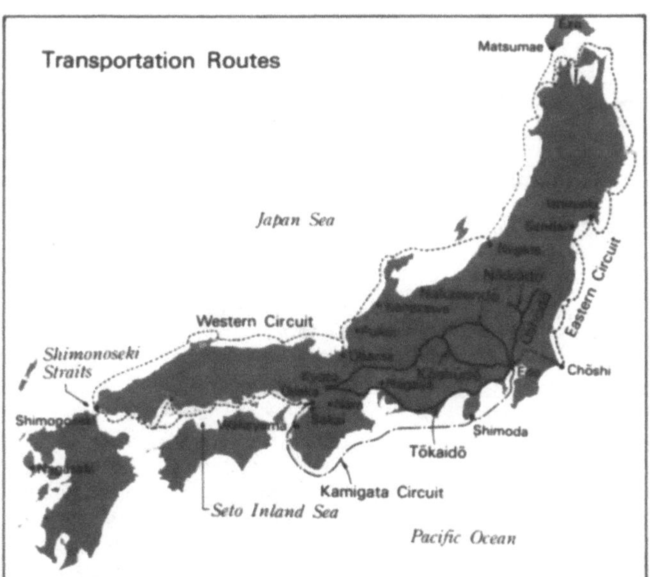

Figure 6.2 Transportation Routes in Tokugawa Japan

Osaka's arrival as a great financial and commercial metropolis was facilitated by the completion of the "Western Circuit" (Nishimawari) shipping route in the mid seventeenth century. This famous sea route, which linked Ezo (Hokkaido), the north and west coasts with the Inland Sea, greatly enhanced Osaka's position as a key trading port and commercial terminal. Serving as the central entrepot of Japan, it was important enough to affect market movements throughout the nation. With its key role of establishing the commercial price of rice, and a population of some 400,000 urban commoners, Osaka gained a secure place in the national economy.

After Ieyasu became shogun in 1603, the capital city function shifted from Kyoto to Edo (later Tokyo). Edo began as a small medieval town during the sixteenth century but was transformed into the country's political capital by the (sankin kotai) system of alternate attendance. The regular, long-term residence of daimyo in Edo resulted in many large mansions appearing rapidly. This also triggered many other land reclamation projects and the ensuing urban growth proceeded at a frantic pace. It has since remained the political centre of Japan.

The Tokugawa period was one in which the bulk of production was distributed by merchants as intermediaries. Towns and cities expanded rapidly. Castle towns became numerous, being built on the plains where transportation facilities were good. Commercial centres grew up around these castles. Many improvements were made to communications, with overland routes spawning travel and itinerant trade and waterways carrying the bulky goods. Free markets and free ports facilitated free distribution of goods. Disputes were settled by the merchants themselves, thus ensuring the development of the first truly free cities. Free trade at home later bred free trade with foreign countries. In a similar manner to the rebirth of Medieval Europe, the network economy of Japan gradually expanded its arena of influence.

6.4 SOME INTERESTING PARALLELS FROM EUROPE AND JAPAN

On closer scrutiny, four important common features of these two evolutionary periods of development in Europe and Japan may be discerned:
(1) improvements to land and sea transportation systems,
(2) simultaneous improvements to the reliability and speed of communications,
(3) the production of written knowledge and the persistent spread of literature, and
(4) the emergence of a merchant class willing to undertake trade over long distances.

These common trends also resulted in some remarkably similar avenues of development, such as
(a) the growth of port cities and post towns,
(b) the rise of merchant guilds, and
(c) the establishment of fairs and periodic markets.

There are some obvious interdependencies between the four key trends mentioned above. The growth in cities and towns, and the interactive activity which knit them together, both fostered and was facilitated by the development of an extensive communication system. In the case of Japan, that system featured major links between Kyoto, Osaka and Edo, but it eventually encompassed towns all over the country. The provision of infrastructure to support overland and maritime transport, which did so much to promote communication and trade over longer distances, occurred at a remarkably fast pace. Within Europe we can speak of the Champagne Route, whereas in Japan we can allude to the Western, the Eastern and the Kamigata Circuits.

Communication systems were evident in both societies during their medieval periods. But communication speed and reliability increased dramatically as the quality of both inland and coastal transport rose. The communication network also expanded to encompass lower-level links between provincial towns.

The rise of an urbanized, mercantile society bound together by a dense communication system required and encouraged the production of written knowledge and the spread of literacy. Kyoto held a dominant role in the production and dissemination of published material during the 17th century. In the 18th century, however, the industry had spread more evenly, with 536 publishers in Kyoto, 564 in Osaka, and 493 in Edo. During the 19th century, Edo's growing primacy was becoming evident with 917 publishers, compared with only 494 in Kyoto and 504 in Osaka. Bookstores also flourished, with about 200 of them serving a Kyoto population of between 350,000 and 400,000. The level of literacy was surprisingly high as a result of the flourishing cultural industries.

In summary, it is worth noting that the spread of knowledge and the expansion of trade both came about as a result of the removal of key barriers hindering long-distance communication. The lifting of these barriers may have served as a catalyst for the sudden revival of towns and cities (see Mees, 1975). This would seem to support an earlier assertion, namely that the emergence of similar networks facilitating movement in each of these regions may not have been entirely coincidental. In both instances, they emanated from a fundamental need for trade in goods and ideas over longer distances - corresponding to the emergence of new and more sophisticated levels of interaction in the network economy.

6.5 CONCLUSIONS

It is worthwhile recalling some statements made in the introductory section of this chapter. The social and economic networks of today are complex, interdependent and hierarchical in character. They do not function according to traditional principles of geographical contiguity. Thus a spatial economy ought to be viewed as a system of networks consisting of (1) the economic activities servicing or located within a set of nodes, (2) the existing links which connect these nodes, and (3) the flows occurring on these links. What is perhaps remarkable about the above statements is that they are just as pertinent to the ancient economies of Medieval Europe and Tokugawa Japan as they are to the international economy of today. Admittedly the spatial scale has altered markedly, but the need for a multilevel network approach seems equally compelling.

In this chapter, I have sketched some of the key elements and stages in the evolutionary paths taken by the European and the Japanese economies. This sketch is far from complete but has nevertheless pointed to some striking similarities. As possible catalysts for the revival of both economies during their medieval periods, I have alluded to four principal factors (1) improvements to land and sea transportation systems, (2) simultaneous improvements to the reliability and speed of communications, (3) the production of written knowledge and the persistent spread of literacy, and (4) the emergence of a merchant class willing to undertake trade over long distances. Undoubtedly other factors were also instrumental during these two eras of innovation and development. Their pursuit is the objective of further research in the near future.

REFERENCES

Batten, D.F. and G. Törnqvist, 1990, "Multilevel Network Barriers: the Methodological Challenge", *The Annals of Regional Science* 24:271-287.

Braudel, F., 1982, *The Wheels of Commerce*, William Collins and Sons, London.

Jacobs, J., 1984, *Cities and the Wealth of Nations*, Penguin Books, Middlesex.

Johansson, B., 1989, "Economic Development and Networks for Spatial Interaction", CERUM Working Paper 1989:28.

Mees, A., 1975, The Revival of Cities in Medieval Europe", *Regional Science and Urban Economics* 5:403-425.

Moriya, K., 1990, "Urban Networks and Information Networks", in C. Nakane and S. Oishi, (eds.), *Tokugawa Japan*, University of Tokyo Press, Tokyo.

Pirenne, H., 1925, *Medieval Cities: their Origins and the Revival of Trade*. (English Translation by F.D. Halsey, Princeton University Press, Princeton, 1952).

Sheldon, C.D., 1958, *The Rise of the Merchant Class in Tokugawa Japan*, J.J. Augustin, New York.

Takekoshi, Y., 1967, *The Economic Aspects of the History of the Civilization of Japan*, Dawsons of Pall Mall, London.

CHAPTER 7

State Market Networks in Japan: The Case of Industrial Policy

T.R. Lakshmanan

Boston University

7.1 INTRODUCTION

Private enterprises and Governments have both played key roles throughout history in the creation of national wealth. In the last two or three centuries business interests in the West have become progressively more autonomous from the state's centralized power, initially taking advantage of new religious bases (Calvinistic) to legitimize their ascendancy (Weber, 1930). The emerging liberal ideology further propelled the business sector into greater autonomy, loosening various feudal fetters. Over time, this autonomy of the business sector created an ambience where resources could be applied to innovate, to produce and to trade - with successful experiments rewarded handsomely. This led to a system of continuous enterprising and continuous innovation (in products, in processes, in inputs, and in organization). The outcome of this system for continuous change was a number of Western economies which have sustained two centuries of continuous wealth creation.

Two interactive parallel developments underlie this evolution in the West. On the one hand is the rise of liberalism, as noted above, with its institutional evolution of an autonomous business sector and its key role in the commercial and industrial transformation of western Europe. The long period required for the business sector to secure its autonomy from the feudal and mercantalist interests left behind a strong distrust of Government and centralized power. This led in turn to the view that the economy is an analytically distinct part of society with its own internal dynamics and the *laissez-faire* view of the undesirable nature of any intervention in the economic realm by the state.

The second key development during the last three centuries is the emergence of the modern state with its legal, political and economic claims and dogmas. These powers and views led to pervasive intervention in the economy by the state during the Mercantalist era in order to promote a favorable balance of trade for European colonial powers in the then emerging World economy with a core (European) - periphery (Non European) structure. As industrialization began in the UK, ushering in a new technical and social environment for production and consumption, the state's participation in the economy was assailed as leading to inefficiencies and eventually became minimal.

Elsewhere in continental Europe (and Japan) the role of the state in economic development was crucial among late industrializers. It coordinated the industrialization through technological borrowing and the creation of the 'functional equivalent' of economic and social institutions necessary for development (Gerschenkron, 1962). While there is thus a clear cleavage between Anglo-American and Continental European experiences, theorizing on the role of the state has been guided by the liberal ideology and the specific economic experiences of the 'lead' developers for much of the last and the present century (namely the UK and the US) leading to a *laissez-faire* view.

In the last decade and a half - characterized by the globalization of business, fierce intercountry economic competition and considerable technological change - the role of the state among successful societies has been noted to be important (Johnson, 1982; Lodge, 1989). As nation states increasingly become "platforms" where multinational corporations play (Porter, 1986), the role of the state as an economic vision setter, as a strategic guide, and as a creator of comparative advantages (e.g. through the creation of relevant infrastructures) is being perceived as increasingly crucial (Lodge, 1989). As a consequence, the issues concerning the role of the state are proceeding beyond simple binary formulations (negative or positive) of the contributions of the state to a more sophisticated discussion of the state-market interface during the course of economic development. This chapter aims to examine these issues related to the state- market interface from a broader perspective which includes the purposes and roles of the state in technology induced development. Such purposes and ideas are not universal but depend on the functional requirements of the development process as well as on historical junctures and the institutional and cultural styles of the countries.

Section 7.2 of the chapter provides an historical summary of government business relationships, first in the preindustrial era and then during industrialization. We present briefly the alternative extant models of government business relationships during industrialization and their drawbacks. We proceed in Section 7.3 to a brief review of the Japanese 'Industrial Policy' experience, which is credited with transforming the Japanese economy - in the quarter century leading up to the oil crisis in the early seventies - into a world class industrial power. This review will suggest that all is not what it seems. The Japanese state has had a strong role in promoting economic development. However, the Japanese state is far more minimalist in the economy than terms such as "Japan, Inc." would suggest. We interpret the Japanese experience in Section 7.4 and highlight the emerging view that the Japanese state does not engage so much in market-displacing policies but rather in market-conforming policies. The 'strength' of the state derives from the 'relational' or 'network' nature of the state-market relationships and from the general national consensus on economic growth. Indeed, the network nature of the state-market interface needs to be explored further. The characteristics of the network and how they vary in different national contexts is a worthy focus for further inquiry.

7.2 STATES AND MARKETS IN THEORY AND PRACTICE: A REVIEW

In the West, political community, as in the Aristotelian notion of the 'organic state', had priority and was used to exert authority over the individual to advance the common good. While this view led in many cases to 'corporatist' and authoritarian states, until early modern times its functional reach was limited to defence, justice, revenue collection, coinage, etc. However, this situation changed dramatically in the two centuries following the Portuguese voyages of discovery. As Europeans secured access to and control of non-European commodities, lands and even peoples, a new world economy was created and developed. This world economy, European in incentive, in organization and in its goals and biases, emerged at a time when the modern nation-state was being established with its claims, loyalties and legal and economic dogmas. With the vast flow of produce, the new labour market in the new world and the colonies, and the dynamics of bullion flow, the nation-state increased its functional reach. The view that "bullion is the sinews of war" in an age of mercenary soldiers led predictably to a concern with the balance of trade - to regulate the trade with colonies in order to capture the growth from finance, insurance and shipping for the mother country and economic intervention known as the *mercantalist* system.

Over the two centuries or more prior to 1800 (in some places longer) and throughout much of Europe, a number of governments, their administrations, and subjects shared a positive view of the regulation of economic life of the community by its rulers. While the community was not initially a nation state, it tried over time to assume that form. The resulting mercantalist system, built by governments - aware of the relationships between their own solvency and the economic welfare of their subjects - had in Adam Smith's words, "twin engines of wealth creation, twin sets of laws to regulate exports and imports in order to protect domestic industry from imports, encourage industrial exports, provide cheap raw material imports and reduce industrial imports". In this effort, trade and industrial interests combined with the fiscal need to create a large apparatus of economic control, some of the elements of which were medieval in origin while others were of an infrastructural nature such as canals, roads and product quality standards. There were predictably national differences, with the English system having more economic purposes such as increasing production and nurturing producers, while the French regulations were more political as regulations were handed down from the state - described as "royal Colbertism". As Wilson (1967) notes, the central century and a half of mercantilism was a period of stable or falling prices, and limited innovation in production and exchange methods, and largely of a great expansion of the scale of activities over large areas. In that era, the mercantalist practices helped greatly in the organized persuit of material gain and wealth, but also attracted influential critics.

To Adam Smith, who analyzed three relatively mature mercantalist economies (England, The Netherlands and France), the ideas behind the system were misguided, its objectives unjust and partial, and a basis for corruption. While the ideas of Smith and others of his kind were influential in England and elsewhere, they had less appeal in countries which were still in the earlier stages of the evolution from an agricultural to mercantalist economy. Smith's criticisms of the practice of producing in one country that which could be bought cheaper abroad had little appeal to those whose choice to import cheaper products meant an indefinite position as a peripheral economy - either in continental Europe or in Alexander Hamilton's US.

The advent of the Industrial Revolution towards the end of the eighteenth century led to a radical departure from the socio-economic and cultural conditions of the mercantalist era. Table 7.1 highlights the major differences between the industrial and mercantalist societies. In the Mercantalist era, the central conflict concerned *liberty*, viewed as the autonomy of business and the freedom of the individual from the feudal restraints on occupational choice, of domicile etc. The earlier arrival of this autonomy of business in England had a lasting effect not only on the organization of the economic institutions but also on the way that organization was conceptualized in the first industrial country. While the society is the locus of centralized power reflecting the balance of power among different economic and social interests and embodying certain ideas of justice and public interest, the economy is viewed as analytically distinct from it. While political and cultural factors may influence economic behavior, they could be treated as exogenous to the operations of the economy. As rational consumers maximize utility and producers maximize profit, a self regulating market harmonizes transactions in the various markets for labour, capital, materials and products. The outcome of such a dynamic equilibrating economy is the efficient use of resources and increased output.

The upshot of this view was not only a down playing of the institutional, cultural and political aspects but also of highlighting *the centrality* of economic factors in development and social evolution - the latter fallacy subscribed to by both liberals and Marxists, in spite of the Marxist assertion of the historical contingency of economic arrangements. If the economy is viewed by neoclassical theory in this fashion as an autonomous sector of society, the logical corollary is to see any government intervention in the market - regulation, welfare, pensions, etc. - as undesirable interference leading to inefficiencies. Even if the economic theorists may differ as to whether a particular intervention is benign or not, they all share the negative view of various states' role in markets and economic development. Such a *laissez-faire* view of the state-market interface gained further ground in the theoretical discussions in the more liberal ambience of the US, and generally in the theories of the development process. As the latter theories were broadened to include the noneconomic dimensions of development and modernization, the role of the state continued to be minimized. This was particularly so in the early post World War II era, when the modernization theories were offered as guides to structuring the economic and social development strategies of the less developed countries in Latin America, Asia and Africa (De Palma, 1978; Lakshmanan, 1990a, 1992b).[1]

Table 7.1: Key attributes of the mercantalist and industrial eras

Attributes	Mercantalist era	Industrial era
1. Central idea	Organize trade	Transform the conditions of production: Division of labour
2. Central conflict	*Liberty*: business autonomy, freedom of the individual from feudal restraints (e.g. arbitrary power, occupational and domicile choice)	*Justice*: Division of social accumulation between labour and capital
3. Central metaphor	Zero sum game	Marginalist positive sum game
4. Key assets	Land and trading assets	Physical capital, transportation
5. Infrastructure	Waterbased; ports, canals etc.	Land transportation, rail, road, power systems
6. Spatial form	Agriculture and market towns	Industrial town

Source: Lakshmanan 1990a

This non-interventionist model of government business relationships is untenable in relation to many theoretical and empirical considerations. Karl Polyanyi (1944) noted that the *laissez-faire* view, which aims to free the market from the state, was itself brought into existence by the state through institutional arrangements such as the law of contract and the law on corporations. The state was not external to but part of 19th century capitalism; in both the UK and the US the state on the one hand helped the

[1]In this model of modernization, modernization was viewed as endogenous in the sense that society is capable of transforming itself from within, without interventions from the state or outside society. Culture (in the sense of knowledge or a set of *situational models* of behavior and thought) changes first, followed in turn by new forces of production; followed by specifically increasing secularization and differentiation of institutions and by greater economic individualism; followed by a mobilization of physical and human capital, all these leading to social transformation. This model was then exalted and presented as a general model of development towards which all countries would progress at various speeds from their different points of departure (Lakshmanan, 1990a).

market by making corrections for its failures by the creation of public goods (e.g. physical, educational and monetary infrastructures) and on the other hand shielded society from the ravages of the market system through government regulation, authorizing labour unions etc. In Polyanyi's view, the first industrial economy in England was the result of interactions between *markets, state and culture*.[2]

Further, this stylized liberal view of the state cannot be sustained to describe the experience of late industrializers. By analyzing the modernization of continental European countries before World War I Gerschenkron (1962) showed that the role of the state was crucial in coordinating the industrialization process through technological borrowing and through the creation of the "functional equivalent" of the economic and social institutions necessary for development. The national state, whether run by a bourgeoisie or by a modernizing elite (e.g. Germany, Japan, Italy, etc.) or Socialist regimes (Russia) remained the modernizing agent. In more recent times, the role of the state in the successful modernization of less developed societies (e.g. Korea, Taiwan, Brazil, etc.) is well known.

The consequence of this diversity of the experience of state-market interactions has been a classification of the government-business relationship into two models in the business and political science literature (Lodge, 1989; Johnson, 1982). Lodge recognizes two models of the role of the government:

* An *Individualistic* society where the role of the government is limited to protect and promote market competition; government intervention is only to protect public health, and welfare (e.g. environmental protection, occupational health, epidemics, etc.)

* A *Communitarian* society, where the government has a role as a vision setter, defining short term and long term needs. It plays a central role in meeting those, not necessarily through its own efforts; it could be democratic or authoritarian. Among the capitalistic societies, Lodge would view the US as the most individualistic state and Japan as the most Communitarian with the European countries in between. Johnson's (1982) terms for a similar grouping of states would be 'regulatory states' and 'developmental states'.

During the last decade, however, some analysts of the state-market interface appear to have swung around to rather extreme views. Some commentators on the Japanese economic successes, particularly in the eighties, have attributed its stellar performance to a government run "industrial policy" (e.g. Johnson, 1982). The state in Japan is viewed as large and ubiquitous, which favours vigorous intervention over a laissez-faire approach in promoting economic development. The Ministry of International Trade and Industry (MITI) is viewed as the Centre of "Japan, Inc." and is said to be instinctively distrustful of laissez-faire capitalism and to favour centralized direction and administrative guidance.

While the bureaucrats at MITI and elsewhere may not regard the market as sacrosanct, since the price system could lead to outcomes that in their view, are not always in the national interest, their interventions are not as massive as often believed and indeed are more often market-conforming rather than market-displacing (Okimoto, 1989; Samuels, 1987). There is considerable evidence that the Japanese state is rather 'minimalist', as measured by key traditional indicators - the tax burden, state ownership of key industries and the degree of regulatory intrusion into the market. Japan ranked 21st among the OECD countries in the late eighties in terms of tax revenues to GDP (27%) as compared to the US (29%) W. Germany (38%), France (48%), and Sweden (52%). State ownership of strategic industries (e.g. steel, shipbuilding, energy, aircraft, telecommunications, etc.) is about as limited as in the US and far lower than in many

[2] In England, the Reform of the Poor Laws facilitated the transfer of immiserized rural labour as necessary streams of labour to the early Industrial towns. Subsequently, as life in those "Satanic mills" became progressively worse, factory laws regulated the work hours which in turn stimulated labour saving technical innovations and growth in the market.

European countries (The Economist, 1978). Finally, in the area of regulatory control the Japanese state is far less intrusive than even the US Federal Government. The latter intervenes much more in areas such as interstate commerce, occupational health, affirmative action, national defence technology contracting etc. than the Japanese Government (Okimoto, 1989).

Which is the correct view of the Japanese state-market interface? A 'minimalist' state or a powerful state that directs the market? To address this question intelligently one needs to go beyond simple binary formulations - such as interventionist-non-interventionist, or weak-strong state etc. We must analyze the underlying process of economic development, its dynamism and requirements, and the respective roles of the state and market in this process in contexts dictated by the stage of development, technology, and institutional and cultural styles. Thus, a useful focus is not the autonomy of the market but the interconnections between the market and the state in different situations over time. To get a handle on these aspects of development we proceed next to a discussion of the Japanese development experience since World War II and the role played by the State.

7.3 INDUSTRIAL POLICY AND THE JAPANESE STATE

Japan's transformation from the humble beginnings of a poor subsistence semifeudal economy to a great economic, military and naval power in six decades is well known (Beasley, 1990). Table 7.2 provides a portrait of Japan's economic transformation over nine decades as compared to other industrial countries. Japan clearly evidences the fastest growth of product per capita over the entire period from the lowest initial per capita income level. Again, since the end of World War II, which resulted in major material losses, the Japanese economic performance has been the envy of the industrial world, moving from less than 1/25 of the US economy in 1950 to nearly 1/3 of the US economy in twenty-five years (Table 7.3). The same period witnessed a major structural change in Japan from a semi industrial rural society to a modern industrial urban society.

Table 7.2: Modern economic growth of Japan and other developed countries

Country	Period		Rates of growth per decade (%): Total product	Population	Product per capita	GNP per capita Inital date 1965 Prices
Great Britain	1765/85	to 1963-7	23.7	10.0	12.8	$ 227
France	1831/40	to 1963-6	21.8	3.2	18.1	242
Netherlands	1860/70	to 1963-7	27.7	13.4	12.6	492
Germany	1850-9	to 1963-7	31.0	10.8	18.3	302
Switzerland	1920	to 1963-7	26.3	8.8	16.1	529
Denmark	1865-9	to 1963-7	32.5	10.2	20.2	370
Norway	1865-9	to 1963-7	31.4	8.3	21.3	287
Sweden	1861-9	to 1963-7	37.4	6.6	28.9	215
Italy	1895-9	to 1963-7	31.4	6.9	22.9	271
United States	1834-43	to 1963-7	42.4	21.2	17.5	474
Canada	1870-?	to 1963-7	41.3	19.0	18.7	508
Australia	1861-9	to 1963-7	36.4	23.7	10.2	760
Japan	1874-9	to 1963-7	48.3	12.1	32.3	74

Table 7.3: International comparison of Japan's production levels

1. Total domestic production

	1950	1960	1965	1970	1975
US	100.0	100.0	100.0	100.0	100.0
Japan	3.9	8.4	12.8	19.8	32.3
West Germany	8.1	14.2	16.6	18.9	28.1
France	10.0	12.0	14.3	15.0	21.6
UK	12.9	14.1	14.4	12.2	15.0
Italy	6.0	6.9	8.5	9.4	11.4

2. Industrial production

	1958	1960	1963	1968	1972
US	100.0	100.0	100.0	100.0	100.0
Japan	15.6	20.5	25.5	36.7	49.0
West Germany	23.0	24.0	24.0	23.1	25.9
France	21.2	20.2	20.6	19.7	23.6
UK	19.5	19.3	17.7	16.0	14.8
Italy	13.6	15.1	17.3	17.5	17.7
USSR	59.4	63.6	71.9	81.6	95.7

Source: Lakshmanan, 1990b

This stellar economic performance in the post World War II era leading up to the oil crisis in the seventies has been attributed by a number of observers to an active "Industrial policy" pursued by the Japanese state (e.g. Johnson, 1982; Lodge, 1989). Industrial policy is one of those terms often used both in professional and popular usage without a clear and rigorous definition of its meaning - variously used to describe the state's efforts to redesign industrial structure or supply side policies, particularly tax proposals. A useful definition of Industrial policy would refer to the state's use of its power and resources to develop and administer policies intended to meet the needs of specific sectors and industries (and in some cases those of individual firms) in order to promote the efficient use of resources. Thus, industrial policy is different from macroeconomic policies, which are a set of public policies dealing with the economy as a whole.

The objectives of industrial policy may vary from country to country. In the Cold War era the aim of industrial policy - to the degree it can be described as such in the US - was to support national security and maintain full employment. For a late industrializer such as Japan, the objectives of industrial policy include technological catch-up, dynamic adjustment to changing international comparative advantage, the provision of infrastructure, and protection against foreign economic and political domination. To attain these goals the typical policy instruments used include Research and Development (R&D) funds, subsidies, tax incentives, restraints on foreign imports etc.

Before we describe briefly these industrial policies and assess the role of the Japanese state in their administration and their successes and failures it will be constructive to highlight the antecedents of the state's role in industrial structuring in the Post World War II era.

The role of the state was pervasive in economic activity in the Tokugawa period as it was in the comparable premodern (mercantalist) era in the West. As noted above, the autonomy of the business sector and its relative freedom from Government control in the West represented an institutional evolution in the early industrializing countries. In Japan, by contrast, the late industrialization in the context of a weak business sector and ill developed social and economic institutions led the Meiji state to play an active role in modernization. Briefly stated, the Japanese state initially played a pervasive role not only in the transfer of new knowledge technologies and physical infrastructure (to promote cheap movement of people, materials and ideas) but also in actively promoting the normative orientations towards a more individualistic exchange economy and by creating new social and economic institutions including a progressive entrepreneurial class (Andersson, et al., 1992; Lakshmanan, 1993a, 1993b). The latter group assumed (within a decade and a half from the start of the Meiji era) a key role in modernization, complementary to that of the state (Hirschmeier and Yvi, 1975). The government played a more important role in the national economy: (a) as a consumer rather than a producer (b) through mercantalist measures such as subsidies, low interest loans, bounties and technical assistance to business in areas such as shipping, heavy industry etc.[3] A specific ministry in the Japanese Government was responsible for this financial and technical support to industries. That ministry's name and organization changed over time - from the Ministry of Industry to the Ministry of Agriculture and Commerce, then to the Ministry of Commerce and Industry and finally to the present Ministry of International Trade and Industry (MITI).

It is this MITI, in particular the vice Minister and top level bureaucrats (along with some senior people at the Ministry of Finance), that is believed by a group of observers to be responsible in the post World War II era for the Industrial Policy and the Japanese miracle (e.g. Johnson, 1982). In contrast to many Western Governments, the Japanese state's (as represented by MITI) approach is anticipatory and aimed at structuring the market positively in ways which can promote desired industry specific goals. This is consistent with the Japanese view that a laissez-faire market, while confering the benefits of competition and efficiency, has several drawbacks (Morishima, 1982). The MITI officials view their industrial policy as compensating for some of the above drawbacks.

The ability of one ministry to carry out an industrial policy (which would have diverse effects on a variety of social groups) over a long period derives from the specific structure of the Japanese system. It is what Okimoto (1989) calls 'segmented political configuration'. The Liberal-Democratic Party (LDP), with its monopoly of power in the postwar period, has assembled a stable coalition of interests whose support is exchanged for favorable legislation, subsidies, tax forgiveness etc. in the areas of agriculture, small scale enterprise, the health professions, construction etc. This clientilistic and pork-barrel relationship combined with financial support of Big Business and generalized voter support (based on fast economic growth) has not only kept the LDP in power for such a long time. In addition, the MITI bureaucrats have also been relatively free from political interference as they develop and implement industrial policies for the manufacturing and high tech sectors (Okimoto and Rohler, 1985).

[3]The state's purchases amounted to about 10 percent of GDP in the 1890-1930 period. The Government's share of gross domestic fixed capital formation averaged over 40%, and often higher in the period 1887-1940 (Ohkawa and Rosowsky, 1973). In the area of technical assistance, the Government developed with great care modern technology in naval ship yards and in Yawato iron and steel mills, where engineers and skilled workers from all over Japan were exposed and trained, so that they could carry back this technology to their work places.

It is in this political context that the Japanese state sought in the post World War II era economic prosperity for its own sake (not to develop military power as in the prewar era). Given the material losses during the war (estimated at over a quarter of the national wealth) the low wages and the resource limitations, MITI's strategy to push for advanced industrial development looked both risky and inattentive to the strictures of the theory of comparative advantage. MITI's rationale lay in a long term perspective, where capital and technology-intensive industries offered big spillover effects, the potential for rapid technological change, a progression towards goods with high income elasticity of demand and the potential for rapid increases in total factor productivity.

This strategy called for an extensive state role - in long term vision setting, the preparation of a set of midrange plans, the mobilization of scarce capital resources facilitating foreign trade but offering protection from foreign suppliers for infant industries, the regulation of technological flows in and out of Japan, and 'investment guidance' and 'administrative guidance'. The outcome was a robust heavy manufacturing base by the late sixties, built on achieving large economies of scale and improving process technology. The combination of captive domestic demand and accessible overseas markets gave Japanese industry a great cost advantage and formal and informal trade barriers protected domestic industry. Other aspects of the Japanese Industrial system - subcontracting networks, lifetime employment, an emphasis on production engineering and quality - also contributed to the success of the industrial policy.

The ability of MITI to rationalize heavy industry in order to gain price and technical advantages for Japan and to lead to a recent shift to knowledge intensive industry, is less impressive (except in the case of NIPPON steel). Indeed, as Okimoto (1989) notes, there have been failures of the MITI industrial policy as in the case of the petrochemical industry - though part of this failure derived from the completely unanticipated oil price increases arriving at the same time that the petrochemical industry matured. However, overall Japan is one of the few countries in the World which seems to have found a relatively effective method for administering a coherent industrial policy.

7.4 STATE MARKET NETWORKS IN JAPAN

A central question asked by many observers of the Japanese scene is: why has industrial policy succeeded relatively well in Japan, while such efforts to upgrade chosen industries or firms in other countries such as the UK, France, and Italy are considered failures? How does MITI have such a broad scope of authority over most manufacturing sectors, with its relative insulation from political pressures and how does it acquire the ability to reach binding consensus with different producer groups? How are typical serious conflicts e.g. infant industry protection versus more efficient production and consumption or antirecassion cartels - versus lower end user prices, etc. - managed without derailing these policies?

One position taken by Johnson (1982), as well as supporters of the notion of 'Japan Inc.', is that the Japanese state is intrinsically powerful. Johnson claims that Japan is a 'developmental' state with a strong concern for promoting the structure of domestic industry which will enhance Japan's international competitiveness. The great degree of state intervention and protection implied in the industrial policy implies a sharp difference in the role of the state in a 'developmental state' Japan as compared with the state role in the US with its market rationality (Johnson, 1982).

A second view holds that the state in Japan is not powerful and it capacity to operate in an effective fashion depends upon (a) the specific segmentation of the political structure noted above (Okimoto, 1989), (b) the general social consensus on economic goals, (c) the network nature of relations among different interest groups leading to Samuels' (1987) "politics of reciprocal consent". Indeed, it has been suggested that the early Meiji period and the two decades after the end of World War II were unusual periods of state dominance reflecting the fear of industrial backwardness and its likely impact on national independence, the perception of national crisis and a late comer's impatience to industrialize and thus provided an unusual opportunity for the exercise of state power. Okimoto (1989) points out that the Japanese state, lacking the universalistic principle of the legitimacy of power, governs more by persuading rather than through confrontation.

Samuels (1987) and Okimoto (1989) have argued that the power of the Japanese state, as emphasized by so many observers, results from the capacity of the state (as embodied in the dominant political party and the high bureaucracy) to work in concert with the private sector - each side considering and accommodating the needs and objectives of the other. Without the legitimacy conferred on it by universalistic principles, the basis for the Japanese state to exercize autonomous power is absent (Okimoto, 1989). Power is exercised in Japan through a complex process of interaction between the dominant political party, the high bureaucracy and the private sector in a process of interaction among groups in the form of a subtle give and take rather than confrontational power relationships. This is what Samuels (1987) calls "The politics of reciprocal consent". If the power wielded by the state in the early Meiji era was so dominant it reflected a confluence of special circumstances, such as the threat of foreign domination, the need for rapid industrial development to offset that threat, and the late industrializer's urgency in a national crisis. The strength of the state comes from the shared national purpose and the good performance record of the Japanese state in delivering on economic growth.

It is therefore useful to view the role of the current Japanese state in this structure of extensive interdependence and voluntary relationships in a 'relational' framework (Dore, 1987). The state and market are in a "network" relationship rather than being two autonomous groups.

Some may argue that the market system with its heavy flow of information is less prone to major errors and possesses several feedback mechanisms. But the Japanese society is so organized as to defuse power and decentralized decision making. Consequently, the information costs of adopting an industrial policy are minimized. By drawing information from a broad cross section of society - academia, mass media, banks, the labour industry etc. - Japan's industrial policy system averts the danger of major errors. Indeed, many observers claim that the MITI's interventions in the market are more often "market-conforming" than market displacing.

In such a network, the state performs many important functions: communications, coordination, the mobilization of resources, the aggregation of interests, conflict resolution, and collective vision development (the last function can be viewed as a weak and modern version of the early Meiji role of the state as a social movement). As Okimoto (1989) notes, such functions can be best exercised in a persuasive (rather than coercive) manner in a state-market network. This same observation is relevant to a considerable degree for effective state-market development relationships in most advanced contemporary societies.

REFERENCES

Andersson, Å.E., T.R. Lakshmanan and Wei-Bin Zhang, 1992, "Technology and Development: Lessons from Success Stories, *Journal of Scientific and Industrial Research*, 51:157-171.

Beasley, W.G., 1990, *The Rise of Modern Japan*, St. Martin's Press, London.

De Palma, G., 1978, "Dependency: A Formal Theory of Underdevelopment?,*World Development*, 6:881-924.

Dore, R., 1987, *Flexible Rigidities: Industrial Policy and Structural Adjustment in the Japanese Economy 1970-80*, Stanford University Press, Stanford, Ca., 106-112.

The Economist, Dec. 30, 1978.

Gerschenkron, A., 1966, *Economic Backwardness in Historical Perspective*, Harvard University Press, Cambridge, Mass.

Hirschmeier, J. and T. Yui, 1975, *The Development of Japanese Business 1600-1973*, Harvard University Press, Cambridge, Mass.

Johnson, C., 1982, *MITI and the Japanese Miracle: The Growth of Industrial Policy 1925-75*, Stanford University Press, Stanford, Ca.

Lakshmanan, T.R., 1990a, "Cultural Contingencies in Technological Transformation", paper presented at the Western Regional Science Association, Hawaii.

Lakshmanan, T.R., 1990b, "The Social Context of Technological Progress in Japan", paper presented at the European Regional Science Congress, Istanbul, August 27-31.

Lakshmanan, T.R., 1992, "Technology-Induced Industrial Development: Implications for Developing Countries", lecture presented at the Second Pacific Summer Institute, Taiwan.

Lakshmanan, T.R., 1993a, "Social Change Induced by Technology: Promotion and Resistance" in Ackermann, N. (ed.) *The Necessity of Friction*, Penguin.

Lakshmanan, T.R., 1993b, "The State and Market in Japanese Development", in Nijkamp, P. and H. Kohno, (eds.), *Festschrift for Prof. Oishi*, Springer Verlag, Berlin.

Lodge, C.G., 1989, *Comparative Business-Government Relations*, Prentice Hall, Englewood, N.J.

Morishima, M., 1982, *Why has Japan Succeeded: Western Technology and the Japanese Ethics*, Cambridge University Press, New York.

Ohkawa, K. and H. Rosowsky, 1973, Japanese Economic Growth: Trend Acceleration in the Twenthieth Century, Stanford University Press, Stanford, Ca.

Okimoto, D.L., 1989, *Between MITI and the Market,* Stanford University Press, Stanford, Ca.

Okimoto, D.L. and T.P. Rohler, 1985, *Inside the Japanese System: Readings on Contemporary Society as Political Economy*, California University Press, Stanford, Ca.

Polyanyi, K., 1944, *The Great Transformation*, Beacon Press, Boston.

Porter, M.E., 1986, "Competition in Global Industries: A Conceptual Framework" in M.E. Porter (ed.), *Competition in Global Industries*, Harvard University Business School Press, Boston, 39-92.

Samuels, R., 1987, *The Business of the Japanese State: Energy Markets in Comparative and Historical Perspective*, Cornell University Press, Ithaca, New York.

Weber, M., 1930, *The Protestant Ethic and the Spirit of Capitalism,* trans. by T. Parsons, Scribner, New York.

Wilson, C.H., 1967, "Trade Society and State" in Rich, E.E. and C.H. Wilson (eds.), *The Cambridge Economic History of Europe*, IV:487-575.

PART II: TRADE AND PRODUCTION NETWORKS

CHAPTER 8

Domestic Demand, Learning and Comparative Advantage

Jan Fagerberg

Norwegian Institute of International Affairs, Oslo

8.1 INTRODUCTION

Traditionally, most attempts to explain the specialization patterns of countries in international trade have focused on supply conditions. According to the standard neoclassical theory of international trade, countries ought to specialize in areas of production that make intensive use of factors of production with which the country is relatively well equipped. However, empirical research has shown that the explanatory power of this type of theory is rather limited (Bowen et al., 1987).

In spite of the dominant role played by the traditional neoclassical theory in this area, there has always been a strand of thought that has emphasized *learning* as a potential source of comparative advantage (for an overview, see Dosi and Soete, 1988). This tradition points to the learning creating effects that relations between firms or sectors within the domestic economy may have, and the impact of this on the development of the international competitiveness of the country and its specialization pattern in international trade. List (1959), in his famous defence of protectionism, pointed to the positive impact that such relations may have in the process of industrialization. Various concepts have been developed to cover (different aspects of) such dynamics. Schumpeter (1934, 1939, 1943) used the concept "clusters of innovations". Other writers in the Schumpeterian tradition use other concepts, such as "growth-poles" (Perroux, 1956) or "development-blocks" (Dahmén, 1970). Starting from a somewhat different perspective, Hirschman (1958) coined the concept "linkages" to cover the positive impact that links between different sectors of the economy may have for economic development. Burenstam-Linder (1961) was the first to discuss the implications of these ideas for trade theory. In the case of developed countries, he suggested that it is demand-induced innovation *within each country*, not supply factors, that determine comparative advantage.

A recent attempt to construct an evolutionary scheme of economic development based on these ideas was made by Porter (1990). Echoing Burenstam-Linder he argues that traditional supply factors, although important in the earlier stages of development, are not among the prime determinants of "competitive advantage" in more advanced countries, where growth is assumed to be innovation-driven. The most competitive industries of an advanced country, he argues, tend to be highly integrated ("clustered"), both vertically and horizontally, with favourable consequences for learning, innovation and "competitive advantage". In Porter's scheme, this typically starts with vertical integration between customers in traditional industries and suppliers of machinery and other types of advanced equipment, and then widens through spill-overs and feed-backs to and from related and supporting industries (Porter, 1990, p. 554-5). In this chapter, we will focus solely on the first of these two mechanisms, emphasized by both Burenstam-Linder and Porter; i.e., *the favourable impact that vertical integration between customers and suppliers (or users and producers) may have on comparative advantage*. Building on earlier work by Andersen et al. (1981), the next section outlines how this can be measured empirically. Then the data are examined, a formal test designed and the results presented.

The Burenstam-Linder-Porter hypothesis, as outlined above, is intuitively appealing and there is a large amount of descriptive evidence that can be used to support it (Porter 1990). However, in spite of the growing popularity of this approach, many would probably still feel that "clustering" is a phenomenon in search of a theoretical explanation. One possible response to this criticism may be found in the modern literature on "networks" (Håkansson, 1987; De Bresson and Amesse, 1991) and "user-producer relationships" (Lundvall, 1988, 1992), combining Schumpeterian insights in the innovation process with analyses of transaction-costs and market behaviour. In this approach, stable relations between users and producers of technology are analysed as a way to minimize costs related to information and communication and to internalize positive external effects. It is argued (Lundvall, 1988) that such relations are especially important in cases where technology is complex (and changing) and the need for close communication and interaction between users and producers of technology is large. Interaction processes of this type involve learning and - in many cases - the modification of an existing technology or the creation of an entirely new technology. When this happens, the competitive position of the firms involved will normally improve.

To the extent that this type of interaction mainly takes place *within country borders*, it should be expected to affect the patterns of export specialization (or comparative advantage) of countries as well. Since, as both Burenstam-Linder and Porter pointed out, the costs associated with communication and interaction increase with distance and differences in culture, language, institutional settings etc., this may be a reasonable assumption to make. Porter even holds that the importance of the domestic market for competitive advantage is growing. However, it may also be argued that the increasing role of multinationals in world production has reduced the costs of communication and interaction significantly, and that the Burenstam-Linder-Porter hypothesis was therefore more relevant in the past than it is presently. Section 8.4 of this chapter considers some of the empirical implications of these two conflicting views.

8.2 INTERPRETATION, DATA AND METHODS

The hypothesis that a high degree of vertical integration between customers and suppliers (or users and producers) affects the competitiveness of the latter positively, needs some further qualification. First, as pointed out in the introduction: it is assumed to hold for advanced countries only. Second, it is of little relevance for suppliers of relatively standardized products. This was pointed out by Burenstam-Linder:

> "... exceptions to our proposition are likely to occur in those cases (1) where it is easy to become aware of the foreign demand in spite of the non-existence of home demand of the product; (2) where the product as such is available without inventive effort; and (3) where no or little product development work is needed." (Burenstam-Linder, 1961, p. 90).

Third, not all cases of vertical integration are equally conducive to innovation and "competitive advantage". As pointed out by Lundvall: "Being closely linked to conservative users having weak technical competence might be a disadvantage for a producer, and vice versa" (Lundvall, 1988, p. 356). Thus, the mere existence of a home market for a particular product or technology is not enough: *A necessary condition is that the domestic users are both sophisticated and demanding.*

To test the hypothesis, a relatively large number of exporting sectors, as well as domestic users of the products from these sectors, must be identified and defined. While this may be easy in theory, it is more difficult in practice, since the available statistics are not collected for this purpose. Most "advanced" products are not classified according to users, and even when this is the case it is not always easy to find internationally comparable data for users on a sufficiently disaggregated level. Furthermore, we have to establish what we mean by an "advanced user sector", and decide how this can be distinguished from a less advanced one empirically.

To the best of our knowledge, the only attempt to face these problems was by Andersen et al. (1981). The empirical hypothesis considered by them was the following: *If internationally competitive producers exist in one sector of the economy, and these producers buy their technology from another sector of the economy, the latter sector should be expected to be internationally competitive too.* For instance, if a country is export specialized in agricultural products, it should also be expected to be export specialized in agricultural machinery. Thus, in the interpretation of Andersen et al. "advanced user sectors" are identified as sectors in which the country has a revealed comparative advantage. This interpretation has the advantage that it enables us to use the same data source, trade statistics, and the same index, revealed comparative advantage (Balassa, 1965), to measure the strength of both "producers" (export products) and "advanced domestic users" (home-market sectors).

A problem with this interpretation is that it introduces a bias towards products where the trade statistics allow a link to be made. For export products this implies that most of them belong to the group "specialized machinery" (SITC, 71), where users in many cases are relatively well specified. On the user side (home market sectors) the consequence is that users in the non-trading sectors of the economy (or in other sectors not covered by the international trade statistics) are excluded from the investigation. For instance, the possible impact on exports of links between technology producers and public-sector users, which is an interesting issue from a policy point of view, cannot be taken into account.

The empirical methodology adopted in this chapter generally follows the work by Andersen et al., but some attempts have been made to overcome the limitations of their data. The most important novelty in this respect is the extension of the analysis to three service sectors, not covered by the trade statistics, and the construction for these sectors of special "home-market indexes". These sectors are health care, telecommunications and shipping (two of which are dominated by public-sector services). For the remaining pairs of export products/home-market sectors there was an attempt to make the definitions more precise by the use of more disaggregated statistics. The resulting sample is larger than that of Andersen et al. (23 pairs compared to 13). However, the group "specialized machinery" (SITC, 71) still accounts for around two thirds of the "export products" included in our sample. Table 8.1 lists the 23 pairs of products included in the test.

For a particular country and product, the index for revealed comparative advantage (S) is the ratio between the market share of the country on the world market for this particular product and the market share of the country on the world market for all products. Letting X denote exports, j the export product and i the exporting country, the index for revealed comparative advantage (S) can be presented as follows:

$$S_{ij} = \frac{\dfrac{X_{ij}}{\sum\limits_{i} X_{ij}}}{\dfrac{\sum\limits_{j} X_{ij}}{\sum\limits_{i}\sum\limits_{j} X_{ij}}}. \qquad (1)$$

Table 8.1: Export products and home market indicators

SITC (REV 1)	EXPORT PRODUCT	SITC (REV 1)	HOME MARKET INDICATOR
54	PHARMACEUTICALS		HEALTH[1)]
6291	RUBBER TYRES AND TUBES	732 - 734	ROAD MOTOR VEHICLES, AIRCRAFT
6951	HAND-TOOLS FOR AGRI-CULTURE AND FORESTRY	04-08(-0814), 24	AGRICULTURAL PRODUCTS, WOOD PRODUCTS
7114	AIRCRAFT ENGINES	734	AIRCRAFT
7115	INTERNAL COMB. ENGINES	732	ROAD MOTOR VEHICLES
7121 7122	AGR. MACHINERY FOR PREPARING SOIL AND HARVESTING	04-08(-0814)	AGRICULTURAL PRODUCTS
7123	MILKING MACHINES	02	DAIRY PRODUCTS
7125	TRACTORS	04-08-(0814)	AGRICULTURAL PRODUCTS
7129	AGR. MACHINERY N.E.S.	04-08(-0814)	AGRICULTURAL PRODUCTS
7151 7152	MACHINE TOOLS FOR WORKING METALS	69	METAL MANUFACTURES
7171	TEXTILE MACHINERY	65	TEXTILES
7172	LEATHER MACHINERY	61	LEATHER
7173	SEWING MACHINERY	84	CLOTHING
7181	PAPER WORKING MACHINERY	25, 64	PULP AND PAPER, PAPER PRODUCTS
7182	PRINTING MACHINERY	829	PRINTED MATTER
7183	FOOD PROCESSING MACHINERY	0 - (00)	FOOD
7184 7185	CONSTRUCTION AND MINING MACHINERY, MACHINERY FOR MINERAL CRUSHING, ETC.	27, 28	CRUDE MINERALS AND METALS
7191	HEATING AND COOLING EQUIPMENT	01 - 03	MEAT, DIARY PRODUCTS, FISH AND EGGS
7249	TELECOMMUNICATIONS		TELE[1)]
726	ELECTROMEDICALS		HEALTH[1)]
7294	AUTOMOTIVE ELECTRICAL EQUIPMENT	732	ROAD MOTOR VEHICLES
735	SHIPS AND BOATS		SHIPPING[1)]
8617	MEDICAL INSTRUMENTS N.E.S.		HEALTH[1)]

1) For the definition of this indicator, see the text and Appendix 1

In principle, this index may vary between zero and infinity, although it seldom takes on very high values. It has the property that the weighted mean is identical to unity for each country across all commodity groups, and for each commodity group across all countries. Thus, a country is said to be specialized (have a revealed comparative advantage) in a product if the RCA index exceeds unity.

The "home-market indexes" for the service sectors were calculated in a way that made their structure as close as possible to the RCA index. For instance, if the index for a specific country for shipping exceeds unity, this implies that the market share of the country for shipping services exceeds the market share of the country for goods and services in general. For health services and telecommunication services, which are not traded on the world market to the same extent, the population was used as deflator. Thus, in these cases, a value larger than one implies that the per capita "quality" of these services in the country is higher than the OECD average. For a more detailed account of the way in which these indicators were constructed, including sources, the reader is referred to the Appendix.

The trade data for the years 1965 and 1987 used in this chapter were collected from OECD Trade Series C. Since the theory is not expected to hold for developing or semi-industrialized countries, we excluded the less industrially developed OECD countries from the sample (Greece, Iceland, Portugal, Turkey and Yugoslavia). Australia and New Zealand were excluded due to lack of data for 1965. The countries included in the sample are: Canada, USA, Japan, Austria, Belgium, Denmark, Finland, France, Germany, Italy, the Netherlands, Norway, Spain, Sweden, Switzerland and the United Kingdom.

8.3 A PREVIEW OF THE DATA (1987)

Before moving to a more formal test, it may be useful to take a closer look at the data. The data set consists of 16 countries and 23 pairs of "export products"/"home-market sectors". For each year this gives us a total of 368 observations about the link between the home market sector and the export product. For exploratory reasons, in this section we will limit the examination to the most recent year (1987) and to the home-market sectors where the specialization is highest, irrespective of the country of origin. Only home-market sectors with a specialization index of 2.0 or higher, i.e. at least twice as large as the mean, are included. Table 8.2 ranks these home-market sectors, in descending order of the specialization index, together with the associated export products, for the year 1987.

The results reported in Table 8.2 appear to be consistent with the hypothesis of a positive impact of the home market on export specialization. In particular, for the top ten pairs in Table 8.2, a strong relation between export specialization and home-market specialization seems to exist, with 7 (out of 10) pairs specialized in both the export product and the home market sector. For the remaining pairs of export products/home-market sectors the evidence is weaker, although still not inconsistent with the underlying hypothesis (12 out of 21 specialized in both). A similar exercise was carried out at the opposite tail of the distribution, i.e. pairs with a value of the home market specialization index below 0.2. Here the evidence was even stronger: of the 30 pairs in this category, 24 were "not specialized" in either (in fact, in all 24 cases, the specialization figure of the export product was below 0.7).

The table reveals several illustrative examples, some of these from the Nordic countries, of high specialization in both the home market sector and the related export product. For Denmark, there appears to be a strong link from the agricultural sector to the machinery sector (milking machinery, food processing machinery and heating and cooling equipment). A similar relation holds for the Netherlands. For Sweden and Fin-

Table 8.2: Export specialization and home-market indicators (1987)

COUNTRY	EXPORT PRODUCT	SPECIALIZATION		HOME MARKET SECTOR	RANK
		EXPORTS	HOME		
FINLAND	PAPERWORKING MACHINERY	8.3	9.8	PULP & PAPER, PAPER PRODUCTS	1
DENMARK	HEATING AND COOLING EQUIPMENT	2.5	8.1	MEAT, DAIRY PRODUCTS, FISH AND EGGS	2
SWEDEN	PAPERWORKING MACHINERY	2.6	4.8	PULP & PAPER, PAPER PRODUCTS	3
NORWAY	SHIPS AND BOATS	8.1	4.6	SHIPPING	4
CANADA	CONSTRUCTION AND MINING MACHINERY	0.7	4.5	CRUDE MINERALS AND METALS	5
NETHER-LANDS	MILKING MACHINERY	1.3	4.4	DAIRY PRODUCTS	6
DENMARK	MILKING MACHINERY	11.8	4.4	DAIRY PRODUCTS	7
SPAIN	LEATHER MACHINERY	1.0	4.2	LEATHER	8
DENMARK	FOOD PROCESSING MACHINERY	3.3	4.1	FOOD	9
USA	AIRCRAFT ENGINES	0.8	3.9	AIRCRAFTS	10
ITALY	LEATHER MACHINERY	8.0	3.9	LEATHER	11
ITALY	SEWING MACHINERY	1.1	3.8	CLOTHING	12
CANADA	PAPERWORKING MACHINERY	0.4	3.6	PULP & PAPER, PAPER PRODUCTS	13
NETHER-LANDS	HEATING AND COOLING EQUIPMENT	0.6	3.3	MEAT, DAIRY PRODUCTS, FISH AND EGGS	14
SPAIN	AGRICULTURAL MACHINERY n.e.s	0.3	3.2	AGRICULTURAL PRODUCTS	15
SPAIN	AGR. MACH. (PREP. SOIL, HARVEST)	0.4	3.2	AGRICULTURAL PRODUCTS	16
SPAIN	TRACTORS	0.4	3.2	AGRICULTURAL PRODUCTS	17
SPAIN	HAND TOOLS (AGR. AND FORESTRY)	2.1	2.7	AGRICULTURAL PRODUCTS, WOOD PRODUCTS	18
NETHER-LANDS	FOOD PROCESSING MACHINERY	2.1	2.6	FOOD	19
NORWAY	HEATING AND COOLING EQUIPMENT	0.4	2.6	MEAT, DAIRY PRODUCTS, FISH AND EGGS	20
CANADA	HAND TOOLS (AGR. AND FORESTRY)	0.2	2.2	AGRICULTURAL PRODUCTS, WOOD PRODUCTS	21
AUSTRIA	PAPERWORKING MACHINERY	1.6	2.1	PULP & PAPER, PAPER PRODUCTS	22
SPAIN	FOOD PROCESSING MACHINERY	0.8	2.1	FOOD	23
ITALY	TEXTILE MACHINERY	1.8	2.1	TEXTILES	24
NETHER-LANDS	TRACTORS	0.1	2.1	AGRICULTURAL PRODUCTS	25
NETHER-LANDS	AGRICULTURAL MACHINERY n.e.s.	1.7	2.1	AGRICULTURAL PRODUCTS	26
NETHER-LANDS	AGR. MACH. (PREP. SOIL, HARVEST)	1.1	2.1	AGRICULTURAL PRODUCTS	27
JAPAN	AUTO-ELECTRICAL EQUIPMENT	1.9	2.0	ROAD MOTOR VEHICLES	28
JAPAN	INTERNAL COMBUSTION ENGINES	1.8	2.0	ROAD MOTOR VEHICLES	29
FRANCE	MILKING MACHINERY	0.5	2.0	DAIRY PRODUCTS	30
AUSTRIA	METALWORKING MACHINERY	1.2	2.0	METAL MANUFACTURES	31

land, a relationship was found between specialization in wood products/pulp and paper and specialization in exports of paperworking machinery. In the Norwegian case, shipping and exports of ships seem to be the most prominent example. Other examples are Italy (leather products/leather machinery, clothing/sewing machinery and textiles/textile machinery) and Japan (cars and auto-electrical equipment).

However, the failures are equally interesting. Table 8.2 includes 13 pairs with a specialization figure of the export product equal to 1.0 or less. These are not evenly distributed across countries: three countries account for 10 failures, with Spain in the lead (5 failures) followed by Canada (3 failures) and the Netherlands (2 failures). As noted above, the observations for the Netherlands are generally not failures: in four of the six pairs included in the table, the outcome was the expected one. In contrast, the three pairs for Canada included in the table are all failures. Although no attempt was made to verify this possibility, it would not be surprising if some of the producers who benefit from the Canadian demand for advanced equipment were situated at the other side of the US-Canadian border. Clearly, in this case, the differences in language, culture etc. - not to mention distance - are quite small. Spain was represented six times on the list and failed in five of them. As in the case of Denmark, Spain has a strong agricultural sector, but in contrast to the Danish example, Spain is not specialized in exports of agricultural machinery. The only exception is exports of handtools for agriculture and forestry, clearly the least advanced class of agricultural machinery included in the test. Arguably, what this shows is that Spain, as "the poorest among the rich" countries of our sample, has not reached the same degree of industrial maturity as the other countries. This does not necessarily imply that the hypothesis of a positive impact of the home market on industrial development has no explanatory power in the Spanish case, but it suggests that there may be important differences across countries in how this mechanism works. However, to explore this issue further, more refined techniques are needed.

8.4. A FORMAL TEST

In a general form, the model to be tested is the following:

$$S_{ij}^{t} = f\left(S_{ik}^{t}\right) \qquad \begin{array}{ll} i = 1...n & \text{(Countries)} \\ j = 1...m & \text{(Commodities)} \\ & \text{("export")} \\ k = 1...m & \text{(Commodities)} \\ & \text{("home-market")} \end{array} \qquad (2)$$

In principle, the choice of functional form should be based on theory, but in this case we have no particular theoretical reasons for preferring one specific functional form. However, the data may give some guidance. Despite its many desirable properties, the index of revealed comparative advantage (S) has one important disadvantage when it comes to statistical work: it has a skew distribution, with a long tail to the right. This creates problems in regression analysis, because it violates the assumption of normality. A logarithmic transformation of the data reduced this problem significantly, but since there were zeros in the data matrix, we had to add a small positive number to all observations to allow the transformation to be made. Thus, the tested model is the following:

$$\log\left(S_{ij}^{t}+0.1\right)=a_0+a_1\log\left(S_{ik}^{t}+0.1\right) \qquad \begin{array}{ll} i=1...n & \text{(Countries)} \\ j=1...m & \text{(Commodities)} \\ & \text{("export")} \\ k=1...m & \text{(Commodities)} \\ & \text{("home-market")} \end{array} \qquad (3)$$

The questions we want to ask are:

1) Is there a positive relationship between the two specialization indexes, as argued by Burenstam-Linder and Porter (i.e. is the coeffecient a_1 positive)? The 5 % level of significance was chosen for the test.

2) Does the model explain past specialization patterns better than present patterns?

3) To what extent does the introduction of a time-lag for the independent variable (S_{ik}) improve the explanatory power of the model?

To answer these questions separate regressions were carried out for 1965, 1987 and 1987 with a lag (the 1965 independent variable). There are alternative ways to group the sample. We can run a cross-sectional regression for each pair of export product/ home-market sectors, using the countries as units or a cross-sectional regression for each country, using the pairs as units. The alternatives partly reflect different aspects or interpretations of the model. For instance, in a cross-sectional regression for each product pair we seek to determine for which pairs the model is most relevant. However, the focus of Burenstam-Linder and Porter was mainly on the specialization pattern of *countries* and the results from the previous section suggest that there may be important differences across countries in how these patterns are shaped. Thus, what will be presented here is *a cross-sectional regression for each country* (for an attempt to explore the other alternative, a cross section for each pair, see Fagerberg, 1992).

Table 8.3 contains a summary of the results with respect to the impact of the home-market variable (s_{ik}). Only the estimate of coeffecient a_1 and its t-value are reported, but with only one independent variable (and a constant term), the relation between the t-value and the fit (F-statistics or restricted R^2) is relatively straightforward. The most important results are the following:

1) For most countries (13 out of 16) there is some support for the hypothesis (a significant positive effect in 1965, 1987 or both years). The three countries with no support at all for the hypothesis under test are Austria, France and the United Kingdom.

2) In most cases the results improve between 1965 and 1987: of the 13 countries where some support was found, 9 report higher absolute t-values and fits in 1987 than in 1965. Thus, if anything, the explanatory power of the model is greater in 1987 than 1965.

3) The introduction of a time-lag for the independent variable did not alter the results very much. The absolute t-value and fit improved in 7 cases and detoriated in 6 cases. Thus, on this question the results are inconclusive.

The results reported here may be compared with those reported by Andersen et al. (1981), although it must be kept in mind that their sample was much smaller (11 countries and 8-10 observations for each country). They reported results for 1954, 1960, 1966 and 1972, using both rank correlation and ordinary least squares (a linear formulation was used). No attempt was made to test for lags. As in the present study, the best results were reported for the most recent years. For these years a significant relationship was reported for approximately one half of the countries included in the sample, compared to around two thirds in the present study. The countries for which they found no support for the hypothesis were Belgium, France, Italy and the United Kingdom (Austria was not included). These results are in accordance with the results for 1965 reported here.

Table 8.3: The hypothesis tested1)

	1965	1987	1987 with lag
	Home2)	Home2)	Home2)
Canada	0.28 (1.45)	0.42 (2.09) *	0.44 (2.52) *
USA	0.43 (2.10) *	0.38 (2.47) *	0.57 (3.46) *
Japan	0.77 (4.03) *	0.51 (2.65) *	0.79 (3.88) *
Austria	0.26 (1.44)	0.16 (0.75)	0.07 (0.36)
Belgium	0.13 (0.40)	0.43 (2.34) *	0.39 (1.34)
Denmark	0.58 (3.33) *	0.77 (4.58) *	0.71 (5.16) *
Finland	0.50 (3.32) *	0.51 (2.01) *	0.52 (3.26) *
France	-0.20 (0.75)	-0.01 (0.01)	-0.34 (1.16)
Germany	0.31 (1.86) *	0.31 (1.98) *	0.10 (1.11)
Italy	0.20 (0.98)	0.54 (2.68) *	0.40 (1.72) *
Netherlands	0.10 (0.36)	0.36 (2.06) *	0.37 (1.91) *
Norway	0.42 (2.77) *	0.60 (3.35) *	0.44 (2.84) *
Spain	0.33 (2.21) *	0.23 (1.18)	-0.03 (0.27)
Sweden	1.00 (4.51) *	0.10 (0.40)	0.35 (1.61)
Switzerland	0.77 (2.95) *	1.12 (4.29) *	0.92 (3.55) *
UK	-0.16 (1.12)	0.39 (0.86)	0.04 (0.20)

1) 23 observations, one for each of export product/home-market sector pair.
2) Estimate of coefficient, absolute t-value in bracket.
*) Positive, 5% level of significance, one-tailed test.

8.5 CONCLUDING REMARKS

The view that domestic demand may have a positive impact on the competitiveness of a country (the Burenstam-Linder-Porter hypothesis) is by no means a new one. Indeed, as pointed out in the introduction, it has been widely held for at least a century. In spite of this, modern (neoclassical) trade theorists normally regard it as "theoretically unsound" and as a cover for protectionism. This chapter has argued that, based on modern innovation and industral organization theory, it is possible to give a plausible micro-economic foundation for the Burenstam-Linder-Porter hypothesis. Basically, the explanation offered is the following: (1) Innovation is the most important competitive factor in advanced industries. (2) Communication and interaction between advanced users and producers of technology play a vital role for innovation. (3) Proximity, a common language, a common education system etc. make the process of communication and interaction much easier. Hence, it is suggested that in the case of advanced products, countries tend to specialize in areas where, by a comparative standard, there are many advanced domestic users.

Based on an empirical methodology developed by Andersen et al. (1981), a formal test of the Burenstam-Linder-Porter hypothesis was made on data for 1965 and 1987. For a large majority of the countries included in the test, the results supported the hypothesis. This holds both for the largest countries in the sample (USA and Japan) and the smallest ones (Denmark, Finland, Norway and Switzerland). The least satisfactory results were reported for Austria, France and the UK. An interesting question for further research is why the factors affecting the trade patterns of these countries appear to diverge from most other developed countries, although it cannot be excluded that these results may be caused by imperfect data or methods. For France and the UK a possible explanation may be that their trade patterns are influenced by their colonial past.

An interesting finding, which is also supported by earlier studies, is that the explanatory power of the Burenstam-Linder-Porter hypothesis improves over time. The only clear counter-examples are Spain and Sweden. This finding suggests that the increasing economic integration in the developed world from the early 1960's onwards has strengthened, rather than weakened, the importance of domestic linkages and learning processes for comparative advantage in advanced products. As pointed out by Porter:

> "While globalization of competition might appear to make the nation less important, instead it seems to make it more so. With fewer impediments to trade to shelter uncompetitive domestic firms and industries, the home nation takes on growing significance because it is the source of the skills and technology that underpins competitive advantage." (Porter, 1990, p. 19).

One implication of this is that, contrary to common belief, the increasing role of multinationals in world trade has not eroded the benefits accruing to comparative advantage from advanced domestic users. Probably, the relationship between multinational activity and country-specific learning capabilities is a complex one. However, it is not possible to resolve this problem within the context of this chapter.

ACKNOWLEDGEMENT

The ideas presented owe much to discussions with the participants of the IKE group, at the University of Aalborg, Denmark - especially Bengt Åke Lundvall, Esben Sloth Andersen and Bent Dalum. A special word of thanks to Bengt Dalum and Vibeke Jakobsen for assistance in data work. I would also like to thank Aadne Cappelen, Central Bureau of Statistics, Margarida Ponte Ferreira, Norwegian Institute of International Affairs and an anonymous referee for valuable comments and suggestions.

Economic support from the Nordic Economic Research Council is gratefully acknowledged.

REFERENCES

Andersen, E.S., B. Dalum and G. Villumsen, 1ß81, *International Specialization and the Home Market*, Aalborg University Press, Aalborg.

Balassa, B., 1965, "Trade Liberalization and 'Revealed' Comparative Advantage", The Manchester School, 33:99-123.

Bowen, H.P., E. Leamer and L. Sveikauskas, 1987, "Multicountry, Multifactor Tests of the Factor Abundance Theory", *American Economic Review*, 77:791-809.

Burenstam-Linder, S., 1961, *An Essay on Trade and Transformation*, Almquist & Wicksell, Uppsala.

Dahmén, E., 1970, Entrepreneurial Activity and the Development of Swedish Industry 1919-1939, *American Economic Association Translation Series*, Homewood.

De Bresson, C. and F. Amesse, 1991, "Networks of Innovators: A Review and Introduction to the Issue", *Research Policy*, 20:363-379.

Dosi, G. and L. Soete, 1988, "Technical Change and International trade", in Dosi, G. et al., (eds), *Technical Change and Economic Theory*, Pinter, London.

Fagerberg, J., 1992, "The 'Home-Market Hypothesis' Reexamined", in Lundvall, B.Å., (ed.), *National Systems of Innovation - Towards a Theory of Innovation and Interactive Learning*, Pinter, London.

Hirschman, A. O., 1958, *The Strategy of Economic Development*, Yale University Press, New Haven, Conn.

Håkansson, H., 1987, *Industrial Technological Development*, Croom Helm, London.

List, F., 1959, *Das Nationale System der politischen Ökonomie*, Kyklos-Verlag, Basel.

Lundvall, B. Å., 1988, "Innovation as an Interactive Process - from User-Producer Interaction to the National System of Innovation", in Dosi, G. et al., (eds.), *Technical Change and Economic Theory*, Pinter, London.

Lundvall, B.Å., 1992, "User-Producer Relationships, National System of Innovation and Internationalisation", in Lundvall, B.Å., (ed.), *National Systems of Innovation - Towards a Theory of Innovation and Interactive Learning*, Pinter, London.

Perroux, F., 1956, "Note sur la Notion de Pôle de Croissance", *Economie Appliquée*, 7:307-320.

Porter, M. E., 1990, *The Competitive Advantage of Nations*, MacMillan, London.

Schumpeter, J., 1934, *The Theory of Economic Development*, Harvard University Press, Cambridge, Mass.

Schumpeter, J., 1939, *Business Cycles I-II*, McGraw-Hill, New York.

Schumpeter, J., 1943, *Capitalism, Socialism and Democracy*, Harper, New York.

Appendix 1

The trade data used in this chapter were calculated from OECD Trade Series C (value data) using the IKE data base on trade statistics at the Aalborg University Centre. Data for health care were taken from OECD: Health Care Systems in Transition, OECD, Paris, 1990; data for merchant fleets and telephone lines were taken from The UN Statistical Yearbook, various editions. Other data were from OECD National Accounts.

Construction of home-market indicators

Tele and Health

T_J = Telephone lines in country j
$j \varepsilon i,\ i=1, ..j\ ..n$

N_j = Number of inhabitants in j

$$I_j = \frac{T_j}{N_j} \frac{\sum_i N_i}{\sum_i T_i}.$$

Similarly for health services, where T_j = health services in country j (in common currency).

Shipping

S_j = Fleet of country, in 1000 tons
X_j = Total exports of country j (goods and services) in common currency

$$I_j = \frac{S_j}{\sum_i S_i} \frac{\sum_i X_i}{X_j}.$$

CHAPTER 9

Revealing Network Properties of Sweden's Trade with Europe

Börje Johansson
Royal Institute of Technology, Stockholm

Lars Westin
University of Umeå

9.1 INTRODUCTION

In this chapter, estimates of the network properties of Swedish exports of commodities to Europe are presented. The standard classical trade models and those generated by Heckscher-Ohlin indicate the direction of trade at the borders of nations but they do not generally tell us with whom a nation trades or the quantity of the flows. The two important factors of destination and quantity are not readily explained in the above models as trade patterns are not revealed by a characterization of the nodes in the trade network alone. It has to be augmented by a representation of link attributes and thus of spatial accessibility. The standard international trade models assume away friction, inertia and barriers in space and thus may not explain bilateral patterns of flows in a network although they do suggest factors in the nodes which are of importance for the creation of inter-industrial trade in homogeneous commodities.

However, the growth of cross hauling between industrialized countries is not, except for the special cases of strategic trade in a bi-national duopoly and various forms of border or cyclical trade, necessarily explained by the standard trade model unless differentiated goods and increasing returns to scale are introduced. The first assumption in particular removes much of the beauty of the traditional models. Since trade in differentiated commodities is less dependent on product prices and relies more on attributes connected with product competition, such a design, customization and service, one may assume that differentiated commodities would require, and be favoured by, links with properties other than those required for more homogeneous goods.

In this study we deal with the relation between node and link related network properties and the export patterns of four commodity aggregates. In order to reveal the properties of the aggregated flows we use a model which is inspired by the tradition of gravity models. The gravity model has been used in the empirical analysis of regional and interregional trade with great success. Generally, such a model may include a characterization of the nodes by variables representing market size, per capita income, population and internal costs of communication, while the links may be described by trade barriers and measures of accessibility such as cultural affinity, membership of trade preference areas, common borders as well as communication costs.

The model described in this chapter was estimated on a panel of Swedish export data for seven years between 1970 and 1987. A classification of commodities which emphasizes common aspects in their market performance, distinguishing between product competitive and price competitive aggregates, is used. Product competition is often associated with markets in the early stage of the product cycle, where product attributes and customization determine the success of sales, while the commodity price is the primary attribute when returns to scale are developed in later stages of the product cycle.

The estimations support our initial assumptions that the pattern of trade is sensitive to a mix of three factors: the accessibility by means of links connecting nodes, the attributes of the nodes and the market performance of the commodities. In particular, our results indicate that successful trade in product competitive commodities is favoured by similarities between Sweden and the import node with respect to cultural affinities, while price competitive commodities are more sensitive to distance and the existence of common borders between the trading nodes. However, we have also observed some contradictions with standard assumptions which indicate possible directions for further research.

The chapter is organized as follows. In Section 9.2 we provide a general introduction to the way that the gravity model can be used in the analysis of trade patterns in a network economy. Section 9.3 is devoted to assumptions regarding the relationship between trade patterns and market performance. Our data set and the Swedish export pattern are introduced in Section 9.4. The estimated model and the econometric results are presented in Section 9.5 while Section 9.6 concludes the chapter.

9.2 CHARACTERIZATION AND ESTIMATION OF LINK AND NODE PROPERTIES IN A NETWORK ECONOMY

Two main features of current trade flows form the focus of the research in this chapter. Firstly, one may observe the growth and dominance of differentiated commodities traded among industrialized countries. This represents a major change away from the trade of bulk and other standardized goods in earlier years. In the case of Sweden, the growing share of intra-industrial trade in relation to inter-industrial trade in its gross trade confirms this observation (Peterson, 1984).

A second feature which may be observed is the increased formation of economic networks among producers and subcontractors and between producers and buyers of final products. The establishment of such networks is connected to investments in links between deliverers and customers. Greater emphasis upon the network aspects of the economy, rather than the traditional price oriented market view, implies that link properties will increase in importance relative to node attributes as explanations of trade patterns, service networks and spatially distributed production systems. The archetypical model of a market economy with independent actors, in which a quantity of a commodity is bought from the seller who offers the lowest prices at the point of delivery, focuses in a myopic way upon production costs in nodes and, rarely, if at all, on transportation costs without observing the dynamic mutualities between actors at the demand and supply sides of a market.

The network economy is thus based on the existence of reciprocity over time and emphasizes mutual investments in the links between customers and deliverers (cf. Teubal and Zuscovitch, 1993; Johansson, 1990). Such investments represent sunk costs and introduce rigidities in customer relations which are reflected by inertia in the trade flows. The amount of investment needed to establish and strengthen a producer/customer connection is obviously negatively related to existing affinities between nodes and a positive function of the friction in mutual links. The dominant flows of a specific product category will thus follow links which have the most appropriate attributes for the products, while the total volume is constrained by barriers, friction and diversities.

The two processes of commodity differentiation and the emergence of economic networks are highly integrated. The demand for differentiated commodities is a function of incomes and social identification while the supply of differentiated products is motivated by the need to form a niche and to obtain a temporary monopoly. Hence, both sellers and buyers have incentives to develop long-standing links of communication and close interaction in order to secure the development of products which fulfil their individual goals. Customer networks involving service, financial assistance and information exchanges are visual responses to this demand. In Johansson and Westin (1991), it was suggested that this behaviour may be modelled by a logit formulation based on discrete choice at the micro level. At the aggregate level such a model would be cast into a gravity formulation of flows in a network.

Various versions of gravity models have successfully been used in the analysis of interregional as well as international trade and barriers to communication analysis. They may capture both dispersions of flows and cross-hauling caused by differentiated commodities and the inertia in flows which is a consequence of strong links between actors at the market. Hence, the models determine with whom, and in what quantities, trade takes place. Early studies may be found in Tinbergen (1962), Leontief and Strout (1963), Linnemann (1966), Gruber and Vernon (1970), Aitken (1973) and Leamer (1974) while more recent studies are Bröcker (1984), Rietveld and Janssen (1990) as well as Bröcker and Rohweder (1990).

The empirical success of the gravity model has not been followed by unequivocal theoretical support. However, recently the relation between discrete choice, multinominal logit, entropy models has been understood more fully.[1] This research has also given the gravity formulation stronger theoretical support, with a foundation in discrete choice theory. Bröcker (1989) also analysed and developed the relation between a model of partial spatial price equilibrium with differentiated goods and various gravity models. As Bröcker also observes, this will improve the possibility to specify the structural form of a trade model and then to transform it in a second stage to a reduced form of gravity model or some other type of model which may be estimated on actual flows. Hitherto, the procedure has been to identify an empirically suitable model first and then, if at all, to try to find a structural model consistent with the theory and the estimated "reduced" form.

In a simple version, the unconstrained gravity model may be given the following well known formulation,

$$x^{rs} = A(Z^r)^{\alpha}(Z^s)^{\beta}f^{rs} . \tag{1}$$

The notation in (1) should be interpreted as follows,

- x^{rs} Commodity flows from node r to node s measured in quantities
- A Intercept coefficient
- Z^r Attributes related to origin node r
- Z^s Attributes related to destination node s
- α, β Parameters
- f^{rs} A function of attributes of the link between r and s

In a specific estimation, a vector of variables may represent node attributes. Variables such as population, market area, income, and educational level may be used. Other attributes of nodes are related to transaction and communication costs which represent

[1] A review of such models may be found in Batten and Boyce (1986).

friction inside each node. The number of telephones, telephone calls, vehicles, passenger miles and other indices of internal infrastructure may, after normalization, be indicators of such internal friction. In Table 9.1 common variables describing such node related attributes are listed.

Table 9.1: Variables describing attributes of node r

GDP^r	Gross domestic product (GDP)
N^r	Population
$GDPC^r$	GDP per capita
$\Delta GDPC^{rs}$	Difference in GDP per capita between r and s as a measure of similarity in income and taste
INF^r	Measures of internal friction inside a nod

The link related function in Equation (1) contains attributes of the links between nodes. In an ordinary linear regression this may either be modelled by an additive, multiplicative or combined function. We may thus write,

$$f^{rs} = \tau^{rs}\, w^{rs} + \exp(\hat{\tau}^{rs}\hat{w}^{rs}). \qquad (2)$$

In Equation (2), w^{rs} and $\hat{w}^{rs}$ are attributes related to the link between r and s while τ^{rs} and $\hat{\tau}^{rs}$ are parameters. Table 9.2 contains variables which characterize link attributes. Distance is generally measured as the crow flies between capitals of nations or between large harbours. The cost of transportation would be an appropriate alternative but it is often difficult to measure. Dummy variables are appropriate to estimate the effects of common borders and membership in trade agreements, while the use of a dummy to represent things such as cultural affinities rather reflects the problems of finding suitable measures in this area.

Table 9.2: Variables which describe link attributes between node r and s

d^{rs}	Distance between r and s as a measure of accessibility
c^{rs}	Transport cost between r and s
D_T	A dummy variable for membership in a trade agreement
D_B	A dummy variable for common borders
D_L	A dummy variable for language and other cultural affinities between r and s

An alternative to the formulation in (1) with actual trade flows as the dependent variable, is to use fractions of total trade flows on links. In panel data, the use of fractions eliminates cycles and trends over time in trade. Instead, processes of substitution and complementarity between flows on different links are stressed. When the fraction of total exports from region r is studied such a variable may be obtained as,

$$\theta^{rs} = \sum_{i} x_i^{rs} / \sum_{si} x_i^{rs}, \qquad (3)$$

while the fraction of an origin-destination relation in the exports from region r of a specific commodity is defined by,

$$\theta_i^{rs} = x_i^{rs} / \sum_s x_i^{rs}. \tag{4}$$

With fractions given by (3) as the dependent variable, equations of the following type may be estimated,

$$\theta^{rs} = A(Z^r)^{\alpha}(Z^s)^{\beta} f^{rs}. \tag{5}$$

The use of fractions, which are constrained to be in the interval between zero and one, suggests estimation with the logistic function as an alternative. The following linearized formulation may then be utilized as an alternative to the gravity formulation in (5).

$$\ln(\theta^{rs} / (1-\theta^{rs})) = g^{rs}(Z^r, Z^s, f^{rs}). \tag{6}$$

The above models may be used in order to test different hypotheses related to trade flows and to find new empirical regularities. In our case, results explaining the pattern of trade for differentiated commodities and the impacts of barriers to trade on various commodities are especially important. The central assumption in relation to differentiated commodities is the assumed correlation with income per capita. This assumption has been discussed and tested in the literature. (See Burenstam-Linder, 1961; Lancaster, 1980; Krugman, 1979; Helpman, 1981; Balassa and Bauwens, 1988.) These studies indicate that the demand and supply of differentiated commodities are positively associated with income. This suggests that,

*Product differentiation is a positive function of the income level in nodes, measured by GDP per capita.

The market size is also assumed to be positively correlated with the number of differentiated commodities produced. Small countries with restricted accessibility to large markets would thus have an advantage in the production of homogeneous commodities since,

*The size of a node - e.g., measured by total GDP or population - constrains the possibility to exploit returns to scale and scope.

The successful development of differentiated commodities is dependent on the possibility to establish close interaction between suppliers and customers. One may thus assume that,

*The production of differentiated commodities is negatively correlated with the amount of internal friction in a node.

Following Balassa and Bauwens (1988), one may furthermore assume that trade in differentiated products between two nodes is positively correlated with link related factors such as,

*Similarities in language and culture.

*Participation in common trade agreements.

*Common borders between nations.

*Similarities in income per capita and thus in taste.

*Accessibility with respect to transportation.

With the exception of similarities per capita, which could also represent similarities in factor proportions, trade in homogeneous commodities may also be positively correlated with those factors. The important question is thus whether significant differences exist between commodities with respect to the above barriers.

9.3 MARKET PERFORMANCE AND TRADE PATTERNS

Increased product differentiation on a market implies that the price of a commodity, although still important, loses dominance as a means of competition. From a product cycle perspective one may also argue that products in the early stage of the cycle in particular are characterized by attributes which are important complements to price as a signal to the customer. The market performance is then characterized by product competition and the trade pattern is similar to that which can be expected for differentiated commodities. In the final phases of the product cycle, where products are standardized, returns to scale exploited and larger volumes sold, the importance of relative prices increases.

In order to test our assumptions regarding market performance and trade patterns a classification of commodities with respect to market performance is needed. Such a classification may be accomplished in a number of ways. Our classification is made from estimated growth rates of export prices relative to import prices and the relative growth of the volumes of exports to imports of commodities during the years 1970-1985. Four categories of commodities were identified by the use of such growth rates and the growth of employment in the industrial sectors producing the respective commodities.[2] The four categories are:

(1) Product competitive commodities with increasing employment (expansive)
(2) Product competitive commodities with decreasing employment (contractive)
(3) Price competitive commodities with increasing employment (expansive)
(4) Price competitive commodities with decreasing employment (contractive)

A commodity for which the relative export to import price has not decreased while the quotient between the export and import volumes has increased is defined as product competitive. In this case price competition has not been necessary in order to increase the market share. The remaining sectors are regarded as price competitive in the sense that they reduce their prices in order to extend or keep their market shares. Sectors which have reduced their prices and their market share are also assigned to the second class and may be regarded as unsuccessful price competitors.

The second division into expansive and contractive categories is made by analysing employment growth rates. Each commodity is then associated with the industrial sector which dominates the production of the commodity. The division is made in order to identify differences in the demand for network properties between growing and contracting sectors on the Swedish labour market in terms of employment. Increasing employment is assumed to be connected with the expansive part of the product cycle. Category one, sectors with an expansive product and a competitive market performance, contains the most successful commodities during this period, and they also have a prosperous future. Employment contraction may in this respect be a first sign of ageing and an outcome of labour saving investments, but it may also be caused by an unsuccessful performance at the market. When it comes to categories three and four it is more dubious to say anything regarding the future of the commodities. Within both categories profitable as well as less profitable commodities may show up. It is a priori difficult to have any specific ideas regarding the differences in their network properties.

To summarize, our first hypothesis is that commodities oriented towards the product development phase need a network with attributes, similar to those which may be expected for differentiated commodities, in contrast to price competitive goods. Secondly, there are also differences between commodities with increasing and decreasing employment within the product competitive segment. The importance of such differences in link attributes can furthermore be revealed by the flows of

---

[2] The classification may be found in Appendix I.

commodities on various links. By comparing differences in demand elasticities between the categories, a further test of our assumptions may be obtained. Thus,

*Differentiated commodities should be less elastic with respect to prices compared to price competitive aggregates.

Our assumptions regarding market performance and the parameters presented above may be summarized in the following way:

*Trade with product competitive commodities is sensitive to affinities such as differences in income per capita and taste, cultural and other barriers between nodes but less sensitive to price changes.

*Commodities in price competition are elastic with respect to price but less influenced by cultural barriers and income differences. However, the cost of transportation and the impact of distance are important.

In this way, we can integrate properties derived from dynamic analyses of product cycle theory and market performance with affinity variables utilized in the analysis of barriers to trade and price elasticities which are used in the estimation of demand functions.

9.4 THE SWEDISH EXPORT OF COMMODITIES TO EUROPE

The foreign trade data used in the following econometric part of the study is based on the CCCN[3] statistics of Sweden's exports of mining and manufacturing commodities to the rest of Europe for the years 1970, 1975, 1980, 1985, 1986, 1987. Commodities produced in sectors under trade protection have been excluded from the material.[4] Unit prices (f.o.b.) on each link may be calculated from the data about the exported quantities and values of each commodity.

Commodities are aggregated with respect to market performance and employment into the four categories discussed above in Section 9.3. The European countries are aggregated into the ten export regions shown in Table 9.3 below. This division reflects the country orientation of Swedish exports with a more detailed representation of the Nordic countries.

Table 9.3: European countries grouped into ten Swedish export-regions. The full name of the country aggregate is given within parentheses.

Export-region	Countries
NOR	Norway
DEN	Denmark
FIN	Finland
ATL (Atlantic countries)	Iceland, Faeroe Islands, Greenland
FRG	Federal Republic of Germany
SWC (Small West and Central European countries)	Netherlands, Belgium, Luxembourg, Ireland, Switzerland, Austria
UK	United Kingdom
FRA	France
WME (West Mediterranean countries)	Spain, Portugal, Italy, Malta
EME (East Mediterranean countries)	Yugoslavia, Greece, Turkey, Israel

[3]CCCCN: Customs Cooperation Council Nomenclature.

[4]Such sectors have been identified by the share of international trade in relation to Swedish production.

Exports in metric tons from Sweden to each of the ten export-regions are given in Table 9.4. The table contains both total flows to each region and a division into the four categories of product and price competitive commodities. Categories with increasing and decreasing employment are indicated by (+) and (-)signs.

In Table 9.4, there is a difference between the sum of the four categories of product and price competitive commodities and the total flows. This is due to the exclusion of agricultural goods and other commodities which are protected from international trade.

Export-regions are ordered according to their importance as receiving nodes of exports from Sweden. In 1985, the export volume to Europe represented 81% of Sweden's global export volume. The regional pattern is dominated by the trade with countries around Germany in the North-West and Central parts of Europe. As one may expect, the total weight of product competitive commodities is substantially below the weight of the two price competitive aggregates which dominate the pattern.

The relative importance of each category of trade with each export-region is given in Table 9.5. In Table 9.5 export-regions are ordered with respect to the share of the expansive product competitive commodities of total trade. The table shows that Sweden's trade with Norway and Finland contains a relatively large share of expanding product competitive commodities, while its trade with Denmark is dominated by successful price competitive commodities.

Table 9.4: Sweden's exports of expansive (+) and contractive (-) product and price competitive commodities, as well as total exports in metric tons (1985).

Export-region	1 Prod(+)	2 Prod(-)	3 Price(+)	4 Price(-)	5 Total
SWC	467	627	1203	7151	9659
FRG	284	629	1189	7283	9653
DEN	196	641	2231	1966	5318
UK	206	1165	701	2532	4726
NOR	801	865	827	1176	4636
FRA	156	319	75	3837	4406
FIN	316	151	150	2005	3016
WME	54	321	152	1034	1668
EME	23	82	4	359	477
ATL	21	25	37	25	125
TOTAL EUROPE	2524	4825	6569	27368	43688
REST OF WORLD	623	916	166	6439	9955
EUR/WORLD	0.80	0.84	0.98	0.81	0.81

Swedish exports to central and southern Europe are, on the other hand, dominated by contractive price competitive commodities. If this aggregate is elastic with respect to distance it provides an argument in favour of increased efficiency in the transport connections with those regions in order to strengthen the competitive situation of a large share of Swedish trade.

As is shown in Table 9.6, a different pattern occurs when flows are measured in values instead of volumes. Still, around 40% of the total trade consists of contracting price competitive commodities. However, in value terms the share of expanding product competitive commodities is more than a quarter of the total export value. One may furthermore observe that exports to the rest of the world are dominated by product competitive commodities.

Table 9.5: Export volumes in metric tons (1985). Percentage of total flows to each export-region

Export-region	1 Prod(+)	2 Prod(-)	3 Price(+)	4 Price(-)
NOR	21.8	23.6	22.5	32.1
ATL	19.4	23.1	34.3	23.1
FIN	12.1	5.8	5.7	76.5
EME	4.9	17.5	0.9	76.7
SWC	4.9	6.6	12.7	75.7
UK	4.5	25.3	15.2	55.0
DEN	3.9	12.7	44.3	39.1
FRA	3.6	7.3	1.7	87.5
WME	3.5	20.6	9.7	66.2
FRG	3.0	6.7	12.7	77.6
TOTAL EUROPE	6.0	12.0	16.0	66.0
REST OF WORLD	7.6	11.2	2.0	79.2

Table 9.6: The value of Sweden's exports of mining and manufacturing products to ten export-regions in Europe (1985). Percentage of the total trade value 156 billion SEK.

Export-region	1 Prod(+)	2 Prod(-)	3 Price(+)	4 Price(-)	1/4
SWC	7.2	2.3	2.2	6.2	1.2
NOR	3.9	2.7	2.1	5.5	0.7
UK	3.8	2.9	1.3	6.4	0.6
FRG	3.0	2.5	2.4	9.0	0.3
FIN	2.8	1.5	1.0	2.7	1.0
DEN	2.6	2.1	3.3	3.9	0.7
WME	1.7	1.3	0.6	3.5	0.5
FRA	1.4	1.3	0.5	4.0	0.4
EME	0.6	0.3	0.2	0.8	0.8
ATL	0.1	0.1	0.1	0.1	1.0
EUROPE	27.2	16.9	13.7	42.2	0.7
REST OF WORLD*	45.7	16.1	4.8	28.4	1.6

*The total export value to the rest of the world is 68 Billion SEK.

The fifth column in Table 9.6 gives the quotient between expanding product competitive trade and contracting price competitive trade in terms of value on each link.

The figures confirm the results from the analysis of volumes and indicate that the Nordic countries and the SWC region have shares above the average, while Sweden's trade with Germany is dominated by commodities in price competition with decreasing employment.

The difference between patterns measured by volumes and values confirms the product cycle related assumption that product competitive commodities are lighter, more valuable and thus more expensive compared with price competitive commodities. The f.o.b. prices of commodities in this study are given in Table 9.7 below.

Table 9.7: Average prices (f.o.b.) of Sweden's exports to European export-regions (1985). SEK/KG.

Export-region	1 Prod(+)	2 Prod(-)	3 Price(+)	4 Price(-)	Average
WME	49.0	6.2	6.2	5.3	7.1
EME	38.0	6.5	64.1	3.6	6.5
UK	28.8	3.9	3.0	3.9	4.9
SWC	24.2	5.7	2.8	1.3	3.0
DEN	20.6	5.1	2.3	3.1	3.8
FRG	16.6	6.1	3.1	1.9	2.9
FRA	14.0	6.4	11.2	1.6	2.6
FIN	13.9	15.7	10.2	2.1	4.5
NOR	7.6	4.8	4.0	7.3	5.3
ATL	6.9	5.6	2.7	8.5	5.4
MEDIAN	18.6	5.9	3.6 (3.1)*	3.4	4.5
ST.DEV	13.5	3.3	19.0 (3.4)*	2.5	1.5

*When the average prices for EME are excluded.

As may be expected, the low weight of expanding product competitive commodities is connected with high unit values. However, one should also observe the relatively low prices of contracting product competitive commodities. This fact may explain why employment in this category is decreasing and indicates that Sweden should specialize even more on the expensive part of the product competitive segments. The small flows to the Mediterranean regions are connected with high unit values which give the regions their top ranking in the table.

In international trade studies the standard deviation of the unit prices to different destinations is often used as a measure of the degree of product differentiation. The standard deviations given in Table 9.7 indicate that the classification of aggregates is reasonable in the context of this test. A comparison with the flows in Table 9.4 indicates that the high standard deviation of the expanding price competitive aggregates is mainly due to the small level of exports to the eastern Mediterranean countries. An exclusion of this flow reduces the standard deviation to 3.4, which is at an expected level. The standard deviation on contracting product competitive commodities has a low value although the average price exceeds the prices of the two price competitive aggregates. This suggests that standard deviation should be used carefully as a measure of product differentiation, at least at aggregated levels.

Interestingly enough, the trade patterns, unit prices and standard deviations are within bounds which may be expected from a priori assumptions. This suggests that the very rough classification which was obtained from the trade and employment statistics alone gives a reasonable picture of a product cycle related trade pattern.

9.5 A MODEL OF NETWORK ATTRIBUTES FOR SWEDISH EXPORTS TO EUROPE

In order to test our hypotheses regarding market performance and trade patterns, the following model has been estimated for each of the four commodity aggregates.

$$x^{SE} = A(GDP^S)^{\alpha 1}(GDP^E)^{\alpha 2}(p^{SE})^{\alpha 3}(\Delta GDP^{SE})^{\alpha 4}\exp\{\alpha_5 t\}f^{SE}, \tag{7}$$

$$f^{SE} = \exp\{\alpha_6 d^{SE} + \alpha_7 D_L + \alpha_8 D_B + \alpha_9 D_A\}. \tag{8}$$

The time dimension of each variable has been suppressed in order to simplify notation. Flows are measured in metric tons, while values are measured in 1980-based prices. The variables in (7)-(8) should be interpreted as follows:

x^{SE}	Flows from Sweden (S) to an export-region (E) in metric tons
A	Intercept coefficient
GDP^S	GDP in Sweden
GDP^E	GDP in the export-region
p^{SE}	Price (f.o.b.) of a commodity produced in Sweden and exported to an export region
t	A time trend
d^{SE}	Distance as the crow flies between S and E
D_L	Dummy for language and cultural similarities
D_B	Dummy for countries with whom Sweden has a common border
D_A	Dummy for the Atlantic export-region

During the estimations the GDP variables have also been substituted for or combined with income per capita variables. But generally multi-colinearity prohibits the simultaneous use of income and income per capita variables. The data set consists of four subsets of observations, one for each commodity aggregate. Each subset consists of 10 trade relations during 7 years; that is, a panel of 70 observations. Differences in accessibility between regions and commodities are estimated through distance and dummy variables. The dummy, D_L, indicating language and cultural similarities, is active for all regions with the exception of France and the two Mediterranean regions. The dummy for a common border, D_B, contains Norway, Denmark, and Finland. A dummy for the Atlantic region, D_A, has been included since the trade with those islands during the residual analysis was shown to contain significant effects compared with the average trade pattern. An alternative to the above procedure would have been to estimate a complete fixed effect model in a first step and to regress the fixed effects against the link specific variables in a second step.

Although the model (7)-(8) has strong similarities with the traditional gravity model, it also contains some specific qualities. Since the data only consist of a row in the European trade matrix, Swedish exports, Sweden does not act as an attractor in the model. Secondly, gravity models are often estimated on a yearly basis over cross-sections while this model is estimated on a panel of data. The latter motivates the inclusion of Swedish GDP and the time trend as variables.

The time trend, t, may be interpreted as an indicator of the correspondence between the preferences among consumers and the attributes of the commodities delivered from

Sweden. It may also capture other trends, for example changes in the overall accessibility to the customers and a development towards lighter commodities. However, the Swedish GDP variable has been included as a measure of the elasticity of exported volumes with respect to increased incomes in Sweden: the hypothesis being that exports become lighter over time.

The price variable captures the static demand price elasticity with respect to Swedish exports. The size of this elasticity reflects the degree of product differentiation and rigidities in the market due to long term contracts and producer/ customer relations. Prices are measured at the Swedish border and thus there should not be a double count for the link costs which are represented by distance and dummies. As discussed by e.g. Linnemann (1966) and Bröcker (1989), prices are generally left out of gravity models as it is argued that the model represents the reduced form of a price equilibrium where prices may be eliminated and the exogenous attributes of nodes should explain the pattern of trade. In this case, the use of a time series of data implies that the notion of a single static equilibrium becomes inapplicable. Prices changes reflecting imbalances caused by devaluations, investments in new technology, demand shifts etc. are then parts of an explanation of trade flows. The fact that we do not estimate a full origin destination matrix provides further support for a formulation with elements from a traditional demand function.

At this stage the model and its results should thus primarily be seen as a model which tests our previous hypotheses and attempts to identify empirical regularities, rather than the reduced form of some consistent structural model.

The results of our estimations are presented in Table 9.8 below. Initially one may observe that all four regressions provide a high degree of explanation and significant estimates of most variables are obtained. However, because of multi-colinearity, income per capita differences have not been found to provide an independent explanation of the trade pattern when GDP variables are included in the regressions.

The coefficient related to the Swedish GDP has a negative elasticity for each category. We interpret this as a confirmation of our assumptions that heavier products are substituted for lighter products when Swedish GDP increases. This process has been considerably faster compared to other categories in the low valued price competitive segment with decreasing employment.

The GDP elasticity with respect to the destination regions may be interpreted as a combined effect of the income in, and the size of, the destination markets. In all four cases this elasticity is positive but inelastic. However, contracting commodities are more elastic than expanding commodities.

Together, the two elasticities with respect to GDP measure the elasticity with respect to relative GDP. One interpretation is that the volume of exports will decrease when the size of the Swedish market grows relative to the export market. Then there are the two largest categories in terms of exported values, categories one and four, associated with positive time trends in their exports.

The language dummy supports our assumption that product competitive commodities are more sensitive to cultural affinities than price competitive products. The first two categories both have a relatively high value on the cultural dummy, while the employment increasing aggregate is considerably more sensitive to such affinities.

As was also assumed, distance is an important factor in explaining the large flows of price competitive commodities with declining employment to central Europe. In the two middle cases of successful price competition and contracting product competition the dummy representing common borders - that is, trade on the Scandinavian markets - is above the average. The low value of the border dummy in the case of the successful product competition category was not expected. The result lends some support to other indications that Swedish intra-industrial trade with Scandinavia has declined.

Table 9.8: Node and link attributes of Sweden's exports in metric tons to Europe. Product and price competitive commodities (1970-1987)*

Export-region	1 Prod(+)	2 Prod(-)	3 Price(+)	4 Price(-)
A	3.97 (2.18)	1.56 (1.37)	2.09 (1.32)	6.15 (5.46)
ln GDP^S	-0.42 (-2.03)	-0.59 (-4.48)	-0.58 (-3.42)	-0.99 (-7.96)
ln GDP^E	0.46 (3.82)	0.87 (12.30)	0.79 (7.54)	0.96 (15.88)
t	0.05 (6.60)			0.03 (4.42)
D_L	1.31 (10.45)	0.81 (10.98)	0.62 (6.62)	0.69 (10.04)
d^{SE}				-0.0002) (-4.71)
D_B	0.74 (2.48)	1.68 (10.28)	1.79 (6.79)	1.18 (6.70)
D_A	-1.57 (-2.45)	0.89 (2.47)	0.77 (1.40)	0.67 (2.18)
ln p^{SE}	-0.93 (-8.04)	-1.13 (-30.00)	-1.21 (-30.60)	-0.78 (-13.06)
R^2	0.95	0.98	0.98	0.98
F-value	176	499	451	645
Observations	70	70	70	70

* Plus and minus signs refer to employment development in the production sectors in Sweden. t-values, corrected for heteroscedasticity are given in parentheses. Heteroscedasticity is taken care of by White's (1980) method.

Finally, the price elasticity was shown to be relatively low for the expansive product competitive category. Although the elasticity was surprisingly high, this was an expected result from a product cycle point of view. However, the elasticity of contracting price competitive commodities is even lower. This was not assumed. One explanation may be that the measured elasticity is static. When the market is characterised by long term contracts, the short term effect of price changes may be small. A dynamic measure of the price elasticity would then be higher. The unequal length of the periods between the years in the data set have prevented such a model. A second explanation to be observed elasticities may be that the demand function is not identified in an appropriate way.

Income per capita variables were not introduced in the estimations presented above. In Table 9.9, estimations with per capita variables are reported in order to study the robustness of estimates with respect to a change of GDP for GDP per capita.

Only the Swedish per capita income was found to provide a significant degree of explanation while the absolute difference in GDP instead became a significant variable in two cases. In all four cases this estimation provides models which are slightly inferior to the previous models.

Table 9.9: Node and link attributes of Sweden's exports in metric tons to Europe. Product and price competitive commodities (1970-1987). Estimates with per capita variables*

Export-region	1 Prod(+)	2 Prod(-)	3 Price(+)	4 Price(-)
A	2.66	4.25	4.75	4.11
	(5.21)	(18.93)	(26.44)	(5.47)
$\ln GDPC^S$		0.89	0.66	
		(11.19)	(8109	
t	0.06	-0.03		0.04
	(8.63)	(-2.86)		(3.92)
$\ln \Delta GDP^{SE}$	0.21			0.33
	(4.97)			(5.95)
D_L	1.30	0.71		
	(14.58)	(8.02)		
d^{SE}		0.0003		-0.0007
		(5.26)		(8.14)
D_B				-0.66
				(-2.98)
D_A	-3.75	-3.95	-3.34	-3.56
	(-26.58)	(-26.09)	(-26.42)	(17.33)
$\ln p^{SE}$	-1.04	-1.31	-1.30	-0.79
	(-15.55)	(-24.14)	(-31.52)	(-10.01)
R^{SE}	0.95	0.98	0.97	0.97
F-value	259	438	293	294
Observations	70	70	70	70

* Plus and minus signs refer to employment development in the production sectors in Sweden. t-values, corrected for heteroscedasticity are given in parentheses. Heteroscedasticity is taken care of by White's (1980) method.

The overall conclusions from the first set of estimations regarding cultural affinities, distance and price elasticities remain. although the positive sign on distance in category two is counter-intuitive. In this second set of estimations, the difference between product competitive and price competitive commodities regarding the importance of cultural affinities is even stronger.

The positive signs on the variable measuring GDP differences in absolute terms indicate, as the first regressions also indicated, increased exports with increased differences in market size between Sweden and the export destination. The estimated regressions did not support our assumption about how differences in GDP per capita influence and explain trade with product competitive commodities. The regression coefficient has the wrong sign, compared with expectations. Neither does the contracting price competitive aggregate display the expected high price elasticity.

9.6 CONCLUDING COMMENTS

In this chapter we have discussed why network properties and the existence of appropriate links and barriers to trade and communication should be important components in the analysis of market performance and patterns of trade. Our assumptions regarding the impacts of cultural affinities and distance on the trade of commodities in price and product competition have gained some support from our estimations of the link and commodity dependent attributes in Sweden's exports to Europe.

Since the accessibility to links between countries reflects investments in infrastructure, our results also have at least one policy implication. Expensive, differentiated products in the early stages of the product cycle may obtain a strong comparative advantage if their advertisements, designs and standards are adjusted to the traditions in the region of destination and if efforts are made to improve the understanding of the foreign culture and language. Such gains may be greater than those obtained from investments in physical infrastructure.

The study has also identified areas where further work may be concentrated. Alternative measures of market performance is one such area. A second area would be to use our approach on less aggregated data both with regard to commodities and regions. In the model we have chosen a set of dummy variables instead of estimating a fixed effect model. As was discussed in the text, an alternative to our approach would be to first estimate the fixed effects and then to explain those effects by other variables. In a future study this approach will be compared with the one chosen here. Other measures of income per capita differences may also be tested to further analyse the Burenstam-Linder hypothesis.

We have not corrected our analysis for autocorrelation in time and space. The lack of regularity in our time series makes the first type of correction difficult while a spatial correction may be included in further work.

ACKNOWLEDGEMENT

The authors wish to thank Lena Sahlin and Jens Tjernström for their assistance in creating the database for the econometric part of this chapter.

REFERENCES

Aitken, N.D., 1973, "The Effect of the EEC and EFTA on European Trade: A Temporal Cross-Section Analysis", *American Economic Review*, 63:881-891.

Balassa, B. and B. Bauwens, 1988, "The Determinants of Intra-European Trade in Manufactured Goods", *European Economic Review* 32:1421-1437.

Batten, D.F. and D. Boyce, 1986, "Spatial Interaction and Interregional Commodity Flow Models" in Nijkamp, P., (ed.), *Handbook of Regional and Urban Economics*, Vol. 1, North-Holland, Amsterdam.

Bröcker, J., 1984, "How do International Trade Barriers Affect Interregional Trade?", in Å.E. Andersson et al., (eds.), *Regional and Industrial Development Theories, Models and Empirical Evidence*, North-Holland, Amsterdam.

Bröcker, J., 1989, "Partial Equilibrium Theory of Interregional Trade and the Gravity Model", *Papers of the Regional Science Ass.*, 66:7-18.

Bröcker, J. and H.C. Rohweder, 1990, "Barriers to International Trade. Methods of Measurement and Empirical Evidence", *Annals of Regional Science*, 24:289-305.

Burenstam-Linder, S. 1961, *An Essay on Trade and Transformation*, John Wiley and Sons, New York.

Deardorff, A.V., 1984, "Testing Theories and Predicting Trade Flows", in Jones and Kenen, (eds.), *Handbook of International Economics*, Elsevier Science Publ.

Gruber, W. and R. Vernon, 1970, The Technology Factor in a World Trade Matrix" in R. Vernon, (ed.), *The Technology Factor in International Trade*, Columbia University Press, New York.

Helpman, E., 1981, "International Trade in the Presence of Product Differentiation, Economies of Scale and Monopolistic Competition: A Chamberlin-Heckscher-Ohlin approach", *Journal of International Economics*, 11:305-340.

Johansson, B., 1990, "Innovation Processes in the Urban Network of Export and Import Nodes: A Swedish Example", in Nijkamp, (ed.), *Sustainability of Urban Systems*, Avebury, Aldershot.

Johansson, B. and L. Westin, 1991, "How Network Properties Influence Trade Patterns", Paper presented at the 31st European RSA congress in Lisbon.

Krugman, P. (1979), "Increasing Returns, Monopolistic Competition, and International Trade", *Journal of International Economics*, 9:469-479.

Lancaster, K, 1980, "Intra-Industry Trade under Perfect Monopolistic Competition", *Journal of International Economics*, 10:151-175.

Leamer, E.E., 1974, "The Commodity Composition of International Trade in Manufactures. An Empirical analysis", *Oxford Economic Papers*, 26:350-374.

Leontief, W. and A. Strout, 1963, "Multiregional input-output Analysis" in T. Barna, (ed.), *Structural Interdependence and Economic Development*, MacMillan, London.

Linnemann, H.J., 1966, *An Econometric Study of International Trade Flows*, North-Holland, Amsterdam.

Peterson, L., 1984, Svensk Utrikeshandel 1871-1980. En Studie i den Intraindustriella Handelns Framväxt, Lund Economic Studies, No. 30.

Reitveld, P. and L. Janssen, 1990, "Telephone Calls and Communication Barriers. The Case of the Netherlands", *Annals of Regional Science* 24:307-318.

Tinbergen, J. 1962, *Shaping the World Economy: Suggestions for an International Economic Policy*, The Twentieth Century Fund, New York.

Teubal, M. and E. Zuscovitch, 1993, "Demand Revealing and Knowledge Differentiation through Network Evolution", Chapter 2, this volume.

White, H., 1980, "A Heteroscedasticity-Consistent Covariance Matrix Estimator and a Direct Test Heteroscedasticity", *Econometrica*, vol. 48, pp 4:817-838.

APPENDIX I. SECTOR CLASSIFICATION

The classification is made according to the standard of Swedish classification of economic activities (SNI) as well as market and product cycle performances. Average export prices (1985) are also given in SEK/Kilo.

PRODUCT COMPETITIVE COMMODITIES:

INCREASING EMPLOYMENT		DECREASING EMPLOYMENT	
SNI	SEK/KILO	SNI	SEK/KILO
2302	2.1	3212	64.9
2909	0.3	3215	46.4
3115	7.2	3219	40.0
3419	9.4	3232	580.1
3511	4.1	3311	2.2
3522	242.2	3319	8.2
3560	30.6	3513	8.7
3720	15.5	3521	8.2
3821	84.8	3829	55.0
3832	397.7	3831	74.3
3843	48.3	3844	82.3
3851	419.8	3845	3318.6
3852	531.9	3853	383.0

PRICE COMPETITIVE COMMODITIES:

INCREASING EMPLOYMENT		DECREASING EMPLOYMENT	
3119	21.5	2301	0.1
3529	6.4	2901	0.1
3530	1.8	3114	15.4
3540	2.8	3211	54.4
3811	150.0	3213	124.2
3824	70.2	3214	32.0
3833	59.4	3220	198.4
3849	29.6	3231	58.9
		3233	98.2
		3240	75.6
		3320	17.6
		3411	3.6
		3512	1.2
		3523	11.9
		3551	22.7
		3559	31.0
		3610	31.8
		3620	5.9
		3691	3.0
		3710	4.8
		3812	19.3
		3819	15.5
		3822	43.9
		3823	76.4
		3825	395.5
		3839	40.1
		3841	145.6
		3901	327.3
		3902	40.1
		3903	49.3
		3909	50.4

CHAPTER 10

Regional Integration as a Vehicle for Microeconomic Disintegration; Some Macroeconomic Implications of the Reorganization of Firms

Pontus Braunerhjelm

The Industrial Institute for Social and Economic Research, Stockholm

10.1 INTRODUCTION

This chapter discusses the impact of the EC program for 1992 on the organization of production in manufacturing firms and the derived macroeconomic consequences for a small open economy. Flexibility is the key concept in this chapter. On the micro-level it is argued that the intensified competition due to "EC 1992" is forcing firms to adopt more flexible production technologies. The type of micro-level flexibility - i.e., whether it can be characterized as being dynamic or static - determines the scope for flexibility at the macro-level.

Turning to the macro-level, allowance for factors of production to move between countries - in contrast to the neo-classical trade model - is an essential part in understanding the adjustment process to policy shocks. The extent to which the reallocation of factors will eventually take place is determined by differences in factor endowments, national policies, international commitments, firm characteristics etc. Such reallocations have a bearing on the distribution of production and welfare, and hence also on trade between regions and countries.

Consequently, the analysis must incorporate the effect of (expected) deviations in economic policies between regions or nations - and/or other relevant differences - from the firms' perspective in order to understand factor flows. It should be noted that such differences are important not only for outsiders to the integration process, but also for insiders, since conditions for firms differ between regions within the integrated area. However, it is particularly relevant for outsiders since integration always contains an element, or potential threat, of discriminatory policies towards outsiders (see Figure 10.1).[1] Statistics reveal that the 1992 proposal has markedly affected the rate of mergers and acquisitions as well as the flows of direct investments (Smith and Walter, 1990).

The analysis is based on traditional trade concepts, with special attention on the firm as a vehicle to transfer factors of production. The way in which the firm's position and role is determined in a network context and the associated effects of factor flows are also considered briefly. Normative conclusions derived from a social welfare function -

[1]Figure 10.1 reveals that a company's net acquisitions, defined as the acquisitions abroad by a nation's firms minus the foreign acquisitions in that nation, are strongly concentrated to non-member countries.

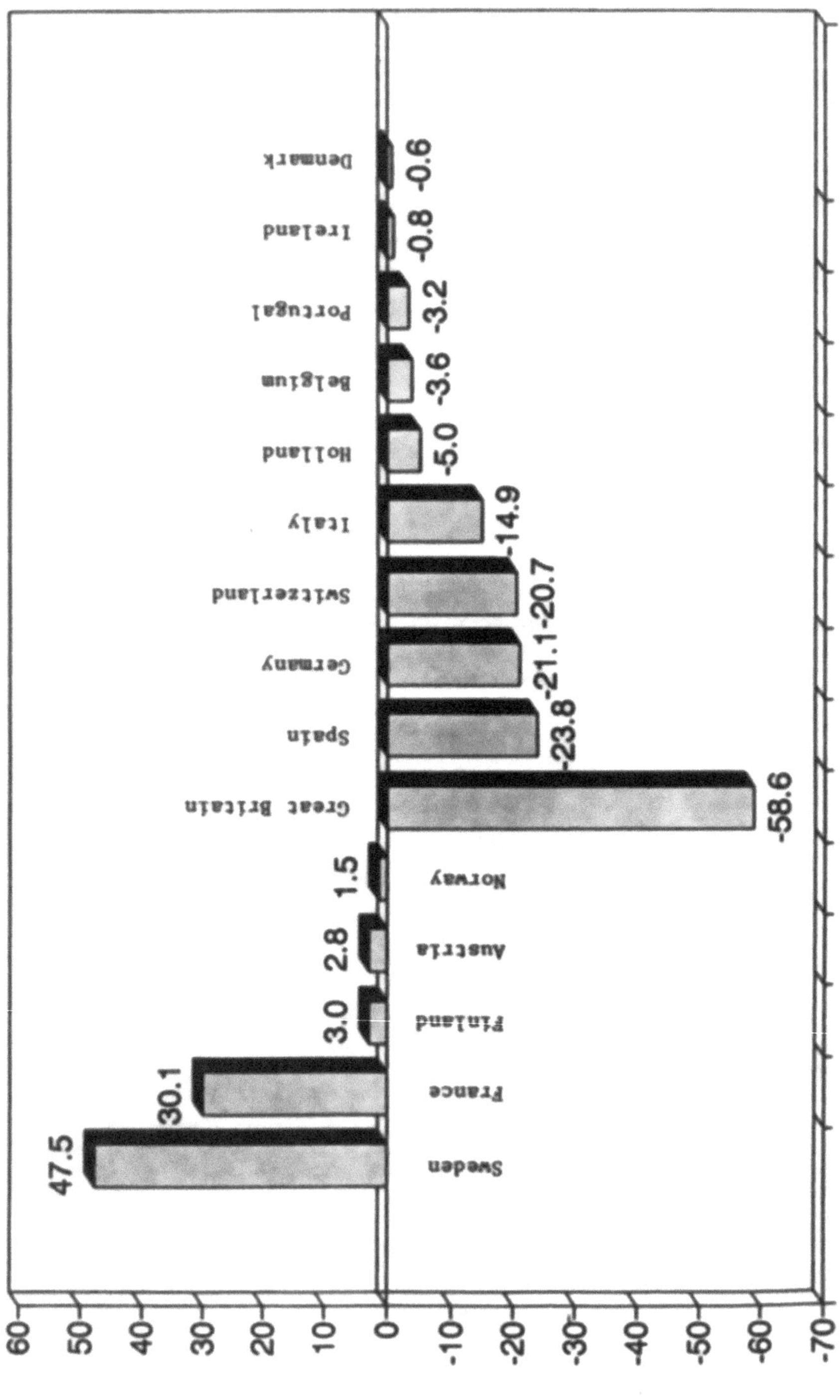

Figure 10.1: Net investment, company acquisitions in Europe, 1990 billion SEK
Source: Translink European Deal Review, April 1990

such as whether policies should be geared towards retaining or attracting firms - are not the purpose of this chapter.[2] Rather, the chapter seeks to describe how, from the firm's view, exogenous policy changes may induce firms to shift production between regions, and examines the related macroeconomic implications of this process.

The chapter is organized as follows. Section 10.2 presents the microeconomic theoretical framework. In Section 10.3 this is linked to the macroeconomic effects on specialization and the ability of countries to cope with shocks on the macro level. Thereafter a conceivable model of the structure of the Swedish economy is discussed as are the adjustments arising from an - as perceived by Swedish firms - exogenous institutional change such as that involved in the EC proposal for 1992. The last section summarizes the main findings of the chapter.

10.2 PRODUCTION FLEXIBILITY AND NETWORKS: SOME MICRO ASPECTS

The term flexibility is quite new in economics. It was probably introduced by Stigler in 1939; he defined it as the slope of the short-run average cost curve, where costs in turn depended on the produced volume (Carlsson, 1989a). A flat curve was interpreted as high flexibility (Figure 10.2). This definition is generally known in economics as *volume flexibility*. That kind of flexibility is restricted to the amount produced and no variability in product characteristics is possible, which renders volume flexibility a quite static feature.

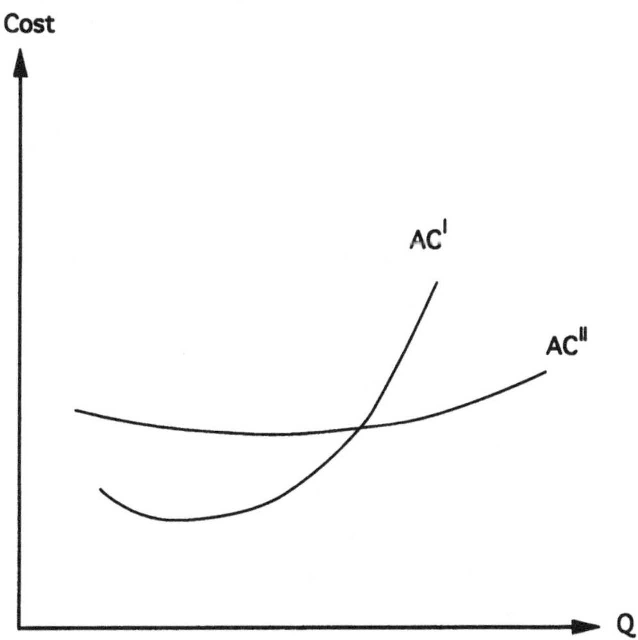

Figure 10.2: Example of average cost curves

[2]More precisely, the effects of industrial policies and strategic trade policies on imperfect markets are not considered here.

Stigler's original definition of flexibility has since been modified and reinterpreted in several ways. In Taymas (1989) different types of flexibilities are unified in a hierarchical structure, relating the traditional volume flexibility to the engineering concept of *machine flexibility*.[3] The term machine flexibility refers to flexibility in products, processes and operations at the plant level. Hence, it relates to flexibility in production technology in general. Further, it grants flexibility a more dynamic character since in addition to volume alteration, product attributes can be varied. Taymas also shows how volume flexibility can be derived from machine flexibility. For our purpose, it is the latter concept which is of particular interest.

The implications of flexibility as defined above are best understood if confronted with the classical models of organizing production. The Fordistic assembly technique and the Tayloristic hierarchic organizing system, both designed to enhance the efficiency of traditional mass production, lose validity if machine flexibility turns out to be a key function in production. This also means that there is a trade-off between static efficiency and flexible production technologies (Mills 1986). In other words, the choice of production technology is made by taking into account the structure of the demand fluctuations. If demand frequently changes with regard to its level, then volume flexibility - or static efficiency - is preferred. On the other hand, if demand is volatile with respect to different varieties within the same product-group, then systems emphasizing machine flexibility are favoured, even if they imply higher average costs in production compared to the traditional system of volume flexibility.

In retrospect it is obvious that among industrialized countries there is considerable demand for varieties of the same product. During the 70s many of the traditionally organized industries were severely hit by the recession. As demand changed and competition increased from "new" countries, not only were these industries sluggish in adjusting with regard to product characteristics, but they were also hit by high fixed costs due to large and inflexible plants. Indeed, as shown by Diwan (1989), flexible production technologies are more conducive in smaller plants.[4]

Hence, production of high quality differentiated products seems to be best accomplished in smaller units, enabling more efficient supervising, monitoring and "customization". World wide the average plant size has also diminished during the last decade (Braunerhjelm, 1993; Carlsson, 1989b). At the same time firms have grown. This illustrates that the importance of economies of scale has been transferred from the traditional production line - or plan level - to the creation of firm specific assets derived from R&D, marketing, organization and management, finance etc. Moreover, the decline of plant size coincides with closer relations between firms (networks), quite dissimilar to the traditional subcontractor-purchaser links. Hence, two parallel paths of development evolve; decentralization and externalization.

Technological progress has paved the way for cost reducing modes such as just-in-time (JIT) delivery systems etc., thereby fostering a closer relationship between firms. Such a close-knit relation between users and producers leads to strong interdependence and favors the establishment of networks. These networks may be national or international - or a mixture of both - and they blur the traditional notion of the unit of production, i.e., the firm. From a theoretical point of view, the existence of transaction costs, while at the same time allowing for more flexibility, justifies the existence of networks (Coase 1937; Williamson 1975, 1985). As asset specificity increases, as there is more uncertainty and more infrequent transactions, the need to internalize

[3] For a more detailed description, see Taymas 1989.

[4] The concept of "flexible manufacturing systems (FMS)" often appears when new production modes are discussed. It is defined as "several automated machine tools of the universal or special type and/or flexible manufacturing cells and, if necessary, further manual or automated workstations. These are interlinked by an automatic workpiece flow system in a way which enables the simultaneous machining of different workpieces which pass through the system along different routes" (Warnecke and Vetting, 1982).

transactions becomes larger. Network organization is one possible strategy to achieve this.

One of the advantages of a network organization is the ability to reinforce the signalling effects of prices compared to alternatives such as structures of vertical integration. Still, the competitive role of prices is often overshadowed by the emphasis on quality aspects in networks. The diffusion of quality enhancing technology and the associated possibilities to exploit externalities are two of the cited superior characteristics of network organizations, especially for high-tech firms. Thus, firms have increased their cooperation in fields traditionally strictly kept to the individual firm. One evident example is R&D, where firms now often share basic expenditures whereas the "pure" product-related R&D is undertaken by the individual firm (two examples in the Swedish economy are the Volvo-Renault agreement and the attempted "outsourcing" of R&D to subcontractors). Such relations are not only developed between firms but - although to a lesser extent - also between firms and institutions such as universities and different governmental bodies.

Simultaneously, in other parts of the economy firms still derive their competitive edge from mass production. In particular, scale economies at the plant level are crucial in the primary industry (forestry, ores, etc.). Within these industries the demand for customized intermediate products carrying certain product characteristics is negligible. Here, conventional subcontracting relations and vertical integration structures of production are much more common.

If the economy is structured in the way described above, exchanges of commodities take place in three different dimensions; trade among the entities belonging to a network, trade between the network and the market and "traditional" transactions at the market between firms. The more standardized products will be traded at the markets whereas the transactions of customized, quality demanding products take place within the network (Ylä-Antilla and Lovio, 1990).

To summarize the micro effects, firms facing relatively frequent fluctuations in demand with regard to production characteristics favour small plants and tighter networks, emphasizing dynamic flexibility. This may well coincide with growth and the internationalization of firms in order to reap the benefits of economies of scale (and/or scope) emanating from non-tangible, firm-specific assets, such as R & D, marketing etc. Internationalization renders market knowledge and increases the possibility of technological spill-over effects that can be internalized by the firms. In other industries, demand mainly varies with respect to the level of outputs. Firms in these industries place priority on traditional, static (volume) flexibility. In the following section we will link this development to macroeconomic performance.

10.3 MACROECONOMIC IMPLICATIONS OF MICROECONOMIC FLEXIBILITY

Adjustments are continuously taking place within firms. The economic, technological and social environments of a firm determine the pace for the adjustment pressure. How these environmental factors interact is far from evident. However, it seems quite indisputable that social change has implications for technological progress. One example is the evolution of the rules for commercial policies between nations. "EC 1992" is such an institutional change that could - for instance, by attracting foreign capital and facilitating intra-regional factor movements - augment competition and thereby spur technological progress (Geroski ,1988). Whether such technological advances are dispersed throughout the whole economy or restricted to certain sectors is, however, not self-evident and may influence the locational decisions of firms.

If the reorganization of production within firms, including the adoption of partly novel production technology, implies higher imports of components, more scattered production etc., such changes will influence the dependence of firms on the outside world. This has ambiguous macroeconomic effects. On the one hand, nations become more sensitive to outside shocks if they become more intertwined with the rest of the world as trade barriers are dismantled. If networks become more international and if firms turn from "multi-sourcing" to "single sourcing" strategies in their delivery systems, their dependence on the world market is likely to increase. On the other hand, the larger proportion of a nation's firms that employs flexible production technologies due to stiffer competition, the higher the aggregate national flexibility. Hence, production can then adjust comparatively smoothly to macroeconomic shocks. Consequently, these two forces work in opposite directions; foreign shocks are transmitted more directly whereas the enhanced industrial flexibility increases the ability for a country to cope with such shocks.

10.3.1 How do the Strategies of Firms Impinge on the Structure and Vulnerability of the Swedish Economy?

The overwhelming part of manufacturing employment, exports, value-added etc., is generated by the 30 largest firms in Sweden. From a macro perspective, the Swedish manufacturing industry could be decomposed into two sub-industries. The machinery, or engineering, industry (which we denote by M) is dominated by comparatively few, but large, multinational firms. Examples are SKF, Atlas Copco and Electrolux. These firms have a long tradition of internationally dispersed production. Predominantly they compete with certain firm-specific assets on highly competitive markets. Marketing, R&D, managerial and organizational know-how are the cornerstones of their competitive edge. Economies of scale are derived on the firm level as several plants take advantage of these intangible assets. Hence, there is no need to concentrate production to one plant or to one country, especially as production technologies must entail a high degree of flexibility in order to enable swift changes in response to alterations in demand, even at the level of national markets.

The other sector, which we denote as the basic industry (called X), contains forestry, ore and other raw material (or process) intensive production.[5] It produces a relatively homogeneous good and production is organized to emphasize volume (static) flexibility. Economies of scale are mainly derived on the *plant* level, i.e., in the actual production process. The technology is characterized by large, process- and capital intensive plants. Compared to the engineering sector, fewer resources are spent on intangibles, especially R&D. Production is generally confined to a particular country since it is tied to the prevalence of some natural resource, whereas firms in the engineering sector may take up production anywhere. Thus, at this level of aggregation the two sectors could also be referred to as being internationally mobile (the engineering industry) and immobile (the basic industry). Sweden's dependence on the respective sectors is illustrated in Figure 10.3. The basic sector is particularly important as a generator of foreign exchange.

Consequently, the basic structure of the manufacturing sector can be captured in a traditional two sector trade model (Jones, 1965). In such models the possible combinations of production are generally illustrated through a production possibility curve (Figure 10.4). Assuming a two-sector economy (engineering and primary industry goods), the flexibility within the economy - i.e. how easily production is switched between the two sectors - is revealed by the shape of the production possibili-

[5]The two sub-industries could just as easily be divided into an assembly sector (engineering, chemicals, etc.) and a process intensive sector (forestry, ore, parts of the chemical and food industry).

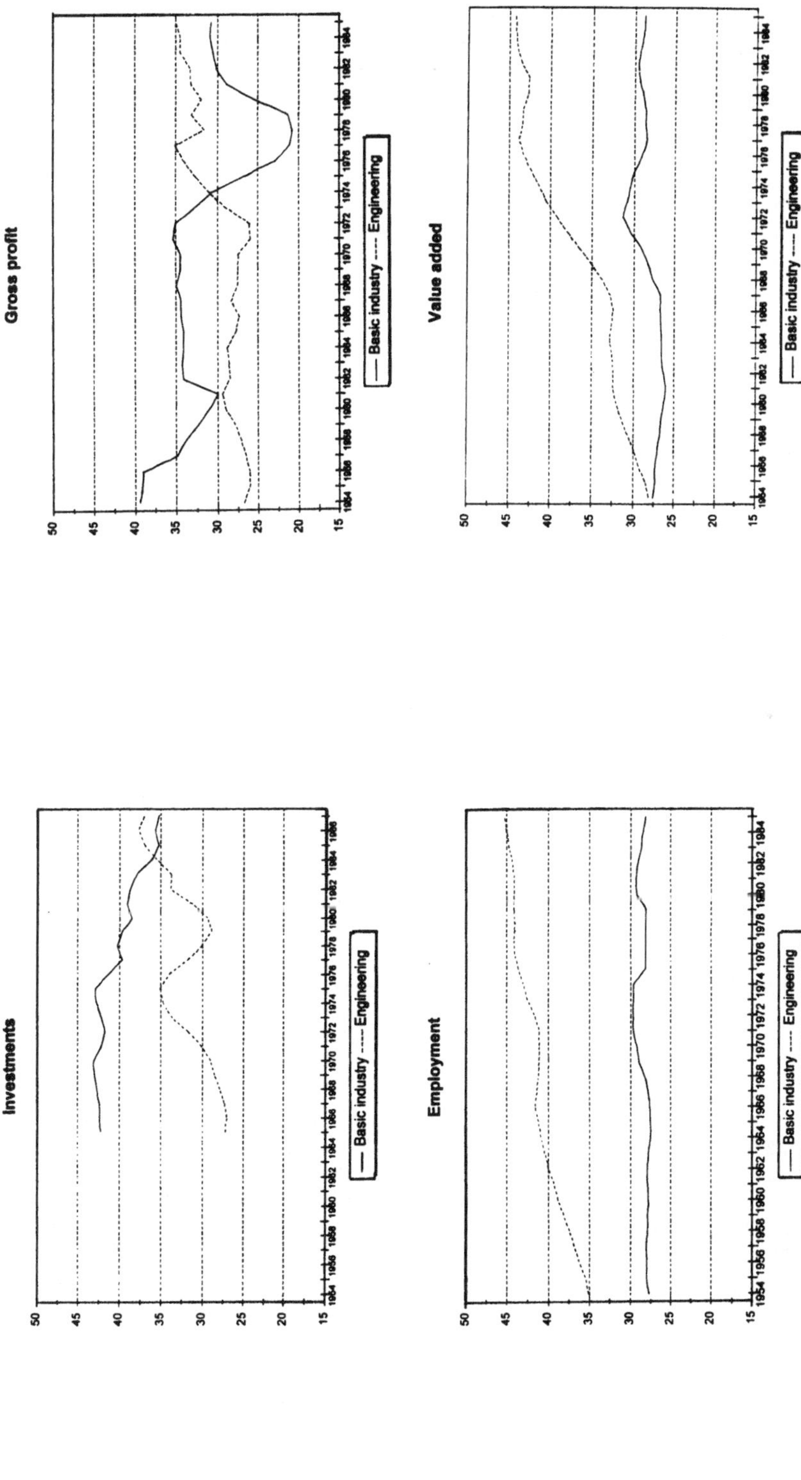

Figure 10.3 The engineering (ISIC 38) and basic (ISIC 2, 34, 37) industries share of total industrial investment, employment, value-added and gross-profit, 1954-1985. (Percentage, 5 year moving averages)

Source: Eliasson 1979, additional calculations from national statistics.

ty curve, which in turn depends on the implemented technology.[6] The lower flexibility is, the more exposed a nation becomes to external (or internal) shocks. Consequently, the duration of an adjustment process due to exogenous shocks will vary according to the way that a country organizes its production.

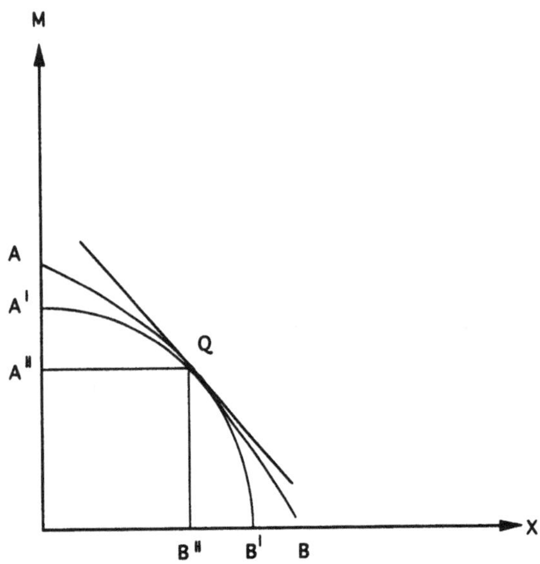

Figure 10.4: Flexibility in a general equilibrium setting

The different curvatures represent different degrees of flexibility. A'' - B'' is completely inflexible and production will take place at Q whatever the price-ratio.

The curves can also be interpreted as the adjustment possibilities in the short-run (A'' - B'') and in the long-run (A-B).

The curvature of the production posibility curve is affected by changes in the endowment of factors.

The role of flexibility (or elasticities) in demand and supply are, on the macro level, perhaps more explicitly elicited in a partial equilibrium context. As Figure 10.5 shows, the more inelastic demand and supply are, the harder the adjustment imposed on the domestic economy to cope with shocks that disturb the foreign supply of goods. The more flexible the production technology adopted in the manufacturing sector (i.e. the larger the engineering industry), the more elastic is supply. If supply and demand are relatively price elastic, a shortfall can be compensated for fairly easy through movements along the respective curves (Figure 10.5). Such adjustment is achieved quite instantly. On the other hand, inelastic supply requires that the curves shift to alleviate the shortfall which takes time and increases vulnerability.

[6] The production possibility curve is superior in illustrating flexibility compared to the integrated equilibrium technique used in the Helpman and Krugman (1985) framework.

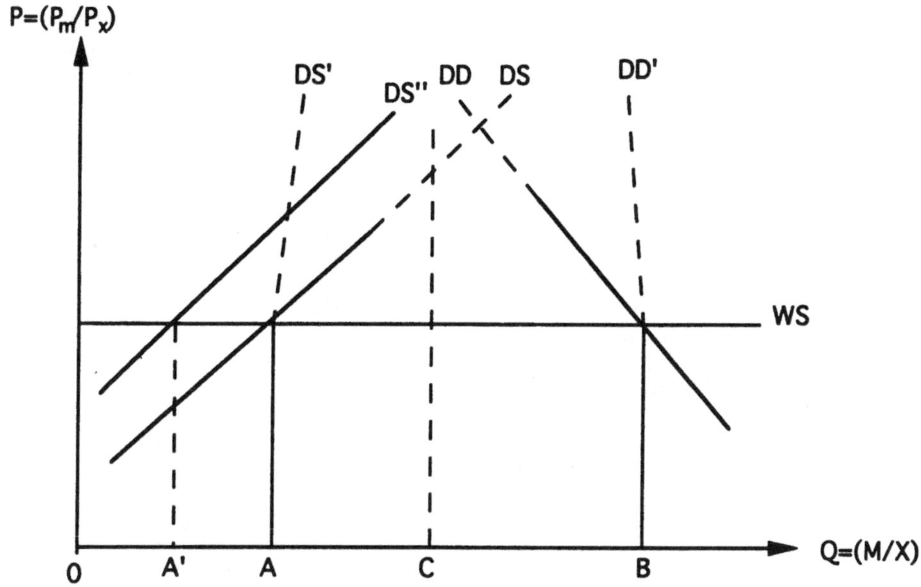

Figure 10.5: A partial equilibrium representation of national vulnerability

DS = domestic supply
DS' = inelastic domestic supply
DS'' = decrease in factors of production
WS = world supply
DD = domestic demand
DD' = inelastic domestic demand
C = estimated minimum level of consumption of Q

Initial import = B-A
Initial consumption = O-B
Import after factor shift = B-A'
Initial domestic production = O-A

The more elastic the domestic supply (DS), the easier it is to attain the minimum level C of consumption of M if a nation is isolated from world market supply (WS). Inelastic curves require shifts of the supply curve. DS'' represents an outflow of factors.

Consider now the role of the international firm in such a setting. The overall objective of the "EC 1992" program is to increase competition between firms by abolishing different protectionistic barriers. Companies formerly protected behind national borders must act in order to prepare for changing market conditions. If competition spurs innovations and technological progress, firms competing in the "high-tech" engineering sector (M) must implement production technologies that allow them to adapt quickly to changing conditions. Further, to become "best performers", they have to concentrate on their core activities (Eliasson and Lundberg, 1989). More peripheral production is either sold out or organized in independent profit accounting entities; i.e., disintegration is embarked upon. Consequently, a competition enhancing integration process at the macro level encourages disintegration at the micro level in the sector producing the "high-tech" good. A change in organization towards machine flexibility changes the curvature, as shown in Figure 10.4. Hence, macroeconomic flexibility increases.

In particular, assume that each firm employs skilled and unskilled labor in addition to machinery and capital of a more physically fixed nature (raw materials, real estate etc.). To simplify we assume that skilled labor and machinery are employed in fixed proportions and likewise for unskilled labor and fixed capital. Further, we assume that skilled labor and machinery can move between different countries whereas unskilled labor and fixed capital may not due to language obstacles etc. In addition, the basic industry predominantly employs the immobile factor (called V) while the engineering sector is intensive in its use of the mobile factor (called h).

Obviously, if conditions change between the outside world and the domestic economy, the adjustment of the engineering sector will be accomplished comparatively more easily. Likewise, if globalization strategies for one or another reason are regarded as important for the companies, this will predominantly affect the engineering sector and could thereby influence the industrial structure of the economy.

External shocks - such as "EC 1992" - and internal exogenous shocks influence the decisions of firms with regard to their location. Looking at the overall production structure of the economy, and assuming that exports take place within both sectors - although the engineering sector is assumed to be a net importer whereas the basic sector is a net exporter - one possible outcome is shown in Figure 10.6.

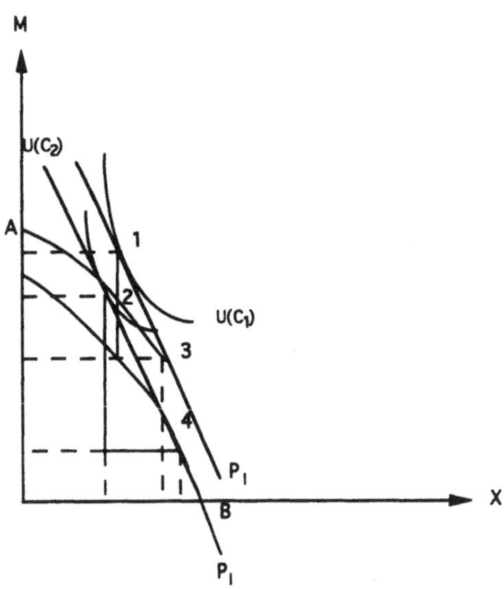

Figure 10.6: Factor flows and specialization in production

U = welfare level
P_I = terms of trade
1 = consumption ratio before integration (C_1)
2 = consumption ratio after integration (C_2)
3 = production ratio before integration (Q_1)
4 = production ratio after integration (Q_2)

An outflow of production factors (or firms), affecting especially the M-sector, shifts the production possibility curve down- and inwards. Such outflows could take place for several reasons. For a country participating in an integration process, closeness to technological centres (Silicon Valley effects, spillovers etc.) may induce location in certain areas. Or, the desire to compensate lost market shares on the previously protected home market could start an internationalization process. For outsiders we could imagine a situation where the engineering goods sector is discriminated against - or expects discrimination - by the integrated area for political, or other, reasons. Or, in general, a situation where divergencies in economic policies favour the production of the engineering goods in the integrated area. To be drastic, assume that the exports of firms in the engineering sector of an outsider country are stopped (or that firms believe that this will happen). Terms of trade remain constant since the impact of the small open economy is neglible in relation to the large, integrated area. However, factors of production may move. It is intuitively appealing that producers of the engineering good in the small open economy with a large export share will move to the integrated area.

Hence, factor flows - as described above - imply that the small open economy will become more specialized and, hence, more dependent on imports from the integrated area. In addition, in this case, it will also become more inflexible (Figure 10.6). In the integrated area the opposite development occurs. This hypothetical situation is of course an exaggeration. However, we can easily conceive other situations involving less severe exclusions from a market (local requirement rules, anti-dumping etc.) that also could set off a transfer of factors of production through direct investment.

An extension of the model is to incorporate networking firms. Assume as before that the manufacturing sector consists of a engineering and a basic industry. A slightly different definition of the production factor is necessary. Skilled labor (called S) produces the production technology (f), while unskilled labor and machinery combine into a factor (L) which is needed in the production of the intermediate product h.[7] V is composed in the same way as above. Thus, firms in both industries employ the fixed factor (V) and S whereas intermediate products are purchased from network firms. In the production of the engineering good (M) the intermediate product is used intensively while the basic sector (X) mainly employs the fixed factor.

Thus, we have a model with four factors, four goods and two countries (the small open economy and the rest of the world),

$$M = f(h,V)$$
$$X = f(h,V)$$
$$f = g(S)$$
$$h = l(L)$$

and for each firm

$$m = g(S)\,(1(L),\,V)$$
$$x = g(S)\,(1(L),\,V)$$

where M and X are tradeable whereas h and f are non-traded.[8] Factors of production are fully mobile between sectors within each nation. However, only S and L are mobile between nations. S can be thought of as a level of knowledge, although it is differentiated between individuals, imposing firm specific characteristics upon each firm which renders a monopolistic market structure. Such knowledge is easily moved

---

[7]This is along the lines of Helpman and Krugman (1985) where it is postulated that some factor in the production process assumes firm-specific features as soon as it is employed by the respective firm.

[8]Assume that the small firms producing h lack all knowledge of foreign markets which limits their export possibilities. There are fixed costs in exports, hence, larger firms have economies of scale advantages in exports. Similarly, f is a firm specific asset which, due to problems of appropriability, is not exported or transferred through arm's length contracts.

across regions and nations. Knowledge for the individual firm is derived from the overall level of knowledge in a country or region (Eliasson 1991b). Thus, by locating in knowledge intensive areas, firms can increase their own knowledge base.[9] Since an increase in S has a positive effect on the marginal productivity of the factors h and V, an emigration of S will stimulate producers of h to relocate and a bandwagon effect will occur (see Appendix).

To summarize, the larger the engineering sector is - characterized by dynamic machine flexibility - the larger is a nation's capacity to cope with macroeconomic shocks. A small engineering sector generates inelastic supply on the macro level, even though supply may exhibit high flexibility in each sector on the micro level. However, a large flexible sector also implies that firms will move if they are exposed to some sort of exogenous shock that alters the prerequisites for production between different regions. Firms' decisions are affected by a multifold of factors, including political factors (Braunerhjelm, 1990b, 1991; Mundell, 1957; see Figure 10.7). As production moves abroad, the only way to achieve the same consumption level is to rely more heavily on imports. In other words, dependence on the international market is increasing and so is national vulnerability. Flexibility on the macro level will also be reduced. If the small open economy takes part in the integration, it will de facto be part of the region and the risks of being discriminated against are eliminated, which influences the expectations of firms and their risk hedging behaviour.

10.4 CONCLUSION

An integration process that enhances competition - which is one of the primary objectives of "EC 1992" - forces firms formerly protected on national markets, to reorganize in order to stay competitive or become best performers. If firms fail to respond to the new market situation, they have to exit. For a small open economy with close ties to the EC, there is no shelter behind national borders. The need to adjust might even be more pronounced for outsiders since the risk of being exposed to discrimination cannot be completely eliminated.

The stiffer competition implies that firms must concentrate on core activities while more peripheral forms of production are dismissed. This induces a disintegration process, encouraging the establishment of smaller, flexible *plants* and network arrangements. Production is decentralized into independent profit accounting, and/or strategic business units. Further, to preserve, or augment, their competitive strength, *firms* must fully exploit scale economies and embark upon internationalization. As quality, defined according to several criteria, becomes more important, networks based on confidence and competence will be established.

The move towards smaller plants, geared by high machine flexibility, decreases the exposure to macroeconomic shocks for a country. The adjustment potential depends on the possibility of swift adjustment, transferring factors and organizing production according to the new terms and conditions. What matters for flexibility on the macroeconomic level is how production is organized on the microeconomic level and the size of the respective sectors. High volume (static) flexibility on the micro-level will never be compatible with high flexibility on the macro-level. The necessary conditions for macroeconomic flexibility are a relatively large engineering sector organized in a dynamically flexible way.

---

[9] I firms believe that their particular knowledge in producing M is enhanced by moving the production to another region - for example because other firms are concentrated in that region (Silicon Valley effects) - they will move to preserve their competitiveness. Or they will move part of their human capital i.e., individuals, to that area.

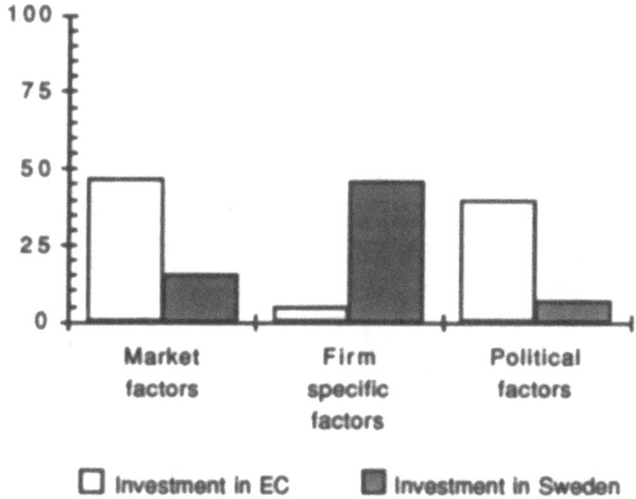

Figure 10.7: Factors influencing the regional distribution of investments and production

The columns indicate the firms in a recent survey (for details, see Braunerhjelm, 1990b) that estimate the importance of the respective factors in the number of investment decisions. The striped column implies that investments within the EC are advocated while the dotted column favours investments in Sweden.

From a policy viewpoint, this also implies that if governments fail to launch distinct signals with regard to their economic policies, in particular when it concerns international commercial relations, expectations may be influenced in such a way that investments of both domestic and foreign firms are discouraged. The massive outflow of investment from Sweden to the EC during the late 80s coincided with a period of extreme uncertainty regarding Sweden's future EC-policy.[10] At the same time foreign investment within Sweden has remained at a very modest level. Since it is particularly the large, inflexible process-intensive firms in the basic industry that stay, the economy has become more vulnerable to external shocks.

Hence, low macroeconomic vulnerability is attained through the complexity and diversity of production. This means that incentives must be appropriately designed to encourage firms to produce and invest within a nation. As integration is undertaken by a small open economy's main trading partners, full membership is probably necessary to keep or restore competitiveness. This argument can of course be generalized to other differences in economic policies.

[10]There are of course numerous reasons why a firm might decide to invest abroad rather than within the domestic economy; these are shown in Figure 10.7. However, the uncertainty concerning the relations towards the EC was one of the factors favouring investment abroad (Braunerhjelm, 1990c).

REFERENCES

Braunerhjelm, P., 1990a, "Effekter på den Svenska Industristrukturen av EG 1992", *Ekonomisk Debatt* 8:713-720.

Braunerhjelm, P., 1990b, "Regional Integration, the International Firm and the Allocation of International Production", *WP No 14-91*, Institute of International Economics and Management, Copenhagen.

Braunerhjelm, P., 1990c, *Svenska Industriföretag inför EG 1992*, IUI, Stockholm.

Brunerhjelm, P., 1991, "Svenska Underleverantörer och Småföretag i det Nya Europa", *Forskningsrapport nr 38*, IUI, Stockholm.

Braunerhjelm, P., 1993, Globalization and the Future Role of Small and Medium-Sized Enterprises, in Oxelheim, L. (ed.), *The Global Race for Foreign Direct Investments. Prospects for the Future*, Springer-Verlag, forthcoming.

Carlsson, B., 1989a, "Flexibility and the Theory of the Firm", *International Journal of Industrial Organization* 7:179-203.

Carlsson, B., 1989b, "The Evolution of Manufacturing Technology and Its Impact on Industrial Structure: An International Study", *Small Business Economics* 1:21-37.

Coase, R., 1937, "The Nature of the Firm", *Economica* 4:13-16.

Diwan, R., 1989, "Small Business and Economics of Flexible Manufacturing", *Small Business Economics* 1:101-110.

Eliasson, G., 1979, Att Välja 80-tal, IUI, Stockholm.

Eliasson, G., 1991a, "Deregulation, Innovative Entry, and Structural Diversity as a Source of Stable and Rapid Growth", *Journal of Evolutionary Economics* 1:49-53.

Eliasson, G., 1991b, "The Firm as a Competent Team", *Journal of Economic Behaviour and Organization* 13:273-89.

Eliasson, G. and L. Lundberg, 1989, "The Creation of the EC Internal Market and Its Effect on the Competitiveness of Producers in Other Industrial Economies" in H. Siebert and J. Mohr, (eds.), *The Completion of the Internal Market Symposium 1989 or IUI Booklet No 263*, IUI, Stockholm.

Geroski, P. 1988, "Competition and Innovation", in *Research on the Cost of Non-Europe* 2, EC Commission.

Helpman, A., and P. Krugman, 1985, *Increasing Returns, Imperfect Competition and International Trade*, MIT Press, Cambridge.

Jones, R., 1965, "A Simple General Equilibrium Model", *Journal of Political Economy*, 73:557-572.

Lundberg, L., 1989, *Svenskt Näringsliv och den Europeiska Integrationen*, bilaga 3, LU-90.

Mills, D., 1986, "Flexibility and Firm Diversity with Demand Fluctuations", *International Journal of Industrial Organization* 4:302-315.

Mundell, R., 1957, "International Trade and Factor Mobility", *American Economic Review* 47:321-325.

Smith, R. and I. Walter, 1990, "The European Market for Corporate Control Structure, Transaction Flow, and Regulation", mimeo, Conference Paper Marstrand, May.

Taymas, E., 1989, "Types of Flexibility in a Single-Machine Production System", *International Journal of Production Research* 27:1891-1899.

Warnecke, H. and G. Vetting, 1982, "Technical Investment Planning of Flexible Manufacturing Systems - The Application of Practice Oriented Methods", *Journal of Manufacturing Systems* 1:89-98.

Williamson, O., 1975, *Market and Hierarchies. Analysis and Antitrust Implications*, Free Press, New York.

Williamson, O., 1985, *The Economic Institutions of Capitalism*, Free Press, New York..

Ylä-Antilla, P., and R. Lovio, 1990, "Flexible Production, Industrial Networks and Company Structure - Some Scandinavian Evidence", ETLA WP, No. 338.

APPENDIX

The production functions in the respective sectors are

$$M = A(1(L), V) = g(S)\,(1(L), V)$$
$$X = A(1(L), V) = g(S)\,(1(L), V)$$

assumed to be homogeneous of degree one in L and V, concave (for given A) and increasing in all arguments.[11] M and X are tradeable while h and f are non-tradeable. However, skilled labor (S), as well as the production factor L used in the production of intermediate goods, can move between nations. The level of the skills within a firm is influenced by the overall amount of S within an economy through spillovers and off market interactions. Therefore, the marginal productivity of S in producing M increases in areas abundant in S. Hence, this increases the reward to S and influences the location of S. Thus,

$$(dm/dS)^* > (dm/dS) \quad \text{if} \quad S^* > S$$

where starred variables represent the foreign region (or country).

The shift in the production function for the firms due to improved or extended knowledge also influences the reward to other factors of production. Assume that technology is captured by a standard Cobb-Douglas function (for a given knowledge level). Then the reward to L, or the price of h, is affected in the following way,

$$m^* = A^* L^a V^{1-a}$$

and the marginal product of L is derived as

$$dm^*/dL = (a/L)\,(A^* L^a V^{1-a}) = (a/L)\,(m^*)$$

where

$$m^* > m \quad \text{since} \quad A^* > A \quad \text{if} \quad S^* > S$$

and consequently the marginal productivity of L is larger in the starred (foreign) country than in the domestic economy. At given prices the reward to the L factor is then also larger. This induces a bandwagon effect where L also locates to the foreign country (region). Alternatively it could be argued that part of the level of skill in M production (call that fraction s) is transferred (spillovers) to subcontractors producing h which increases the price relative to domestically produced h (this could also be interpreted as a quality difference). As S and L flow out of the outsider country, the production possibility curve shifts downwards. The shift will continue until factor rewards are equalized or regions (nations) become completely specialized.

---

[11] Let small x and m denote firms in the respective sectors.

CHAPTER 11

From Trade Flows to Corporate Networks

Roberta Capello
Bocconi University, Milan

Tomaso Pompili
University of Pavia

11.1 INTRODUCTION[1]

The aim of the present chapter is to analyse the growing phenomenon of corporate networks in relation to the evolution of industrial markets towards greater levels of internationalisation.

The interest for such a study stems from the fact that a growing literature is developing in both fields, but what is still overlooked is the interrelation between internationalisation and networking, these two phenomena being strictly related in their dynamics through strong and circular synergies.

The analysis of this linkage can be based on the profound evolution in industrial systems, which has brought about a high complexity of markets and, more generally, of economic transactions governing markets. These deep changes have generated the need for a different form of industrial organisation both at national and international levels.

In particular, our interest is to analyse the interrelation between company behaviour and internationalisation. The literature seems to suggest that all forms of cooperation agreements (joint ventures, equity or non equity agreements) are a *new* and *substitutable* form of company behaviour. Our question is whether this new organisational form can also be thought of as a new and substitutable form for company internationalisation. We reply by linking the three major international research issues (trade, multinationals and agreements) together.

This issue is analysed from a conceptual perspective in Section 11.2 where we present a taxonomy of forms of internationalisation and their dynamics. The theoretical results stressing the importance of the content of international flows and of the nature of industrial relationships will be validated by an empirical analysis of the Italian/Swedish case, dealing, in sequence, with aggregate quantitative data on trade flows (Section 11.3); on Italian inward and outward foreign direct investments (Section 11.4); and with qualitative case studies about the more recent dynamics of cooperation agreements - namely, on their nature, organisation and evolution, in selected Italian subsidiaries of Swedish corporations (Section 11.5).

---

[1] Though the chapter is the result of common work by both authors, R. Capello has written Sections 11.1, 11.2.1, 11.2.3, 11.5 and 11.6, while T. Pompili has written Sections 11.2.2, 11.2.4, 11.3 and 11.4.

11.2 FORMS OF INTERNATIONALISATION: TOWARDS A DYNAMIC TYPOLOGY

11.2.1 Internationalisation in Network Theory: A Still Undefined Relationship

The relatively recent phenomenon of corporate networks between independent companies (in its different forms of equity and non equity agreements, and cooperation agreements), has become an interesting field of analysis for experts on firms' behaviour, as is witnessed by a growing literature.

Numerous studies try to explain the phenomenon from a quantitative viewpoint, through the use of empirical analyses (Cainarca, Colombo and Mariotti, 1989; Camagni, 1989a; Camagni and Gambarotto, 1988; Chesnais, 1988; Foresti, 1986). At the same time, a series of theoretical explanations has been developed, in order to conceptualise this new phenomenon in a more general framework on company behaviour. These attempts have been very fruitful, as is demonstrated by the relevance of this concept to the understanding of firms' behaviour (Camagni, 1989b; Mariotti, 1989; Rullani, 1989; Teece, 1986 and 1989; Vaccà and Zanfei, 1989).

A stimulating, though very general, definition of network is the following: "A corporate network is an organisational structure allowing horizontal, lateral, circular relations; a structure capable of having multiple centres, each of them playing an active and propositive role, even if limited" (Lorenzoni, 1986, p.70).

Although a debate still exists on the theory of corporate networks, the present chapter benefits from recently established conclusions on the concept of "network" reached by some authors:

a) The definition of this term, suffering from undefined and unclear boundaries, is in this chapter the following: "corporate networks are *formalised, selected and long distance* relationships" (Camagni, 1989b);

b) we refer to the term "network" only in the case of *explicit* linkages among selected partners (Camagni and Pompili, 1993; Freeman, 1990);

c) all kinds of organisational forms, *intermediate* between market and hierarchy (Williamson, 1973), are corporate networks, regardless of their legal and institutional form (equity agreements, non-equity agreements, joint-ventures) (Camagni and Pompili, 1993).

During a conference of the Groupe de Recherche Européen sur les Milieux Innovateurs (proceedings published in Camagni, 1991), corporate networks were analysed from a different perspective. Their existence was explained through their capacity of generating external synergies and know-how to local economies. "Through formalised and selective linkages with the external world, local companies may attract the complementary assets they need to proceed in the economic and technological race. Internal, mainly informal and tacit linkages may not be sufficient to achieve the main goal, especially in times of rapid economic and technological changes" (Camagni, 1991).

Thus, in economic theory, corporate networks generate synergies, hence *external economies*, among participants (corporate networks as club goods) and also act as a driving force to overcome the *constraints* imposed by less dynamic (and better studied) infrastructure networks.

These brief but efficient snapshots of more recent theories on corporate networks show that this phenomenon has been widely analysed in a multitude of ways. What the literature still does not emphasise is the fact that most of these cooperative agreements among firms take place at an international level. In the following paragraphs we try to explain this fact.

The first, and simpler, explanation stems from the definition of network itself, presented by Camagni (1991). If a corporate network is a way of bringing external know-how to a local economy or firm, this is more evident and true if external inputs come from outside national boundaries. In this way, national constraints, in terms of both economic and institutional rules, are overcome and local economies and firms take advantage of non-country specific complementary assets.

Alternatively, international cooperation agreements can be explained through the hypothesis of a territorial life-cycle theory of cooperation agreements. On the basis of empirical studies, Hansen (quoted in Camagni and Pompili, 1993) elaborated the idea that corporate networks initially develop locally, through a spin-off effect from large firms or research laboratories. In a second stage, linkages expand spatially and intensively between firms located in different regions. The last stage is characterised by the proliferation of inter-firm synergies outside the region. The logical consequent stage is the proliferation of international networks, in order to exploit synergies among partners coming from more differentiated spatial locations.

Neither interpretation satisfies our purposes, as both explain internationalisation only as a consequence of the geographical diffusion of firms' networks over national boundaries.

A more satisfying explanation can be found outside network theory, and stems from the analysis of the evolutionary path of industrial structure and of internationalisation.

11.2.2 Corporate Networks in Trade Theory: A Step Forward

Internationalisation has always developed through two traditional forms: namely, physical exchanges between countries and direct investments abroad by multinational corporations.

The characteristic feature of the first form was the transfer, through national barriers, of physical goods and production factors - namely capital and labor - where the former made unnecessary the latter, as the classic theory and the Swedish school on international trade point out.

The second form consists of the coordination of multi-country production and sales by a single strategic decision-making center, and is related, according to Dunning's (1981) eclectic approach, to three sequential necessary conditions regarding the existence of company specific ownership advantages (R&D, advertising, skills), compensating the costs of not being local (Kindleberger, 1969); internationalisation advantages deriving from transaction costs among independent plants and units (Williamson, 1975); and country-specific locational advantages, as defined by the modern life-cycle theory (Vernon, 1966).

We suggest that in recent decades international trade has been increasingly characterised by a new object of exchange different from physical goods and production factors: *information*, both as property rights and knowledge, transferred thanks to multinational companies and corporate networks. This statement is coherent with the idea that *dynamic* comparative advantages matter, in so far as countries tend to create their own comparative advantages by accumulating technology, knowledge and human capital (Krugman, 1986). In this sense, foreign direct investments and cooperation agreements create comparative advantages rather than exploiting them (Onida, 1990).

Thus, a definition of these cooperation agreements can be reached through an analysis of the transferred object through international exchanges.

When the object of transfer is physical (goods or production factors) we are confronted by simple exchanges of international trade (demand-led internationalisation); this is true even when they are embodied in long-term contracts. These can be sub-divided according to their characteristics, as being either internal to the firm, through commercial subsidiaries, or external, through commercial licenses to foreign agents.

Only when the object of transfer is information (knowledge or property rights) we have a proper (supply-led) case of internationalisation. Again, a further sub-division is feasible, between an internal transfer, as in the case of multinational corporations with production subsidiaries, and an external transfer, as in the case of cooperation agreements (production licenses, R&D agreements) with foreign firms.

According to this definition, we can create a taxonomy of different forms of internationalisation, on the basis of both the type of relation taking place (external or internal to the firm) and the objects of international transfer (goods and production inputs or knowledge and property rights). (See Table 11.1).

Table 11.1: Taxonomy of forms of internationalisation

Flow relations	Goods	Knowledge
Internal	Commercial subsidiaries	Foreign direct investment (production subsidiaries)
External	Commercial licenses	Cooperation agreements (production licenses, R&D agreements)

One peculiar result of our taxonomy is that corporate networks are characterised only by production, research and strategic agreements between companies, whereas commercial cooperation agreements are part of traditional international trade exchanges, having physical goods as their object of exchange.

The symmetric and asymmetric relationships between firms engaged in agreements can be better focused by drawing a distinction between property rights and knowledge as objects of agreements. Property rights in relation to technology, whose nature is to be in the form of private goods, can be exchanged on the basis of asymmetric agreements (such as production licenses), depending on the characteristic of the destination market. The smaller the total size of demand and the lower the degree of supply concentration in the destination market, the more asymmetric cooperation agreements are in favour of the holder of property rights. On the contrary, knowledge, a typical club good, is difficult to exchange on the basis of asymmetric agreements, the property of the exchanged object being irreversible, after it has been shared among partners. Moreover, in a dynamic sense the accumulation of knowledge is faster with cooperation. Besides, symmetric approaches are easier in R&D agreements since often the agreements are stipulated between firms and non-profit institutions, i.e., between organisations whose goals are not directly competing.

11.2.3 Determinants of Company Behaviour in Multi-National and Global Systems

Cooperation agreements represent the optimal behaviour of firms confronted by the trade-off between market resort and internal development, when both alternatives present high costs (high use-costs of the market and high costs for creating an internal know-how) (Camagni and Pompili, 1993). However, this explanation is not sufficient once we try to interpret company behaviour at an international level.

In a *multi-national system*, like the one characterising the 1950s, 1960s and 1970s, the constraints on internationalisation were being released not only by the creation of free trade areas and by the proliferation of trade fairs, but also by the development of motorway and flight infrastructure; in this context, company strategies for the most appropriate method of internationalisation (whether internally through subsidiaries or externally through licenses and agreements) depended on the results of the trade-off between strategic and managerial autonomy and flexibility, as defined by the following four factors:

a) the characteristics of the two markets. The more different, the less well known or the smaller the national markets, the higher the costs of external firms to enter the national market, in terms of appropriate marketing, commercial, personnel and strategic policies. Thus, the greater the possibility of finding agreements with local partners;

b) the needs for country-specific complementary assets. The more necessary country-specific complementary assets are, the more interested a firm will be in finding an agreement with local partners. This is, for instance, the case for the AT&T-Italtel agreement in the telecommunications sector. AT&T gained shares of the Italian switching equipment market through an agreement with the large Italian telecommunications firm Italtel, at the expense of sharing technological and scientific know-how with the Italian partner;

c) restrictions upon direct investments formulated by the destination country (Ozawa, 1985). Finally, the larger the restriction are, the more widespread cooperative agreements with local firms will be;

d) traditional locational advantages. The less significant the traditional locational advantages are (in terms of the local labor force, financial and fiscal rules, etc.) the more efficient cooperation agreements with local firms will be (Dunning, 1981).

It is widely recognised that since 1970 a deep and radical transformation in the characteristics of the industrial structure has been taking place, pushing the system towards greater levels of *globalisation* (Cozzi and Vaccà, 1986; Vaccà, 1986).

What changes is the spatial context in which corporate strategies and economic choices have to be contemplated, becoming one global context rather than a sum of national contexts. Economic and industrial behaviour must refer to all possible resources, markets and forms of technological know-how, regardless of national or continental boundaries. This change has been synthesised in the fact that we are moving towards a global system. This is especially true for R&D and technology, but it is also increasingly so for markets and demand.

These changes in markets and in economic transactions can be summarised as being characteristic of increasing *complexity*, a complexity emerging from the accumulation of information itself and resulting in a greater *variety* of economic situations firms have to cope with at any moment (static complexity) and a greater *variability* of economic situations over time (dynamic complexity) representing a much more difficult task for firms to cope with (Cozzi and Vaccà, 1986).

The spatial transformation from multi-national to global markets (the 1970s being a decade of transition) reinforces the ongoing process of the internationalisation of national economies. However, faced by the choice of the form of internationalisation, company behaviour changes and *multi-national corporations become trans-national corporations*. The difference between the two forms lies in their internal organisational structure. In principle, although in both cases we observe a multi-located structure, the former is organised according to a *geographical* hierarchical principle: the corporation subsidiaries in each country are small-scale replicas of the corporation in the mother country, all of them performing the full array of company functions (R&D, production, marketing, administration, finance, strategy, ...) at their appropriate geographical levels, with few or no flows of intra-company international trade. On the contrary, the latter is organised according to a *functional* hierarchical principle: then, the different functions of the typical subsidiary in one country may well be loosely inter-connected, whereas international linkages are much more strongly forged and perceivable as international trade flows (e.g. R&D or specialised production units perform a global or at least continental role as suppliers of national marketing units). The economic logic underpinning this organisational change is the simple exploitation of economies of scale due to the globalisation of markets.

Thus, in a global system, the most effective and suitable form of internationalisation seems to be a hierarchical system, a trans-national organisational structure capable of exploiting economies of scale and of avoiding market use-costs. Companies are expected to grow by internal processes, reducing (at least in relative terms) their external linkages.

11.2.4 Corporate Networks in a Global System: Testable Expectations

Despite theoretical expectations, empirical analyses continue to point out the relatively increased importance of cooperation agreements in the 1980s and beyond, somehow implying the non-optimal organisational structure of trans-national companies.

The greater efficiency demonstrated by cooperation agreements in comparison with purely trans-national organisational forms can be explained by two crucial events which characterise this period: namely, a globalisation process which has only been partly achieved to date (imperfect globalisation) and a need for an increasingly pluralistic and differentiated form of technological and scientific know-how.

Although it is widely recognised that economies are moving towards a global system, this process is still incomplete. Language, and more generally cultural barriers and national regulations are not disappearing and country-specific complementary assets still play a fundamental role in market penetration strategies. Cooperation agreements with local firms can assure an easier and more specific strategy to enter the market at the production level, keeping preferences and demand volatility under control. Moreover, demand becomes more and more capable of developing its own preferences, of expressing a self-containing dynamic, this feature providing local customers with an active and direct role in innovative processes.

The development of some crucial innovative products is facilitated by, and in some cases dependent upon, the *bridging between supply and local demand*, between technological potentialities on the one hand, and customer preferences and needs on the other. Innovative policies ignoring this crucial linkage are characterised by a high risk of failure. The strategic importance of this linkage has been analysed in a recent work on telecommunications, relating it to the adoption of computer networks (Camagni and Capello, 1991). The adoption trajectories of these innovative physical infrastructures are strictly related to the inter-relation policies between supply, offering innovative and customised products, and demand, expressing its preferences through an active role in innovative processes. The well known dichotomy between technology push and demand pull innovations loses its significance in relation to what we have just said. The propulsive force for innovation does not entirely depend on technological development and scientific potentiality; neither does it depend solely on changes in demand needs and preferences. It is strictly linked to the capability of firms and adopters to create a bridge between them, in order to link their technological potentialities with their real needs and preferences.

From this point of view, networks survive in the trans-national era as a necessary and appropriate *response to "the past that will not go away"*, given that these frictions are not strong enough to justify a return to the multi-national organisational pattern.

The growing static and dynamic complexity of technology explains why even in largely globalised functions, such as R&D, cooperation agreements are preferred to the classical multi-national form of internationalisation. The strategic resource for the competitiveness of firms and for the comparative advantage of countries becomes *pluralistic and differentiated technological and scientific know-how*, allowing for greater levels of flexibility and integration both at the level of the work force, and at the functional level (Capello and Williams, 1992).

The variety of cases, circumstances, and knowledge that firms have to deal with at any given time (i.e. static complexity) can be afforded through a trans-national organisational form. In fact, a greater variety of products demanded on the market, requires a number of initial changes in the organisational structure (from production to strategies). Once these changes have been internalised, firms can also cope with the variety of technologies and markets through a hierarchical structure.

What makes the hierarchical structure totally inefficient is the *continuing changes* in technology and in the market, this variability being manageable only through a highly flexible organisational structure: the only form of internationalisation allowing for greater levels of flexibility is a cooperative structure. To be able to afford this dynamic technological know-how, firms have to coalesce with technologically complementary partners: this form of organisation allows companies to afford the increasing variability over time of cases, circumstances, and continuous product and process innovation (i.e. dynamic complexity), thanks to knowledge accumulation.

In empirical terms the existence of greater degrees of dynamic complexity can be inferred by the presence of additional necessary linkages between a firm and its external environment, by the need for more information for strategic decision making processes and by the need for a more aggressive marketing policy to face the increasing international competition.

From this point of view, networks are typical of the trans-national era as the appropriate *response to "the challenge of tomorrow"*, where internal growth may prove to be too risky a strategy.

The analysis of the economic determinants in the choice of the most suitable form of internationalisation allows for a definition of a dynamic typology of these forms. In the second half of the 20th Century two overlapping periods can be identified in which these three forms of internationalisation - trade, foreign direct investment and cooperation agreements - assume a different relative weight.

During the multi-nationalisation period (1950-1980), faced by the choice between the different forms of internationalisation, company behaviour moved from commercial licenses and commercial contracts to multi-national corporations; at the aggregate level, the pattern of trade changed, as the reciprocal opening of the two economic systems encouraged a dynamic re-definition of comparative advantages. Depending on the economic and legal characteristics of both the origin and the destination country (characteristics of the two markets, the needs for country-specific complementary assets, restrictions upon direct investments to the destination country, traditional locational advantages of the destination country), a cooperation agreement was occasionally preferred.

In the globalisation period (1970 onward) company behaviour predominantly favoured cooperation agreements and multi-national corporations, though expanding and trans-nationalising predominantly occurred through purchases of existing companies. Moreover, in juxtaposition to expected theoretical interpretations, cooperation agreements have not been increasingly displaced by trans-national corporations, this form of internationalisation still being the most efficient organisational structure because of the two processes explained above (imperfect globalisation and increasingly pluralistic and differentiated technological know-how).

In the following sections our intention is to test the dynamics of international corporate behaviour at an empirical level, namely in the Italian/Swedish case. We expect economic relationships between the two countries to evolve following different forms of internationalisation in different periods, also taking into account the lead of Sweden over Italy in terms of its development stage. In particular, *we expect*:

* a temporal lag in the diffusion of the three forms of internationalisation between Sweden and Italy, owing to their different economic development stages;
* a growing importance of international trade relationships in quantitative terms between Italy and Sweden in the 1950s and 1960s;
* a decrease, or at least a slow-down, of international trade, accompanied by a re-definition of comparative advantages, after the establishment and during the subsequent (1970s) expansion of Swedish multi-nationals in Italy;

* a greater recourse to asymmetric (property rights) cooperation agreements in the 1970s and in those industries where regulatory barriers are higher and where products are often customised;
* a new increase of international trade during the development of Swedish trans-nationals in Italy as part of globalisation trends in the 1980s;
* a greater recourse to symmetric (knowledge) cooperation agreements in the 1980s and in those industries where the economic and technological environment of companies is more complex (i.e. varied at any moment and variable over time) and where R&D and other globalised functions are more strategic;
* a substitution of trade specialisation by industry with direct investment in the 1970s and 1980s;
* a bias towards cooperation agreements rather than other forms of industrial organisation as a means of international development, even for multi-nationals and trans-nationals in the 1970s and 1980s;
* a tendency for multi-national investment, and then agreements, to favour areas with more advanced economic development.

11.3 TRADE LINKAGES BETWEEN SWEDEN AND ITALY

11.3.1 Main Trends of Sweden and Italy's International Trade

The creation of free trade areas at the end of the 1950s spurred a period of fast expansion of national production and of international trade, which was to last until the 1970s oil crises; in recent years, the economic upturn has shown its beneficial effects on international trade flows as well.

Taking Sweden as the benchmark, the structural relationship between Italy and Sweden has been a dynamic one: in fact, while in 1962 the two countries were positioned at different stages of development, in 1987 they were both considered to be most advanced countries.

GDP per capita differentials in Table 11.2 synthetise this evolution: in 1962 GDP per capita was twice as high in Sweden (7000 USD) than in Italy (3500 USD); however, over time Italy has continuously improved its relative welfare, with an acceleration between 1972 and 1977, and the differential had shrunk to three fifths by 1987 (12600 USD to 7800 USD), being even lower in purchasing power terms.

The hypothesis that trade increases before the stage of multinationals, during which it decreases to increase again during the stage of transnational corporations, is apparently supported by the data, although the macro-economic and structural explanation may be more appropriate at this aggregate level.

Our expectations about the evolution of trade flows are supported by the data for Italy. In fact, Italian development was largely linked to Italy's increasing insertion into the world economy: its export and import ratios both increased impressively, by about 8 points, during the 1960s (1962-72). Thanks to a sharply diverging trend (a hint of Italy's answer to the first oil crisis), in the mid-1970s the two ratios converged to the same value, which they retained in 1982. Finally, in the mid-1980s the two ratios both picked up again, albeit at a different pace.

On balance, our expectations about trade dynamics are also supported by the data for Sweden. Perhaps more surprisingly, Sweden's ratios also increased, and they did so throughout the period, with the one exception of imports during the 1977-82 period. So a slightly slower rate of growth during the 1960s (7 points) was made up for in the following years, when, however, the Swedish response to the oil crisis was slower than the Italian. The level of export and import ratios has been consistently higher in Sweden, which is obviously due partly to its more advanced stage of development and partly to the well-known small country effect.

Table 11.2: National accounts data at 1985 exchange rates and prices (billions USD)

	1962	1967	1972	1977	1982	1987
ITALY						
Exports	20,18	34,04	53,71	74,02	85,46	102,47
Imports	26,71	40,54	66,71	74,77	86,11	111,54
Gross domestic product	179,11	228,09	283,34	343,17	398,14	449,73
Population (millions)	50,84	52,67	54,40	55,73	56,64	57,33
GDP per capita (USD)	3523,01	4330,55	5208,46	6157,72	7029,31	7844,58
Exports/GDP (%)	11,27	14,92	18,96	21,57	21,46	22,78
Imports/GDP (%)	14,91	17,77	23,54	21,79	21,63	25,47
Exchange rate (ITL/USD)	625,00	625,00	583,20	882,40	1352,50	1296,10
SWEDEN						
Exports	10,46	14,68	21,22	24,40	29,91	38,22
Imports	12,71	17,75	24,07	28,66	29,58	37,98
Gross domestic product	53,03	65,37	78,21	85,58	92,99	105,93
Population (millions)	7,56	7,87	8,12	8,25	8,33	8,40
GDP per capita (USD)	7014,55	8306,23	9631,77	10373,33	11163,27	12610,71
Exports/GDP (%)	19,72	22,46	27,13	28,51	32,16	36,08
Imports/GDP (%)	23,97	27,15	30,78	33,49	31,81	35,85
Exchange rate (SEK/USD)	5,17	5,17	4,76	4,48	6,28	6,34
Cross rate (ITL/SEK)	120,82	120,12	122,47	196,87	215,26	204,43

Source: OECD, National accounts - main aggregates 1960-1989, OECD, Paris

Exchange rate policies have also played a role in these developments: a fixed rate environment encouraged a more stable pattern of trade growth whereas in the 1970s flexible rates were part of a more turbulent environment detrimental to trade; note that Italy used devaluation as a means to defend and balance its openness ratios, an example later followed by Sweden.

In fact, the stability of the exchange rates of both currencies with the US dollar ended in the early 1970s, producing, after a few "weak dollar" years, a devaluation trend for both until the mid 1980s. As an aside in relation to the next section, the exchange rate between the two currencies experienced only one major re-alignment during the period in discussion, taking place in the mid 1970s in favour of the weaker Italian economy.

11.3.2 Main Trends of Trade between Sweden and Italy

Sweden and Italy are, by and large, important commercial partners. For Italian imports Sweden has consistently occupied the 9th or 10th place, whereas for exports its position was a little worse, being around 11th or 12th. For Swedish exports Italy's ranking was consistently 10th, and for imports its position has improved over the last ten years, but still within the 8th to 12th range. Finally, the total net commercial balance does not show a consistent pattern over time, although 1972 and 1977 show a net gain for Sweden and 1982 and 1987 a net gain for Italy.

At an aggregate level, expectations meet a broadly positive response, namely that trade is substituted by multi-national corporate investment and, subsequently, it is associated with corporate reorganisation along trans-national lines (Table 11.3)

Table 11.3: Trade flows of Sweden (million SEK) and Italy (billion ITL)

	1962	1967	1972	1977	1982	1987
SWEDEN						
Imports from Italy	590	916	1085	2453	5407	10678
Exports to Italy	608	837	1264	2590	5258	10593
Total imports	16119	24329	38390	90148	173525	257408
Total exports	15127	23425	41652	85527	167975	281433
Imports from Italy (%)	3,66	3,77	2,83	2,72	3,12	4,15
Exports to Italy (%)	4,02	3,57	3,03	3,03	3,13	3,76
ITALY						
Imports from Sweden	76	97	129	469	1234	2273
Exports to Sweden	63	105	156	414	1036	2044
Total imports	3797	6142	11265	42429	116216	162353
Total exports	2918	5441	10849	39968	99231	150879
Imports from Sweden (%)	2,00	1,58	1,15	1,11	1,06	1,4
Exports to Sweden (%)	2,16	1,93	1,44	1,04	1,04	1,35

Source: Statistiska Centralbyrån, Statistisk årsbok för Sverige, SCB, Stockholm
Istat, Statistica annuale del commercio con l'estero, INS, Roma

In fact, employing Swedish statistics, the share of Swedish imports from Italy has increased since 1977, after falling between 1967 and 1977; the share of Swedish exports to Italy also increased after 1977, but its negative growth was pre-dated to 1962-72. However, taking into account the general trend of openness ratios, despite the counter-trend dynamics of the early decades the shares of trade with Italy over Swedish *GDP* only show a decline for exports in the 1967-72 period.

When employing Italian statistics, the share of Italian imports from Sweden halved, falling continuously until 1982; it only increased in the mid 1980s; the share of Italian exports to Sweden also increased only in the final period, but its decline stopped in 1977. Even taking into account the general trend of openness ratios the shares of trade with Sweden over Italian *GDP* show an increase only in the mid 1980s and, for exports, in the mid 1960s.

In sum, whereas Italian data show mostly the barrier effect of EEC membership from the Italian viewpoint, balances and Swedish data do suggest a time mismatch in multi-national development in the two countries: again two effects are detectable, not just the corporate one.

11.3.3 The Structural Division of Labor between Italy and Sweden

Turning again to Swedish statistics, and focussing upon the specialisation between countries, *the expectation of a re-definition of comparative advantages with foreign direct investment is broadly supported.* Net commercial balances at the industry level (Table 11.4.a) do show some consistent patterns confirming the traditional structural economic image of the two countries: trade in mining and modern manufacturing is in favour of Sweden, whereas trade in agribusiness and traditional manufacturing is in favour of Italy. However, balance reversals do appear, too, affecting transport means (pro-Sweden after 1972), chemicals (pro-Sweden after 1977), machinery (pro-Italy after 1982) and electric/electronics (pro-Italy after 1972); these patterns contradict traditional images, showing the two countries vying for comparative advantages in industries with a higher and/or more dynamic technological content, a reminder of Italy's international specialisation trend since the early 1970s towards skill intensive industries rather than capital- and R&D-intensive ones (Onida, 1989).

Table 11.4.a: Trade flows between Sweden and Italy at current prices (million SEK)

	1962	1967	1972	1977	1982	1987
TO SWEDEN FROM ITALY						
Mining-oil	49	146	146	193	626	131
Modern manufacturing (*)	41	91	136	372	729	1519
Chemic.-pharmac.-petroch.	23	34	56	141	212	468
Electric-electronics (**)	14	84	100	210	382	938
Machinery	77	125	150	422	1072	2746
Transport means	37	41	103	222	303	789
Trad. manufacturing (***)	233	281	259	656	1676	3371
Agriculture-food products	116	124	136	236	407	716
TOTAL	590	926	1086	2452	5407	10678
TO ITALY FROM SWEDEN						
Mining-oil	222	310	351	548	1202	1814
Modern manufacturing (*)	120	149	210	481	893	1894
Chemic.-pharmac.-petroch.	16	20	29	80	280	702
Electric-electronics (**)	36	24	151	142	269	801
Machinery	113	144	283	573	1121	2370
Transport means	13	8	11	393	572	1123
Trad. manufacturing (***)	61	63	113	284	671	1706
Agriculture-food products	26	120	116	89	250	183
TOTAL	608	837	1264	2590	5258	10593

(*) non-metal minerals, metal products, photo-phono-cine, etc.
(**) cables, components, apparel, machinery, electronics, telecommunic., computers
(***) textiles, clothing, leather, shoes, paper-printing-publish., rubber, plastics

Source: Statistiska Centralbyrån, Statistisk årsbok för Sverige, SCB, Stockholm

From a complementary viewpoint, based on relative weights on total trade (Table 11.4.b), as the two countries become reciprocally more open and structurally more similar, *expectations are fulfilled*, in that comparative advantages are indeed re-defined and become increasingly based on specific intra-industry niches and the strategic behaviour of companies therein.

Italy-to-Sweden patterns confirm the structural evolution of Italian comparative advantages as they are perceived in Sweden, albeit in selective ways. In fact, Swedish imports from Italy specialise in traditional manufacturing throughout the period. Initial specialisations in agribusiness and, shortly afterwards, electric/electronics disappeared respectively and were replaced by machinery. Over time, modern manufacturing has reversed its under-specialisation, whereas mining and chemicals have maintained theirs. Finally, transport means and electric/electronics show a broadly neutral position.

The general picture shows that trade flows between Sweden and Italy are structurally more similar than expected, and the more so in recent years. In fact, Swedish exports to Italy specialise, relative to total exports, in agribusiness, mining and transport means throughout the period, and in machinery after the 1960s, while chemicals and traditional manufacturing recover in the last decade from their under-specialisation. Finally, modern manufacturing is in a neutral position and electric machinery shows a fluctuating pattern.

Table 11.4.b: Trade flows between Sweden and Italy at current prices (%)

	1962	1967	1972	1977	1982	1987
TO SWEDEN FROM ITALY						
Mining-oil	8.31	15.94	13.46	7.87	11.58	1.23
Modern manufacturing (*)	6.95	9.93	12.53	15.17	13.48	14.23
Chemic.-pharmac.-petroch.	3.90	3.71	5.16	5.75	3.92	4.38
Electric-electronics (**)	2.37	9.17	9.22	8.56	7.06	8.78
Machinery	13.05	13.65	13.82	17.20	19.83	25.72
Transport means	6.27	4.48	9.49	9.05	5.60	7.39
Trad. manufacturing (***)	39.49	30.68	23.87	26.74	31.00	31.57
Agriculture-food products	19.66	13.54	12.53	9.62	7.53	6.71
TOTAL	100	100	100	100	100	100
TO ITALY FROM SWEDEN						
Mining-oil	36.51	37.04	27.77	21.16	22.86	17.12
Modern manufacturing (*)	19.74	17.80	16.61	18.57	16.98	17.88
Chemic.-pharmac.-petroch.	2.63	2.39	2.29	3.09	5.33	6.63
Electric-electronics (**)	5.92	2.87	11.95	5.48	5.12	7.56
Machinery	18.59	17.20	22.39	22.12	21.32	22.37
Transport means	2.14	.96	.87	15.17	10.88	10.60
Trad. manufacturing (***)	10.03	7.53	8.94	10.97	12.76	16.10
Agriculture-food products	4.28	14.34	9.18	3.44	4.75	1.73
TOTAL	100	100	100	100	100	100

(*) non-metal minerals, metal products, photo-phono-cine, etc.
(**) cables, components, apparel, machinery, electronics, telecommunic., computers
(***) textiles, clothing, leather, shoes, paper-printing-publish., rubber, plastics

11.4 THE ITALIAN EXPERIENCE IN FOREIGN DIRECT INVESTMENT

11.4.1 Global Trends in Foreign Direct Investment

Since 1970 the growth rate of foreign direct investment in the world has exceeded that of international trade, investment and GDP. However, since 1980 investment has overwhelmingly taken the form of acquisitions, as opposed to the previous stress on greenfield development. On the one hand, and initially, this shift often depended on market saturation leading to excess capacity and a trend towards oligopolistic markets. On the other hand, and more recently, it has depended on the increasing interest of corporations in local know-how and competence as strategic assets for long-run survival, especially since product cycles (hence monopoly rents) have shortened to a few years and require high investments and vast markets to repay them quickly.

Between 1960 and 1980 (Cominotti and Mariotti, 1986), traditional investment-originating countries (the USA above all, but also the UK, the Netherlands and France) lost weight to other advanced countries (most notably Japan and Germany, but also Switzerland and Sweden). Conversely, at the receiving end advanced countries (except Canada) gained ground at the expense of less developed countries. These trends were reinforced during the 1980s: the US confirmed their reversal into a destination country (for Japan and the UK above all) whereas Europe became a major area of origin as well as staying on as a recipient.

According to theory, *Italy should be an interesting recipient country for foreign direct investment* for three major reasons:

* a macro-economic reason: Italy sports an increasingly large market in terms both of final and of intermediate products;
* a technology/production reason: Italy has been shifting from an intermediate to an advanced country, where frontier technologies can be exported and applied;
* a political reason: Italy's legal provisions regulating foreign direct investment have always been extremely liberal, also with respect to almost any other developed country.

However, *Italy should be expected to be a scarce and selective originator of investment flows* for the following three reasons (Onida, 1989), which it shares with Japan:
* a macro-economic reason: Italy is a new-comer, i.e. national development is relatively lower and more recent, so foreign orientation is still mostly in terms of export markets and foreign investment, if any, will be mostly market-oriented;
* a structural reason: industry mix is tilted away from high-tech industries, in favour of scale-intensive and traditional ones, which would produce investment towards LDCs and a preference for low-cost non-traditional forms of investment;
* a property reason: ownership patterns are biased towards State-held companies and medium-to-small-size unquoted private companies, which have a relatively negligible foreign orientation.

This quantitative analysis is based on data collected for the National Council for the Economy and Labour (Cominotti and Mariotti, 1986, 1989 and 1991) considering only *investment in companies with production plants*, i.e. excluding commercial subsidiaries and financial investments; further, we focus on majority-controlled companies, and in any case this form dominates inward investment and, to a lesser extent but increasingly so (as expected), outward investment too. From these data, *our expectations are confirmed at the aggregate level.*

In Italy, the dominance of inward over outward investment is unparalleled by other indicators, as expected, and increased during the 1960s and 1970s. However, during the mid 1980s the gap halved (Table 11.5 or 11.6): the company ratio went from 27% to 56%, the turnover ratio from 29% to 55%, the employment ratio from 32% to 66%. This happened because, while the level of externally-controlled employment remained stable (-3%) and companies increased slightly (+9%), Italian companies abroad more than doubled (+122%) and controlled employment doubled (+100%). The weight of foreign-controlled companies over total Italian manufacturing employment was as low as 9% in the early 1980s. As for active internationalisation, this was even lower by international standards (Cominotti and Mariotti, 1986).

11.4.2 Italian Inward and Outward Investment by Industry

The data (Cominotti and Mariotti, 1986, 1991) have been aggregated for the purpose of this study into eight industries, approximately the same as in the previous section (Table 11.5.a). *Our hypothesis is supported: namely that Italy, being a large but recently advanced country, is a diversified but stabilised recipient of direct investment but a more dynamic originator.*

Structurally, Italian outward investment is in a state of flux, in terms of both mix and concentration. Thus, it is possible to detect both a trend towards cost-reducing outward investment (food, textiles, some sections of other industries...) and an increasing ability to manage investment in industries with a more complex content: for this reason, Pavitt's (1984) five-way industrial classification does not provide useful guidelines on the composition of Italian outward investment.

Table 11.5.a: Foreign direct investment Italy-abroad by industry

	Companies		Turnover		Employment	
	1985	1989	1983	1988	1983	1988
TO ITALY FROM ABROAD						
Mining-oil-I°transform.	52	34	16309	7990	16905	10004
Modern manufacturing (*)	122	130	4143	5898	36166	27101
Chemicals-pharmaceuticals	252	289	13777	28194	79137	90907
Electric-electronics (**)	150	175	13845	25414	128315	124815
Machinery	191	205	3847	7677	44294	49714
Transport means	44	56	1642	2833	22105	18926
Trad. manufacturing (***)	127	136	4338	7341	45393	41966
Food, drinks, tobacco	91	101	5723	8685	32131	30064
TOTAL	1029	1126	63624	94032	404446	393497
ABROAD FROM ITALY						
Mining-oil-I°transform.	24	52	6138	8133	6476	12267
Modern manufacturing (*)	30	56	641	1378	6015	12359
Chemicals-pharmaceuticals	28	70	742	3569	5203	10618
Electric-electronics (**)	40	107	2042	11023	16119	89476
Machinery	40	80	1078	3168	7530	18558
Transport means	24	46	2767	8762	29540	49000
Trad. manufacturing (***)	56	116	3706	5828	49968	44277
Food, drinks, tobacco	40	101	1609	9448	8608	23179
TOTAL	282	628	18723	51309	129459	259734

Source: Cominotti - Mariotti (1986, 1991)

Table 11.5.b: Foreign direct investment Italy-abroad by industry (%)

	Companies		Turnover		Employment	
	1985	1989	1983	1988	1983	1988
TO ITALY FROM ABROAD						
Mining-oil-I°transform.	5.05	3.02	25.63	8.50	4.18	2.54
Modern manufacturing (*)	11.86	11.55	6.51	6.27	8.94	6.89
Chemicals-pharmaceuticals	24.49	25.67	21.65	29.98	19.57	23.10
Electric-electronics (**)	14.58	15.54	21.76	27.03	31.73	31.72
Machinery	18.56	18.21	6.05	8.16	10.95	12.63
Transport means	4.28	4.97	2.58	3.01	5.47	4.81
Trad. manufacturing (***)	12.34	12.08	6.82	7.81	11.23	10.66
Food. drinks. tobacco	8.84	8.97	9.00	9.24	7.95	7.64
TOTAL	100	100	100	100	100	100
ABROAD FROM ITALY						
Mining-oil-I°transform	8.51	8.28	32.78	15.85	5.00	4.72
Modern manufacturing (*)	10.64	8.92	3.42	2.69	4.65	4.76
Chemicals-pharmaceuticals	9.93	11.15	3.96	6.96	4.02	4.09
Electric-electronics (**)	14.18	17.04	10.91	21.48	12.45	34.45
Machinery	14.18	12.74	5.76	6.17	5.82	7.15
Transport means	8.51	7.32	14.78	17.08	22.82	18.87
Trad. manufacturing (***)	19.86	18.47	19.79	11.36	38.60	17.05
Food. drinks. tobacco	14.18	16.08	8.59	18.41	6.65	8.92
TOTAL	100	100	100	100	100	100

(*) non-metal minerals, metal products, photo-phono-cine, etc.
(**) cables, components, apparel, machinery, electronics, telecommunic., computers
(***) textiles, clothing, leather, shoes, paper-printing-publish., rubber, plastics

Table 11.6.a: Foreign direct investment Italy-abroad by country

	Companies		Turnover		Employment	
	1985	1989	1983	1988	1983	1988
TO ITALY FROM ABROAD						
USA	362	338	30039	34423	163772	127444
France	118	151	7995	12323	45585	60209
Sweden	48	83	3281	7725	42039	52941
Germany	134	152	4917	10789	34757	49304
Switzerland	140	108	4998	6745	42352	31013
United Kingdom	108	128	3515	6023	26741	28979
Netherlands	60	63	4951	8466	30847	20484
Other countries	59	103	3928	7538	18263	23123
TOTAL	1029	1126	63624	94032	404356	393497
ABROAD FROM ITALY						
France	30	114	1817	12671	14399	61787
Brazil	36	52	1961	4138	36762	43490
USA	32	103	2086	8477	6579	32000
Spain	40	95	1253	5408	14459	31681
Germany	10	32	2345	5107	18303	28728
United Kingdom	15	33	1540	3098	10339	17492
Other countries	119	199	7721	12410	28618	44556
TOTAL	282	628	18723	51309	129459	259734

Source: Cominotti - Mariotti (1986, 1991)

Table 11.6.b: Foreign direct investment Italy-abroad by country (%)

	Companies		Turnover		Employment	
	1985	1989	1983	1988	1983	1988
TO ITALY FROM ABROAD						
USA	35.18	30.02	47.21	36.61	40.50	32.39
France	11.47	13.41	12.57	13.11	11.27	15.30
Sweden	4.66	7.37	5.16	8.22	10.40	13.45
Germany	13.02	13.50	7.73	11.47	8.60	12.53
Switzerland	13.61	9.59	7.86	7.17	10.47	7.88
United Kingdom	10.50	11.37	5.52	6.41	6.61	7.36
Netherlands	5.83	5.60	7.78	9.00	7.63	5.21
Other countries	5.73	9.15	6.17	8.02	4.52	5.88
TOTAL	100	100	100	100	100	100
ABROAD FROM ITALY						
France	10.64	18.15	9.70	24.70	11.12	23.79
Brazil	12.77	8.28	10.47	8.06	28.40	16.74
USA	11.35	16.40	11.14	16.52	5.08	12.32
Spain	14.18	15.13	6.69	10.54	11.17	12.20
Germany	3.55	5.10	12.52	9.95	14.14	11.06
United Kingdom	5.32	5.25	8.23	6.04	7.99	6.73
Other countries	42.20	31.69	41.24	24.19	22.11	17.15
TOTAL	100	100	100	100	100	100

In fact, the previous priority placed upon traditional manufacturing (mostly rubber) and transport means (39% of employment in 1983), followed by transport means (23%) and electric/electronics (12%), has been reversed into a stress on electric/ electronics (34%

of employment in 1988), followed by transport means (19%) and traditional manufacturing (17%). The other five industries in terms of employment together amounted to 26% of the total in 1983 (all between 4% and 7%) and to 30% in 1988 (all between 4% and 9%). The electric/electronic sector was by far the most dynamic industry, followed (in relative terms) by food and machinery; traditional manufacturing was, because of rubber, the only industry to decline even in absolute terms.

Italian inward investment is relatively more stable, less concentrated, and more technologically-advanced from a structural viewpoint. Thus, there is a relative trend towards replacing consumer industries with technologically dynamic and information-rich industries, the exception being chemicals; here Pavitt's typology works, in that science-based industries are much more involved in inward investment (this is even clearer at a less aggregate level: Cominotti and Mariotti, 1991).

In fact, the first two industries are electric/electronics and chemicals/pharmaceuticals; they remained the same during the 1980s, and with a similar share (from 51% to 55%). Medium importance, above 10% employment, can be assigned to machinery and traditional manufacturing (aggregate share 22% to 23%), whereas the remaining four industries have a total share declining during these years from 26% to 22%. Given a stable total aggregate, relative decline often means absolute decline as well. Gains and losses during the mid-1980s have been less impressive than for outward investment: we highlight machinery and especially chemicals/pharmaceuticals on the plus side and modern manufacturing and mining/oil on the minus side.

Comparing inward and outward investment, expectations are fulfilled, as multinational corporations are increasingly active in industries characterised by dynamic complexity.

In fact, we see clear differences in relative specialisations: transport means, traditional manufacturing and mining/oil (all assembly or raw materials industries) concentrate in outward investment whereas chemicals, machinery and modern manufacturing concentrate in inward investment; interestingly enough, food and electric/electronics show a reversal of relative specialisation in favour of Italian investors (the latter being concentrated on cables, components and computing). The employment balance, in contrast, is positive for Italian control only in traditional manufacturing (just slightly, and with internal re-distributions), transport means and recently mining/oil (slightly): the other five industries show a positive balance for inward investment, although only chemicals/pharmaceuticals are not decreasing; and in fact horizontal investment grew in all industries except chemicals.

11.4.3 Italian Inward and Outward Investment by Country

Our expectation is confirmed, that this activity is the preserve of corporations in very advanced countries, a club which Italy has joined only in very recent years.

Outward investment is concentrated to six countries (Table 11.6), the remaining ones having only a 17% employment share and a 24% turnover share in 1988, at the end of five years of increasing concentration. In the first half of the 1980s, the two intermediate countries of Brazil and Spain showed the highest number of companies and ranked first and third for the number of employees. In their trend towards alignment with other advanced countries, Italian investors, too, started in the 1980s to privilege OECD countries at the expense of LDCs: we detect here a sort of horizontal investment which runs parallel to horizontal trade. Thus, in 1988-89 France became the first recipient country in terms of companies, turnover and employment (+329% for employment) and the USA has also proved very attractive, jumping to third place for employment (+386%) and to second place for turnover and companies. Overall, Italian-controlled employment in the four most attractive advanced countries passed from 38% to 54%.

Turning to the other side of the coin, inward investment is even more concentrated, the first seven countries leaving only a mere 6% of employment, 8% of turnover and 9% of companies to other countries. It is significant for Italy's international standing that the two large intermediate countries (Brazil and Spain) are replaced here by three small advanced countries (Sweden, Switzerland and the Netherlands). As in world trends, the 1980s were marked by significant exits from the Italian market by US corporations, selling to European corporations (mostly French and Swedish, but also German and British); the number of Swiss corporations also declined. Despite the decline, the US remains by large the main foreign investor, with around one third of the total. During the mid 1980s France, Sweden, Germany and, to a lesser extent, the UK gained significant ground in companies and employment, whereas Switzerland and the Netherlands followed the US trend of decline through disinvestment.

Comparing inward and outward investment, the employment balance is usually unfavourable to Italy with advanced countries and favourable with less developed countries, although the trend is towards a greater balance with the first group, which has already been reached in the case of France.

The uninterrupted penetration of Swedish companies into Italy is by now a consolidated phenomenon: in 1981 Italy attracted less than 1% of Swedish foreign direct investment, but in 1985 that portion was closer to 5% and it has certainly further increased to date, and at a faster pace than any other recipient country (Svenska Handelsbanken, 1986). Swedish controlled employment increased rapidly in the 1980s to 13.5% of the total, and thanks mainly to a few huge operations Sweden climbed from fourth to third place as an investor in terms of employment and jumped from seventh to fifth place in terms of turnover. Swedish-controlled companies show a markedly above-average employment size, while their turnover size is still above average but less so, possibly indicating the effect of industry mix.

11.4.4 Italian Inward and Outward Investors

Our hypothesis concerning a greater diversification of inward rather than outward Italian investment is supported at the general level, whereas in the case of Sweden it is counterbalanced by a small country effect.

The outward investor structure is heavily concentrated (Table 11.7), since the four major Italian private groups, led by the Agnelli, Pirelli, DeBenedetti and Ferruzzi families, controlled 73% of employees abroad under Italian control (1988). If the two main State holdings (IRI and ENI) are added, the share reaches 78%: thus less than one quarter of the employees abroad are controlled by small groups or medium-sized companies. This pattern has not softened despite the rapid increase in internationalisation, since the number of investors has remained stable and the identity of small investors has proved rather volatile (the turnover involved about one quarter of them in just 5 years).

In particular, Italian investment in Sweden was nil until 1985; after that date, 4 companies have been bought by four different investors, for an aggregate turnover (1988) of 277 billion ITL (0.5% of total Italian outward investment) and 1931 employees (0.7% of total). Why investing in Sweden? In the most recent case the acquisition depended on the Swedish company being a subsidiary of a multinational company being purchased; but in all cases we can detect the attraction not so much of the Swedish domestic market, but of a larger share on the world market through the acquisition of competitors or rather of companies occupying related niches.

Not surprisingly, the inward investor structure is much less concentrated than the outward one, since the fifteen major employers made up 35% (1983) and 37% (1988) of the total. This depends on the much larger company base from which the investors are drawn. The trend towards further concentration is very weak, although mobility even at the top echelons is in order: ITT and GTE sold their activities to CGE-controlled

Alcatel and to Siemens respectively, and Philips quit domestic appliances in favour of Whirlpool. Thus, the US presence has fallen from 6 to 3 top investors, whereas the French rose from 1 to 3.

Table 11.7: Foreign direct investment by main investors (number of employees and per cent)

1983				1988			
TO ITALY FROM ABROAD							
Electrolux	- S	19932	4.93	ABB	- S	21527	5.47
ITT	- US	14637	3.62	Electrolux	- S	15114	3.84
Philips	- NL	14475	3.58	IBM	- US	13809	3.51
Michelin	- F	13565	3.35	CGE	- F	11578	2.94
IBM	- US	12649	3.13	Siemens	- D	10621	2.70
Ericsson	- S	10732	2.65	Michelin	- F	10084	2.56
Unilever	- NL	8023	1.98	Whirlpool	- US	8972	2.28
BBC	- CH	7280	1.80	Ericsson	- S	8869	2.25
SKF	- S	6372	1.58	Nestlè	- CH	8557	2.17
General Electric	- US	6089	1.51	Unilever	- NL	7416	1.88
Utd. Technologies	- US	5932	1.47	SKF	- S	6355	1.62
GTE	- US	5619	1.39	Utd. Technologies	- US	6317	1.61
Hoechst	- D	5538	1.37	Saint Gobain	- F	5871	1.49
Siemens	- D	5236	1.29	Hoechst	- D	5623	1.43
3M	- US	5081	1.26	Philips	- NL	5066	1.29
Other investors		263196	65.09	Other investors		247718	62.95
ABROAD FROM ITALY							
				IFI (Agnelli)		62373	24.01
				CIR (DeBenedetti)		54547	21.00
				Pirelli		51057	19.66
				Feruzzi		21170	8.15
				ENI		6246	2.40
				IRI		6182	2.38
				Saffa (Bonomi)		4334	1.67
				GFT		3417	1.32
				Ferrero		3210	1.24
				GIM (Orlando)		3014	1.16
				Other investors		44204	17.02

Source: Cominotti - Mariotti (1986, 1991)

The number of Swedish major investors rose in the 1980s due to two huge operations: the purchase of the Zanussi group by Electrolux in 1984 and the fusion of Asea with the Brown Boveri corporation (which has been historically active in Italy) to create ABB in 1988. It has to be noted that, as in similar cases, ABB was considered Swedish due to complementary indicators, such as representation on the Board of Directors (Cominotti and Mariotti, 1991). The top four Swedish investors totalled 11% of 1983 employment and 13% of 1988 employment, due to energetic acquisition policies, of which the most active was the Wallenberg group. In both reference years the largest employer was a Swedish investor, and in 1988 the second largest was also Swedish. However, it should not be forgotten that several other major and medium-sized Swedish groups are active in Italy.

Why investing in Italy? The last major operation, the creation of ABB, was a consequence of the international merger and acquisition market (in which, however,

Italian activities were not marginal to decisions), but in most cases the attraction of the Italian domestic market to Swedish investors is apparent, unlike in the opposite direction, and it is perceived as a market quickly responding to products that rank high in technology and/or quality (and image); of course a footing in the EEC market is not overlooked, but it does not play the main role, and labor costs are not really an attraction (Svenska Handelsbanken, 1986). The primary strategic mission assigned to subsidiaries seldom concerns markets other than the Italian one (22%) or functions differently from the production/sale of goods/services to the final market (13%), suggesting that the multi-national organisational structure has not yet fully turned into a trans-national structure (Gualini and Pompili, 1990; Pompili, 1991).

11.4.5 Italian Inward and Outward Investment by Location

In locational terms, large outward investors are concentrated in the North-West (Milan-Turin-Genoa), the home of the longest established manufacturing tradition which, incidentally, is also the largest consumer and intermediate market and the destination of most inward investment.

Our expectation of a concentration of investment in most advanced areas is broadly confirmed, together with the increasing importance of this phenomenon.

As to the time pattern and to the location of inward investment (Table 11.8), 30% of foreign-controlled companies acquired this status after 1981, and at an accelerating rate (for Swedish-controlled companies the recent share is over 50%). We do not have detailed data before 1981, except that it is unlikely that more Swedish subsidiaries opened in 1981-83 (Svenska Handelsbanken, 1986).

Table 11.8: Foreign companies in Italy (1989) by year of acquisition and HQ location

	-1981	1982-83	1984-85	1986-87	1988-89	Total
FROM SWEDEN						
Milan (province)	6	0	2	6	8	32
Other Lombardy	4	1	0	0	2	7
Turin (province)	5	1	0	1	1	8
Other Piedmont, Liguria	1	0	0	1	0	2
Triveneto	1	0	7	1	4	13
Emilia, Tuscany, Umb., Mar.	4	1	0	1	1	7
Rome (province)	3	0	0	1	0	4
Other Latium and South	3	2	0	1	2	8
TOTAL	27	5	9	12	18	81
FROM ABROAD (ALL)						
Milan (province)	302	11	31	30	34	408
Other Lombardy	97	5	12	16	222	152
Turin (province)	50	6	2	6	11	75
Other Piedmont, Liguria	38	5	5	11	14	73
Triveneto	56	2	21	8	25	112
Emilia, Tuscany, Umb., Mar.	72	3	9	16	38	138
Rome (province)	47	1	3	1	4	56
Other Latium and South	90	8	2	6	13	119
TOTAL	752	41	85	94	161	1133

Source: Cominotti - Mariotti (1986 - 1991)

The location of the 1133 foreign-controlled headquarters shows a strong and persistent concentration in the province of Milan (42%); adding to it the rest of Lombardy, this area accounts for 55% of the companies. Consistently, 40% of the 81 Swedish headquarters, managing 127 plants, are in the province of Milan, increasing to 50% for the whole of Lombardy.

The 32 companies headquartered in Milan manage 43 plants (35 of the 36 located in the province), and the 9 Pordenone companies manage 28 plants (among these, all the 13 located in the province); at a lower level, the 8 Turin companies manage 13 plants (8 of the 9 located in the province) and the 4 Rome companies manage 7 plants (4 of the 7 located in the province); finally, the remaining 28 companies manage 36 plants, of which only 3 are outside the headquarter province.

Finally, the locational distribution of the Swedish-controlled employment (Table 8.9), shows the one-investor poles of Electrolux in Pordenone (in Triveneto), Ericsson in Rome and SKF in Turin. Apart from these, only the province of Milan has a significant employment strength, due to a large number of medium-sized subsidiaries, which was further heightened by the subsequent acquisition of the Brown Boveri Corporation, whose Italian companies are mostly headquartered in Milan, so that now Milan's share for Sweden is less distant from the 39% value for all foreign-controlled companies (Gualini and Pompili, 1990).

This pattern provides the basis of the sample selection for the qualitative analysis reported in Section 11.5

Table 11.9: Foreign companies' Italian employment (1986) by industry and HQ location

	Base	Chemic.	Electr.	Machin.	Tradit.	Total
FROM SWEDEN						
Milan (province)	219	1469	1613	1459	52	4812
Other Lombardy	0	0	273	180	67	520
Turin (province)	170	0	0	6169	631	6970
Other Piedmont, Liguria	0	0	418	0	0	418
Triveneto	916	0	15551	118	562	17147
Emilia, Tuscany, Umb., Mar.	494	77	545	137	0	1253
Rome (province)	0	0	7641	0	0	7641
Other Latium and South	0	0	433	80	115	628
TOTAL	1799	1546	26474	8143	1427	39389
FROM ABROAD (ALL)						
Milan (province)	12488	40301	56372	15699	19968	144828
Other Lombardy	4007	8364	15505	4159	6718	38753
Turin (province)	396	1512	4113	14552	14022	34595
Other Piedmont, Liguria	2335	1862	6385	13220	3356	17158
Triveneto	2236	3693	18015	5092	4667	33703
Emilia, Tuscany, Umb., Mar.	4711	3848	5988	8347	5268	28162
Rome (province)	5216	7268	12143	587	3364	28578
Other Latium and South	5019	11421	9251	4850	10220	40761
TOTAL	36408	78269	127772	56506	67583	366538

Base: energy, gas, water, mining, steel, metallurgy, construction
Chemicals: chemicals, pharmaceuticals, artificial and synthetic fibers
Electric: computers, electric, electronics, precision instruments
Machinery: machinery, transport means
Traditional: food, text., cloth., shoes, timber, furnit., paper, print, publish., rubber, plast.

Source: Cominotti - Mariotti (1989), Gualini - Pompili (1990)

11.5 INTERNATIONAL CORPORATE NETWORKS: EMPIRICAL EVIDENCE

11.5.1 Methodology and Sample

An analysis of international corporate networks has been carried out through direct interviews with a sample of eight significant firms[2] in order to test our expectations about the form of internationalisation presented in Section 11.2.

Firms have been chosen among the largest Swedish corporations operating in Italy: five of these firms are located in Milan, one in Rome, one in Turin and one in Pordenone (the North-Eastern part of Italy). Only two of the ten firms contacted had no cooperation agreements in Italy, their external relationship being run only by the holding company in Sweden.

For the interviewed firms, the range of cooperation agreements goes from companies with few networks (2-3) to others with a large number of them.

The interviewed firms operate in different industries, both science-based and other industries: chemical colouring agents, pharmaceuticals, mechanics, steel making, electrical-appliances, defence, telecommunications, electronics.

Our basic expectation that cooperation agreements have been an emerging form of internationalisation during the seventies and eighties has been confirmed.

In fact, in our sample, despite the great difference in the features of these sectors, a common attitude characterises corporate strategies with regards to cooperation agreements: they are interpreted as the unique corporate strategy for internationalisation, allowing for high levels of flexibility, in terms of both internal organisation and external linkages, in the face of markets characterised by great volatility and uncertainty. This result supports the conceptual idea developed by Camagni (1991) that corporate networks are a mechanism for supplying local economies with the scientific, technological, managerial and organisational know-how missing at the local level.

For science-based industries (defence, telecommunications and electronics, pharmaceuticals) this uncertainty stems from the technological process, from its rapidly changing dynamics, from the consequent high costs in R&D and from the continuing process and product innovations required to survive in the market. The reasons for the choice of this corporate behaviour lie in the exploitation of traditional complementary assets, both at the technical and scientific level.

For other industries (electrical-appliances, chemical colouring agents, mechanics and steel making), cooperation agreements are generally not used and, if used, are a way of reducing competition and the risks of losing market shares; in other words, a way of keeping competition and risk under control. In this case, cooperation agreements develop mostly among competing firms, whose incentives in developing such strategies are both a greater product specialisation, allowing for the exploitation of a higher level of economies of scale and a higher level of agreements with competitive firms, allowing for lower risks of market share losses.

The analysis allowed for a deep interpretation of this corporate behaviour, and in particular the study looked at the origin, the organisation, the nature and the dynamics of international corporate networks.

---

[2]The authors would like to acknowledge the fundamental contribution kindly given by the firms to the study. The ten companies contacted are the following:

a) Milan-based, no agreements: Alfa Laval, Sandvik;
b) Milan-based, interviewed: Asea Brown Boveri, Casco Nobel, Esab, Fiar Ericsson, Pierrel, Fermanta;
c) Non-Milan-based, interviewed: Fatme Ericsson (Rome), RIV-SKF (Turin), Zanussi Electrolus (Pordenone).

11.5.2 The Origin of Networks

The strategy behind the choice of cooperation, the reason for the identification of partners and the characteristics of the initial phase of the agreement are described in this section. This helps to create an understanding of the reasons for the increase in the (relative) importance of this form of internationalisation.

Even from an analysis of the origin of networks, the *preference for cooperation agreements rather than foreign direct investments (trans-national) emerges clearly, as expected.*

A clarification is in this respect important. More than a new form of behaviour, cooperation agreements have to be thought of as a growing phenomenon, developing both among industries and among countries. Their way of developing among firms, their nature, their organisation and their dynamics have been widely discussed and the results confirm entirely a similar analysis undertaken for different industries and different firms by the same research group (Camagni and Pompili, 1993). This coincidence of results stresses that these characteristics are endogenous to the phenomenon, rather than dependent on sectors or country-specific features.

The choice between this form of internationalisation and foreign direct investments stems from the relative flexibility towards change characterising this industrial structure in comparison to the other forms. Developing through this kind of structure, firms can remain independent and can thus follow the high volatility of demand and of technology itself.

While both forms of internationalisation agree in underlining the strategic interest in this kind of industrial organisation, the reasons for this choice differ in the two cases.

In the case of science-based industries the reasons for such behaviour are more related to effectiveness issues (complementary assets, including fragmented market exploitation), and competitive advantages issues (niche innovation, know-how to enter a new market).

The most common reasons explaining such behaviour are the high costs in R&D characterising this new "techno-economic paradigm" and the new wave of product and process innovations taking place - especially in electronic and telecommunications sectors - which are related to the profound technical discoveries in micro-electronics.

The area in which these cooperation agreements most frequently take place is R&D, and the object of the agreements is usually technological and scientific know-how.

The great volatility of markets and of customers' preferences explain why co-operation agreements are preferred to internal development through acquisitions: the flexibility of this organisational form is a strategic feature governing external relationships, thus facilitating these contacts by assuring a high level of independence and autonomy.

In juxtaposition to science-based sectors, whose interest in cooperation agreements stems from effectiveness issues and competitive advantages issues, in other industries where product and process innovations do not occur very often, the most quoted reasons for such behaviour with both external firms and firms internal to the group are related to an efficiency strategy (rationalisation, economies of scale).

In the case of public contracts in the electro-mechanic industry, for example, the division of labor among competing firms is governed by cooperation agreements. The complex product is produced by a number of companies, each of them specialising in a specific part: in this way, economies of scale are exploited and competition is kept under control.

In these industries, cooperation agreements among partners belonging to the same group are developed to pursue greater levels of rationalisation in the group's division of labor. In more than one case, firms underline a process of production rationalisation among the group consisting of an integrative division of labor at the production level.

The choice of partners is a crucial phase upon which the success of agreements depends. In most cases partners were previously known by the firm as technological and scientific leaders in the market. The time for defining an agreement is generally very long, more than one year; it lasts for at least two years and is generally very expensive, in terms of analyses of potential partners and of the transaction itself (the division of costs and benefits deriving from the partnership).

In very few cases agreements are developed with institutions, and in general the object of these agreements is a technological and scientific complementary asset, carried out by universities and by research centres. However, the difficulty of developing this sort of agreement has been stressed: strong differences in managerial and organisational mentality are pointed out as the major problems and obstacles for a completely successful complementarity among partners.

Thus, the relatively greater instability of public-private networks and of small-large size firms can be explained by asymmetries in incentives to cooperate and by the difficulties which arise during cooperation.

11.5.3 The Organisation of Networks

Another interesting element in the analysis of this form of internationalisation is the organisation of corporate networks: by this term we mean the study of the legal structure of agreements, the division between costs and benefits, the reasons for failure, the definition of a leadership. *This analysis, once again, reinforces our basic expectation that a bias exists towards cooperation agreements as an emerging form of internationalisation, this form being a stable and low-risk strategy even from the point of view of the organisation of the network.*

In fact, the organisation of the network does not reflect sectoral differences. These partnerships are always formalised through a written contract, both within the group and with external firms, this fact explaining once again the long time required to stipulate an agreement. The legal features of agreements vary from firm to firm and depend on the object of the agreement: large firms usually develop joint-venture and, in general, equity agreements, while smaller firms cooperate through production licenses and contracts, i.e. non equity agreements. Moreover, when the object of the agreement is complex and strategic, as in the exchange of scientific and organisational know-how, agreements take the form of equity agreements.

As the partnership is based on a written contract, benefits and costs issues are discussed and decided ex ante. For this reason the risk of obtaining lower benefits than expected does not exist. For example, opportunistic behaviour of partners does not occur once the preliminary phase of the partnership is developed.

According to interviewed firms, agreements only fail because of external forces changing the external environment in which the agreements were previously discussed. Another possibility for "failure" is the miscalculation of the object of agreements, such as know-how and scientific capacity, revealing a higher strategic value after the agreement and thus demonstrating a profound financial underestimation of the object of agreements.

Integration among partners is developed at the lowest possible level. Firms try to keep independent as much as possible, avoiding in this way the organisational restructuring which would follow corporate integration. Thus, cooperation takes place through the creation of cross-company teams, whose role is to be responsible for carrying out the partnership and achieving its goals. The characteristic flexibility of this industrial organisation would be reduced, and even destroyed, by creating common units: this would increase the costs of the agreements from the organisation point of view and would not allow the corporate capacity to behave autonomously.

Contrary to what might be expected, the leadership issue does not play a crucial role for the sample firms. Apart from one case, where a leader is always defined, and becomes responsible for achieving goals, in the other cases it has always been underlined that contracts are not based on a hierarchical structure but on integration for cooperation at the same level.

Thus the control over partners is very limited, though possible, and the only variable kept under control is the time allowed for the partner to develop his work. This is because when cooperation involves integration in R&D or in production, time is a crucial factor for enhancing the partners' work.

11.5.4 The Evolution of Networks

Another crucial point is the dynamic characteristics of corporate networks, i.e. the evolution of the phenomenon in itself, of the structure, of "network culture".

Our expectation about the dynamics of this form of internationalisation is a greater recourse to symmetric (knowledge) cooperation agreements in the 1980s and in those industries where the economic and technological environment of companies is more complex and where R&D and other globalised functions are more strategic.

In fact, cooperation agreements manifested a clear expansion in the eighties: in all industries there was an evident tendency towards the development of this kind of industrial organisation and of internationalisation, involving new functions, new objects and new partners by creating new networks.

Moreover, a clear dynamic in the form of agreements emerges. In most cases agreements start with simple formal relationships, i.e. property rights (asymmetric agreements), and move towards more complex and stable partnerships (symmetric agreements), once the scientific, technological, managerial and organisational capacity of the partners have been tested. The dynamic of networks also manifests itself in backward and forward phases and functions with respect to phases and functions decided upon in the original contract.

During the contract a network "culture" develops and becomes strategic in two ways: in the development of a common managerial style, i.e. a common organisational and strategic approach to the market, and in the development of a cooperation capacity, this being a very crucial element for the success of agreements.

Partners already know each other from previous joint collaborations. Many agreements develop between firms previously linked by commercial licenses: this is an interesting case of a shift from traditional forms of internationalisation to new emerging forms of internationalisation.

The structure of the networks tends to be stable until the end of the contract, in that they are fixed ex ante, especially in terms of leaders and the number of partners. A change of partners during the contract is impossible. Moreover, if one of the partners changes its legal form, or is merged with a new firm, the agreement is immediately interrupted, thus minimising uncertainty.

Finally, all companies know that competing firms behave in this way, this being another demonstration that cooperation agreements are the strategy for developing in this period. Another expectation we had was about *the territorial dynamics of cooperation agreements, which were expected to take place more easily in more advanced economic areas*. This has proved to be true in the case of our sample. In fact, cooperation agreements take place more easily between more developed countries (USA, Japan and Europe) where a highly consolidated complementary technological, managerial and organisational know-how exists and can easily become the object of exchange.

11.6 CONCLUSIONS

Our analysis has focused on the linkage between internationalisation and networking, on the basis that these two theoretical strands have insufficient explicative power when applied separately. We believe that the national dimension is important enough to warrant the use of international theories and that the organisational dimension is important enough to require the adoption of network theory.

We have searched for support for our hypotheses by adding to the body of knowledge about Swedish multinationals (Swedenborg, 1979): the results confirm the existence of this linkage, and its importance in firms' strategies.

Our empirical analysis of the Italian/Swedish case shows, in detail, in accordance with our conceptual expectations, that:

* overall, in each form of internationalisation we notice the same time lag effects in the different stages of development experienced in Sweden and subsequently in Italy;
* there is a long history of intense trade between the two countries, despite the trade diversion effects of the EEC and of EFTA. This trade relationship shows a greater dynamics in the 1960s and 1980s than in the 1970s;
* trade between the two countries does not follow historical specialisations as comparative advantages were redefined in the 1970s following the development of foreign direct investments. This is especially evident when comparative advantages in industries depend on high technology and/or skills, i.e. on complexity issues;
* during the 1980s Italian/Swedish linkages have intensified. This is shown by the acceleration of Swedish direct investment, reaching an extremely important position in the Italian industrial structure, and by direct Italian investment making its first appearance;
* foreign direct investment parallels trade patterns, in that Swedish corporations are more attracted towards Italy than Italian ones are towards Sweden. However, the operational logic is still in the transition from multi-national to trans-national, as investment is motivated both by quality market orientation and by local know-how, but not so much by the hierarchical intra-group international division of labor. This logic is supported by a concentration of Swedish investment in science-based and specialised supplier industries;
* cooperation agreements initiated by Swedish multinationals present a constantly increasing trend, in accordance with the persistence of extensive trade linkages and the intensification of direct investment linkages. This result reflects our expectation that cooperation agreements are an emerging phenomenon, begun in recent decades and destined to increase their relative weight in comparison with the other two forms of internationalisation. The reasons for the growing importance of this organisational form are the need for common know-how and the desire to preserve companies' identity and flexibility while aiming at reducing costs;
* the dynamic of cooperation agreements follows a specific evolutionary pattern, going from simple relationships to more complex and yet stable linkages, once the capacity of the partners has been tested: knowledge about partners is an important sunk cost. In this case, our expectation about the temporal development of first asymmetric and then symmetric agreements has also proved to be true;
* finally, the territorial pattern of foreign direct investment is coherent with trade patterns in that it privileges advanced countries and - within Italy - advanced city-regions. The development of cooperation agreements also seems to privilege more developed areas. In fact, most cooperation agreements amongst our case-study firms take place with America and Japan, these areas representing the places where more complementary technical, scientific and managerial know-how exists. However, this cannot be confirmed at the intra-national level in the case of Swedish subsidiaries.

The thesis of a gradual substitution of trade with foreign direct investments has been proved to be true only in the short to medium term, whereas in the long run organisational change causes both forms of internationalisation to increase. On the basis of this, and considering the growing complexity of economic and technological environments, we speculate that cooperation agreements will expand further and will be a useful and efficient organisational form of internationalisation. However, they will only partially substitute trade and foreign direct investment.

REFERENCES

Cainarca, G., M. Colombo and S. Mariotti, 1989, "Accordi tra Imprese nel Sistema Industriale dell'Informazione e della Communicazione", in Cainarca, G., M. Colombo, S. Mariotti, C. Ciborra, G. De Michelis and M. Losanno, (eds.), *Tecnologie dell'Informazione ed Accordi tra Imprese*. Comunità, Milano.

Camagni, R. (ed.), 1989a, "Accordi di Cooperazione e Alleanze Strategiche: Motivazioni, Fattori di Successo e Elementi di Rischio", *Rassegna Economica IBM*, 4: Supplement.

Camagni, R., 1989b, "Cambiamento Technologico, 'Milieu' Locale e Reti di Imprese: Verso una Teoria Dinamica dello Spazio Economico", *Economia e Politica Industriale*, 64:209-237.

Camagni, R. (ed.), 1991, *Innovation Networks*, Bellhaven, London.

Camagni, R. and R. Capello, 1991, "Le Caratteristiche delle Nuove Technologie di Comunicazione e Loro Interazione con la Domanda", in Camagni, R., (ed.), *Computer Network: Mercati e Prospettive delle Technologie di Comunicazione*, Etaslibri, Milano, 1-46.

Camagni, R. and F. Gambarotto, "Gli Accordi di Cooperazione come Nuove Forme di Sviluppo Esterno delle Imprese", *Economia e Politica Industriale*, 58:93-128.

Camagni, R. and T. Pompili, 1993, "Irreversible Investment and Internal Evolution of Firms Networks: Two Case Studies in the Lombardy Region", in D. Maillat and L. Senn (eds.), *Reseaux d'Innovation et Milieux Locaux*, EDES, Neuchatel, 235-258.

Capello, R. and H. Williams, 1992, "Computer Networks Trajectories and Organisational Dynamics: a Cross-National review", in C. Antonelli, (ed.), *The Economics of Information Networks*, Elsevier, Amsterdam, 326-342.

Chesnais, F., 1988, "Technical Cooperation Agreements between Firms", *STI Review*, OECD, 4.

Cominotti, R. and S. Mariotti (eds.), 1986, *Italia Multinazionale: l'Internazionalizzazione dell'Industria ilaliana*, I R&P-Ospri report to CNEL, Il Sole - 24 Ore, Milano.

Cominotti, R. and S. Mariotti (eds.), 1989, *Italia Multinazionale: Gli Investimenti Esteri in Italia e dell Italia Versi i Paesi Esteri*, II R&P report to CNEL, Il Sole - 24 Ore, Milano.

Cominotti, R. and S. Mariotti (eds.), 1991, *Italia Multinazionale1990: l'Integrazione Internazionale e le Prospettive del Mercato Unico Europeo*, III R&P report to CNEL, Angeli, Milano.

Cozzi, G. and S. Vaccà, 1986, "Esiste Veramente il Consumatore Universale?", *Economia e Politica Industriale*, 52:3-32.

Dunning, J.H., 1981, *International Production and the Multinational Enterprise*, Allen & Unwin, London.

Foresti, J., 1986, *Une Toile d'Araignée se Referme Autour de l'Industrie Européenne: Les Accords de Coopération Industrielle*, Commission des Communautés Européennes, Bruxelles, mimeo.

Freeman, Ch. (ed.), 1990, *The Economics of Innovation*, Edward Elgar, London.

Gualini, E. and T. Pompili, 1990, "Le Attività Estere nell'Area Milanese", in I. Re.R. (ed.), *Ricerche Oetamm* 15, OETAMM, Milano.

Kindleberger, C.P., 1969, *American Business Abroad*, Yale U.P., New Haven.

Krugman, P. (ed.), 1986, *Strategic Trade Policy and the New International Economics*, MIT, Boston.

Lorenzoni, G., 1986, "Strategia e Flessibilità Organizzativa", *Economia e Politica Industriale*, 52:63-72.

Mariotti, S., 1989, "Efficienza Dinamica e Sistemi di Imprese", *Economia e Politica Industriale*, 64:91-124.

Onida, F. (ed.), 1990, *Specializzazione e Integrazione Internazionale dell'Industria italiana*, CNR Finalised Project "Structure and Evolution of the Italian Economy", Angeli, Milano.

Ozawa, T., 1985, "On New Trends in Internationalisation: a Synthesis towards a General Model", *Economic Notes*, 1:35-52.

Pavitt, K., 1984, "Sectoral Patterns of Technical Change: Towards a Taxonomy and a Theory", *Research Policy*, 13:343-373.

Pompili, T., 1991, "Le Relazioni tra Imprese nell'Area Metropolitana Milanese", in I. Re.R. (ed.), *Ricerche Oetamm* 21, OETAMM, Milano.

Rullani, E., 1989, "Economia delle Reti: I Linguaggi come Mezzi di Produzione", *Economia e Politica Industriale*, 64:125-164.

Svenska Handelsbanken International Division, 1986, *Direct Investment in Italy: the Swedish Experience*, Svenska Handelsbanken, Stockholm.

Swedenborg, R., 1979, *The Multinational Operation of Swedish Firms*, Almqvist & Wicksell, Stockholm.

Teece, D., 1986, "Profiting from Technological Innovation", *Research Policy*, 15:286-305.

Teece, D., 1989, "Concorrenza e Cooperazione nelle Strategie di Sviluppo Tecnologico", *Economia e Politica Industriale*, 64:17-45.

Vaccà, S., 1986, "L'Economia delle Relazioni tra Imprese: Dall'Espansione Dimensionale allo Sviluppo per Reti Esterne", *Economia e Politica Industriale*, 51:3-42.

Vaccà, S. and A. Zanfei, 1989, "L'Impresa Globale come Sistema aperto a Rapporti di Cooperazione", *Economia e Politica Industriale*, 64:47-89.

Vernon, R., 1966, "International Investment and International Trade in the Product Cycle", *Quarterly Journal of Economics*, 80:198-207.

Williamson, O.E., 1975, *Markets and Hierarchies: Analysis and Anti-Trust Implications*, Free Press, New York.

CHAPTER 12

Measuring Horizontal Inter-Industrial Linkages

Makoto Okumura
Kyoto University

Kazuhiro Yoshikawa
Kyoto University

12.1 INTRODUCTION

During the 1980's, Japanese industries groped for a way to regain their comparative advantage in the international market; their solution was a restructuring into technology-intensive and high value-added industries. This choice not only led to the emergence of new "high technology industries", but also to the reformation of "old industries". Elaborate adaptations to changes in demand and the spread of demand are now vital in order to produce high profits. Therefore, the product cycle goes faster, and high-variety small-scale manufacturing production gains greater importance than mass-production. In this situation, Research and Development (R&D) activities become very important for the continuous development of new products.

R&D activities seem more foot-loose than mass-production type activities, because they do not require accessibility to raw materials or energy. Therefore, many remote regions have emphasized the advantages of their peaceful living environment and expected the attractions of R&D activities to act as a generator of industrial development. This is the reason for the intensive competition among numerous local governments to be assigned the "Technopolis Plan". However, the reality is much more severe for remote regions. In the last decade research activities have gradually been concentrated to the Greater Tokyo Metropolis. The lack of highly-educated manpower in remote regions provides a poor base for R&D activities.

In remote regions it is very difficult to introduce new industries requiring an influx of manpower. The utilization of existing manpower is more feasible. It is important to select those industries that will utilize existing technologies or market information in the region. However, such inter-industrial linkages based on technological or local market information have not been adequately discussed or analyzed to date. Existing analyses only focus on "Vertical Linkages" which directly correspond to flows of materials or goods among industries. For contemporary industrial policy, "Horizontal Linkages" are the key concept. New indexes for measuring horizontal linkages are proposed in this study. This chapter also tries to discuss horizontal linkages and their political implications for industrial development in remote regions.

In Section 12.2 we discuss vertical linkages and the drawbacks of industrial policies which are based on them. We show why horizontal linkages are important in industrial development policy making. The following section calculates the measuring indexes of horizontal linkages. In Section 12.4 empirical calculations of these indexes illustrate the effectiveness of the proposed indexes and demonstrate their political implications.

12.2 VERTICAL LINKAGES AND HORIZONTAL LINKAGES

12.2.1 Vertical Industrial Linkages and Growth Pole Policy

Flows of materials or goods have been described as "Vertical Linkages" among industries. They provide the main basis for the "growth pole theory". The industrial complex was planned on the basis of vertical linkages in order to reduce the transportation costs of goods or materials. A growth pole industry is another way of describing an industry which has strong vertical linkages with many other industries. A raw material based industry, such as the iron and steel industry, sells its products to numerous machinery producing industries; down stream vertical linkages from such an industry have a strong power to attract other industries. Upstream vertical linkages also produce a growth pole. For example, in order to provide materials, parts or energy, steel industries, rubber industries and energy plants came to locate near automobile plants.

During the high rate of economic growth which occurred until the oil-crises, the Japanese plan for regional development advocated the dispersion of industrial activities and proposed the adoption of a "growth pole" strategy in which efforts for regional development were concentrated in certain strategic points in various parts of the country. To implement this strategy, fifteen "New Industrial Cities" and six "Special Areas for Industrial Development" were designated by 1966. Raw material based plants such as those for iron and steel, petroleum refining, pulp or chemical processing were settled in coastal areas. In 1969 the second National Development Plan proposed the construction of extensive network systems for transportation and communications to increase the accessibility of remote regions.

Against this background of high economic growth and transportation improvements, the dispersion of industries - especially heavy industry and primary industry - was promoted. These policies contributed to a weakening of the migration flows from remote regions to metropolises to some extent.

12.2.2 Drawbacks of Vertically Introduced Complexes

The oil-crises of 1973 greatly altered the economic situation in Japan. Heavy industries and primary industries which had generated economic growth suffered from cost inflation and the loss of their competitive advantages. The recession in these key industries resulted in an overall slump of the regional economy and the out-migration of young members of the labor force from remote regions. It is very ironical that this situation was more severe in areas where the growth pole policy had succeeded and created a mono-central industrial structure. Actually, most of the New Industrial Cities came to suffer from the complete slump of the regional economy.

Growth pole industries with their strong vertical linkages have a great capacity to gather related industries. However, if the economic environment around the pole industry becomes worse, all of the industries are severely damaged; a decrease of production in the growth pole industry compresses demand to related industries in the upper stream; in addition, higher prices or the lower quality of key products result in the loss of competitive advantages of the down stream industries. In other words, vertical linkages make the industrial structure less substitutable and more susceptible to structural slump.

The other criticism is that growth pole effects based on transportation cost reduction have become less important, because transportation costs represent an increasingly smaller share of total costs. Increasing labor costs in Japan have caused process decomposition and the subrogation of labor-intensive activities to the developing countries. The standardization of production and the remarkable improvements in processing machinery have greatly accelerated this international division of labor. As a result, the importance of vertical linkages between domestic industries has been reduced significantly in the last two decades.

12.2.3 Recent Industrial Policy in Relation to R&D Activities

The trend towards technology-intensive and high value-added industries was the path chosen by the Japanese economy for revival from the oil crises. This tide was strengthened by the Technopolis Program proposed by Ministry of International Trade and Industry (MITI) in 1983. The Technopolis concept was formulated in order to create a high technology complex like Silicon Valley or The North Carolina Research Triangle, where technological innovations in high technology industries such as semi-conductors, new materials and bio-technological advances are underway. This also involves a regional development component: the promotion of economic development and industrial restructuring in remote regions. Since 1983, twenty-six Technopolises have been designated by MITI. The Technopolis must be an area where industries are not excessively concentrated, but MITI set a pre-condition that there must be enough existing local business to provide the nucleus of entrepreneurial skills for the project. The area must have at least one university providing courses in high technology and it must have good access to high speed transportation networks.

Although it is too early to evaluate the Technopolis program, it can be said that the program emphasized the importance of R&D activities in remote regions. However, two major problems can be noted. One is that the policy is no more than a confirmation of existing conditions and it was only applied to a few cities with such pre-conditions. The other problem is that all Technopolises are expected to develop so-called high technology industries such as the production of semi-conductors, new materials and bio-technological advances. They are undoubtedly promising growth industries at present, but they have little relationship with existing industries in the region. This situation is much like that when heavy industries and primary industries were selected as promising industries for "New Industrial Cities" about one quarter of a century ago. Technopolises can not be guaranteed to be free from the problems of "structural slump" in the future.

12.2.4 Importance of Horizontal Linkages

Remote regions do not possess extensive resources or manpower, so it is most likely that R&D activities will only grow in a small number of industries. Therefore, target industries must be selected so that they can utilize regional characteristics and strengthen them. R&D activities can also be expected to receive workers who are engaged in existing industries; current regional industries cannot always catch up with the demand saturation or decline along the progress of the product-life cycle. Therefore, strong linkages to existing industries as well as a high growth potential are desired characteristics of target industries.

Industries have vertical and horizontal linkages. As mentioned in the previous section, the importance of vertical linkages has declined in the last two decades, and there are new fears of a "structural slump". Thus we must focus on horizontal linkages.

Technological similarity and market information commonality are considered to be the most important components of horizontal linkages. Local technologies of current production processes or existing market channels may be utilized for some other products. Even if the original local products experience setbacks, technological marketing information can create the seeds for new production. If a variety of industries or products is expanded according to horizontal linkages, the regional economy can become more resistant to structural slump. Local technologies are usually accumulated according to the skills of local artisans. To use local forms of technology for new products makes suitable use of these existing skills. The discovery of new fields for existing forms of technology provides new fields of activity for people in the region, and enables them to remain there.

12.2.5 Related Studies about Horizontal Linkages

An input-output table is a table of values corresponding to transactions of goods or services. It is no more than a table of vertical linkages, though it can be used to analyze horizontal linkages. The similarity of the rows in an input-output table shows the commonality of selling markets of selected industries, while the similarity of the columns shows the commonality of input markets. The number of industries that have non-zero purchases from the two industries under consideration gives an index of output market similarity. The number of industries that have non-zero transactions to the two industries under consideration provides the input market similarity index. Roepke et al. (1974) suggested a market commonality index which is the sum of these two similarity indexes. Leontief (1965), Streit (1969), and Slater (1977) suggested other similarity indexes from the same viewpoint which were calculated by input coefficients instead of transaction values. The main drawback of these indexes is that no statistical test is possible and there is no theoretical criteria for significant transactions.

Correlation coefficients enable statistical testing as shown by Czamanski (1971). Though the original input-output table gives only direct transactions between industries, Leontief's inverse matrix provides the total relationship considering successive vertical linkages. The row vector of the inverse matrix shows the total backward influence induced by the industry under consideration; the column vector shows the total forward influence based on the demand to the industry. Blin et al (1977) proposed the use of correlation coefficients of the two columns or two laws in order to estimate the similarity of the total influence of the mentioned industries. In any case, the horizontal linkages based on input-output tables are based on market structure, not on a technological relationship.

The similarity of geographical location gives another measure of horizontal linkages. Numbers of employees are usually used. Florence (1944) developed one of the earliest measurements of geographic association between industries. It is obtained by summing the absolute values of the difference between the regional employment proportions of any two industries over the sample regions. That is,

$$G_{jk} = \sum_{j=1}^{N} \left| \frac{x_{ij}}{x_{.j}} - \frac{x_{ik}}{x_{.k}} \right| \qquad (1)$$

where, G_{jk}: the coefficient of geographic association between industries j and k,
x_{ij}: the number of employees of the jth industry in the ith region,
x_{ik}: the number of employees of the kth industry in the ith region.

This coefficient was severely criticized because it is only sensitive for large zones and it gives large values even though the geographical patterns differ in the other smaller zones. The geographical employment dispersion pattern was investigated by Roepke et al (1974), but their methods have the same problem as Florence's measurement.

Correlation coefficients came to be widely used in order to enable statistical testing; see Richter (1969) and Streit (1969) amongst others. However, the use of geographical employment data is not free from the problem of arbitrary zoning or the difficulty of extracting industrial linkages from the effects of population scale. Geographical analyses have not succeeded in the identification of technological linkages.

Other researchers have tried to grasp the technological structure more directly. Isard et al (1959) decomposed processes for chemical industries and picked up the technological relationships in each unit's process. This microscopic analysis can take local conditions or characteristics into consideration, but it requires very detailed data. It is only applicable to short-run assessments, and only for limited industries in a limited area. Patent analysis has become very popular recently. The Invention Association (1985) analyzed the technological structure of the field of electronics. However, not all of the used technologies are registered as patents, therefore this method is only applicable to so-called high-technology industries such as the electronics or bio-chemical industries. Any increase in the overseas transaction of patents makes the analysis more difficult.

These suggested methods for the analysis of technological linkages demand much data and their applicability is limited; a broader method covering a wider range of industries is strongly desired for regional technological policy making.

12.3 MEASUREMENT OF HORIZONTAL LINKAGES

12.3.1 Diversification of Products in Multi-product Firms

Here, let us pay attention to the behavior of firms. Every firm tries to choose profitable products according to its resources, labor force and technology. Many firms produce more than two products; they are called multi-product firms. Bi-production, such as the production of sulfuric acid during steel refinement, is a perfect technical case of multi-production. However, the concept of multi-production covers a much wider area than that indicated by such a trivial bi-production case. Many firms actively select new products which can possibly yield them more profit. This diversification has attracted much attention from researchers in marketing. According to the Japanese Census of Manufacturers (JCM) in 1987, almost half of the manufacturing firms in Japan (43.6%) produce products which are categorized in more than two Standard Industrial Codes (2-digit SIC's). Product diversification becomes an effective and important strategy for firms, especially for large ones. Rumelt (1974) surveyed the product diversification patterns of the U.S. manufacturing firms and their correlation to economic results, such as profit. The ratio of the main product's production value to the total production of the firm provides a criterion for classifying the firm's strategy into Single, Dominant, and Diversified. A diversified strategy was subdivided into Vertical, Related and Unrelated, based on the width of products. Rumelt found that the Related-Diversification strategy, where products were selected so that market information or technology would be applicable to them, gave more profit than the other strategies. Another study, in which Hakoda (1987) interviewed 171 major firms in Japan, shows an increase in the Related-Diversification strategy at the expense of the Single strategy. (See Table 12.1). In many firms, diversification is promoted on the basis of horizontal linkages. Therefore, diversification patterns may provide a nice index of horizontal linkages. We will suggest an index based on the diversification data of JCM.

Table 12.1: Diversification strategy of major Japanese industrial firms

STRATEGY	1960	(%)	1965	(%)	1970	(%)	1975	(%)	1980	(%)	1983	(%)
Single	73	42.6	60	35.0	50	29.2	40	23.3	41	23.9	39	22.8
Dominant	44	25.7	49	28.6	52	30.4	54	31.5	42	24.5	43	25.1
Diversified	32	18.7	49	28.6	47	27.4	59	34.5	74	43.3	74	43.2
(Vertical)	10	5.8	18	10.5	18	10.5	21	12.2	24	14.0	23	13.4
(Related)	22	12.8	18	10.5	18	10.5	22	12.8	30	17.5	31	18.1
(Unrelated)	22	12.8	13	7.6	11	6.4	16	9.3	20	11.6	20	11.6
TOTAL	171	100	171	100	171	100	171	100	171	100	171	100

Note:
Single: Main product gives more than 95% of the total production of the firm.
Dominant: Main product gives between 70%-95% of the total production.
Diversified: Main product covers less than 70% of the total production.

Quoted from Hakoda (1987): Diversification Strategies and Industrial Structural Change

12.3.2 Co-Production Rate (CPR) as an Index of Horizontal Linkages

Although many manufacturing firms expand their activities as extensively as non-manufacturing service industries, concentrating the discussion to the manufacturing industry improves the availability of data. The volume about Products in the JCM series presents annual production data according to categories of products. We can obtain the number of workplaces categorized in each industry group that produce each kind of product. Based on this data, each product is checked to see if it is produced by a certain industrial group. If two different products are produced by the same industrial group, it is probable that these two products are actually produced at the same workplace. When there is no direct vertical linkage nor any technological relationship between the two products, there is no reason to produce them in the same workplace. In this context, the "Co-Production" is proposed as an index of linkages, and defined as follows:

$$C_{jk} = \frac{\sum_{i=1}^{M} \delta_{ik} N_{ij}}{N_{.j}} \qquad (2)$$

where, C_{jk}: the co-production rate of the *j*th product and the *k*th product,
N_{ij}: the number of works in the *i*th industry that produces the *j*th product,
δ_{ik}: the index variable that takes the value of 1 when the *k*th product is produced in the *i*th industry (if $N_{ik}>0$), otherwise it takes the value of 0 (if $N_{ik}=0$).

The index takes a non-negative value. When all parts of the *j*th product are produced with the *k*th product, it takes the maximum value unit, as is the case of perfect bi-production. This Co-Production Rate is a new measurement index of diversification strength at the workshop level, not the firm level. Therefore, it is expected to reflect technological linkages.

Diversification intensity, as measured by the co-production rate, may be influenced by many technological factors. Here we consider that the influences of such technological factors are additive; each Co-Production Rate can be explained as the summation of the effects of the following factors:

$$C_{jk} = \sum_{l=1}^{L} \zeta_{jl} \zeta_{kl} D_l$$

where, ζ_{jl}: the index variable that takes the value of 1 when the *k*th product possesses the *l*th technological factor, otherwise it takes the value of zero,
D_l: the effect of the *l*th factor.

Technological factors are not as numerous as non-zero co-production pairs; some residuals remain. Here we consider a non-linear program, choosing such combinations of ζ_{jl}s and D_ls as to minimize the squared residuals as follows:

$$\textit{minimize} \sum_{j=1}^{M} \sum_{k=1}^{M} \left(C_{jk} - \sum_{l=1}^{L} \zeta_{jl} \zeta_{kl} D_l \right)^2 . \qquad (4)$$

An algorithm of this non-linear problem is available as an "Overlapping Clustering Method" in major statistical packages.

This procedure is repeated by changing the number of clusters, and any decline of residuals is investigated. The number of clusters will be determined according to the inflection of the squared residuals.

After these analyses at the national level, we can develop a procedure for assessing regional technological characteristics. In order to evaluate the technological linkages between existing industries in the region and a new candidate *j*th industry, the Technological Linkage Index T_j is defined. It is the summation of the Co-Production Rate over existing industries as follows:

$$T_j = \sum_{k=1}^{M} \eta_k C_{jk} \tag{5}$$

where, η_k: the index variable that takes the value of 1 when the *k*th product is produced in the region in question.

This index can also be defined for the *j*th industry already existing in the region. It demonstrates the technological linkages of existing industries with other industries in the region. We can define Technological Compactness as the sum of the technological linkage index over existing industries, which means the total technological linkages in the region.

$$TC = \sum_{k=1}^{M} \eta_k T_k. \tag{6}$$

12.3.3 Analytical Indexes for Market Commonality

In order to ascertain marketing commonality, the other facet of horizontal linkages, two indexes are defined on the basis of input-output data; one is the Co-purchasing Rate and the other is the Co-selling Rate. The Co-purchasing Rate (Cp_{jk}) indicates the applicability of information about the input market of the *k*th industry when inputs for goods of the *j*th industry are purchased. It is defined as the proportion of input purchases from the industry which also sells goods to the *k*th industry. That is,

$$Cp_{jk} = \frac{\sum_{m=1}^{M} \lambda_{mk} t_{mj}}{t_{.j}} \tag{7}$$

where, λ_{mk}: the index variable that takes the value of 1 when the *k*th product uses inputs of the *m*th goods ($t_{mk}>0$), otherwise it takes the value of zero,

t_{mj}: the transaction value from the *m*th industry (goods) to the *j*th industry (goods).

This is an index for marketing commonality, though it also has certain significance from the point of view of technological linkages. Let us consider two different industries using the same semi-conductor devices as parts. Because heat, electricity and magnetic lays would damage them, such physical forces must be well controlled in the processing routines of both of these products. There is much technological commonality in the processing skills for two such products.

The Co-selling Rate (Cs_{jk}) is similarly defined as the portion of sales to such industries that purchase the *k*th product to the total sales of the *j*th goods. It represents the applicability of selling market information about the *k*th goods when the *j*th good is sold, and is defined as follows:

$$Cs_{jk} = \frac{\sum_{m=1}^{M} \lambda_{km} t_{jm}}{t_{j.}} \tag{8}$$

It is also true that this Co-selling Rate has a technological meaning as well as market linkages. We can image two different mechanical goods used as inputs for a small sized video recording machine, for example. In order to enable the down-sizing of the VTR machine, both of the parts are required to be compact; fine and careful processing is indispensable for these two products.

It is worth noting that these two indexes evaluate the horizontal linkages at the industrial level, not the firm level as the Co-Production Rate does. However, when both of these indexes take a high value, technological linkages are strongly expected between the two products.

In order to analyze horizontal market information accessibility in the region, we define another index by summing up the Co-purchasing Rate and the Co-selling Rate. The index H_j is a Horizontal Market Linkage Index which evaluates the accessibility to market information in the region.

$$H_j = \sum_{k=1}^{M} \eta_k (Cp_{jk} + Cs_{jk}). \tag{9}$$

The summation of this index over existing industries gives Horizontal Market Compactness. This index evaluates the total horizontal market commonalities in the region.

$$HC = \sum_{k=1}^{M} \eta_k H_k. \tag{10}$$

12.3.4 Vertical Industrial Structure

In order to test vertical linkages, the following two indexes based on Input-output data are calculated; one is the Purchasing Rate: p_{jk} which means the relative importance of the *k*th goods to total inputs of the *j*th industry, and the other is the Selling Rate: s_{jk} which indicates the portion of transactions to the *k*th industry in the total sales of the *j*th goods. Those are:

$$p_{jk} = \frac{t_{kj}}{t_{.j}}, \quad s_{jk} = \frac{t_{jk}}{t_{j.}}, \tag{11}$$

where, t_{jk}: the transaction value from the *j*th industry (goods) to the *k*th industry (goods).

If either of these two indexes takes a high value, there are vertical linkages between the two products.

When new industries decide their location, the existence of vertically related industries in the region may increase the availability of inputs or their accessibility to output markets. In order to estimate the accessibility for the new *j*th industry, a Vertical Linkage Index V_j is defined as the summation of the Purchasing Rate and the Selling Rate over existing industries.

$$V_j = \sum_{k=1}^{M} \eta_k (p_{jk} + s_{jk}). \tag{12}$$

This index can be calculated for existing industries in the region, too. It allows the possibility of the self-reliance of the *j*th industry based on intra-regional vertical linkages. Summing up the index over existing industries makes a Vertical Compactness Index, VC, as follows:

$$VC = \sum_{k=1}^{M} \eta_k V_k. \tag{13}$$

The indexes stated above are calculated in three steps. At first, based on national level data, the strength of inter-industrial linkages is investigated: the Co-Production Rate, the Co-purchasing Rate, the Co-selling Rate, the Purchasing Rate and the Selling Rate are evaluated. These basic indexes provide the basis for the following regional analyses. The second step is the evaluation of the possibility of introducing new industries into the region. This is tested by ascertaining how many linkages there are between candidate industries and existing industries, using three linkage indexes: T_j, H_j and V_j. The last step will exhibit how the regional industries are connected to each other, using three compactness indexes. This compactness displays the strength of the regional characteristics.

12.4 EMPIRICAL STUDY OF LOCAL INDUSTRIAL CHARACTERISTICS

12.4.1 National Level Co-Production Patterns

First we show the results of the calculations at the national level in 1988. Because there is an extensive body of literature analyzing the vertical industrial structure (see, for example, Czamanski et al, 1977, Campbell, 1975), we focus on co-production patterns.

Table 12. 2 shows the distribution of the Co-Production Rate calculated for pairs of 3-digit goods (149 categorized). 1,294 of the 22,052 pairs (149 x 148) have non-zero values. They seem too small considering the remarkable inclination towards a diversified strategy of firms. It is because the Co-Production Rate evaluates the linkages at the workplace level. A firm can diversify its product lineup by operating numerous workplaces with mono-production.

Table 12.3 shows the overall co-production pattern based on an aggregation to 2-digit industries (23 x 23) pairs. This aggregated pattern shows a block-triangulate structure corresponding to vertical linkages to some extent. There are two apparent blocks; light material industries (SIC.14-18, 22, 23), and heavier industries - metal-machinery-electrical (SIC. 25-32). The chemical industry (SIC. 20) connects these two blocks. There are weak linkages in regard to commonalities of marketing information or design concepts between final products such as furniture, plastic and miscellaneous (SIC. 17, 22, 34).

The overlapping clustering procedure revealed basic factors based on these co-production patterns. At the inflection point of the squared residuals, the number of clusters was set at 75, about half of the number of industries. Products using each type of technology are shown in Table 12.4. Many technological factors are only shared by similar industries with the same two-digit code, included in the diagonal entries in Table 12.3. Other combinations correspond to the linkages analyzed by Table 12.3; food-beverages, chemical-fabrics-leather, chemical-metal, metal-machinery, and machinery-electric machinery.

Table 12.2: Distribution of Co-Production Rate (1986)

Range of Co-Production Rate		Frequency	Cumulative Frequency
1.0		1	1
1.0	- 0.5	5	6
0.5	- 0.2	12	18
0.2	- 0.1	23	41
0.1	- 0.05	28	69
0.05	- 0.02	72	141
0.2	- 0.01	86	227
0.01	- 0.005	135	362
0.005	- 0.002	285	647
0.002	- 0.001	255	902
0.001		392	1,294
0		20,758	22,052

Table 12.3: Co-Production Pattern at 2-Digit Level

Industry i / Industry j	12	13	14	15	16	17	18	19	20	21	22	23	24	25	26	27	28	29	30	31	32	33	34
12 Food	◉	◉					△		○														
13 Beverage and tobacco	◉	●							○														
14 Fiber textile			●	◉	○		○		○		○				○		○	○		○			○
15 Apparel products			◉	●	○	○	△	△	△			○	○										○
16 Wood and wooden product			○	○	●	○	○				○									○			△
17 Furniture and fixture				○	◉	●	○				○					○	○	△	△	△			○
18 Pulp and paper product	◉		○	○	○	○	●	○	◉		○	○		○									○
19 Printing, publishing and allied				○			○	●															
20 Chemical Product	○	○	◉	○			○		●	○	○			○	○	◉		○	○	△		○	○
21 Petroleum and coal product									◉	●													
22 Plastic products			○		○	○	○		○		●	○		○			○		○				○
23 Rubber product			◉	○			○				○	●	○	◉		○							○
24 Leather, leather product and fur			○	◉								◉	●										
25 Earthen and related product					△		○		○		○	◉		●	○	○	○	○	△	○	○		
26 Iron and steel material and product		○						○					○	●	●	●	○		○				
27 Non-ferrous metal material and product						○			◉			○		◉	◉	●	◉	○	○	○	○		○
28 Fabricated metal product			○		△	○					○			○	◉	○	●	○	○	○	△		○
29 Ordinary machinery			△			△			○						○	○	○	●	△	○	△		△
30 Electrical machinery and equipment						○			△		○			△		△	○	△	●	△	○	○	○
31 Transport equipment					△	▽			◉					○	○		○	○	○	●	○	○	○
32 Precision equipment														○		○	○	○	○	○	●		◉
33 Ordnance									●									◉	◉	◉	◉	●	
34 Other manufacturing product			○	○	○	○	○		○		○	○				○	○	○	○	○	○		●

● ◉ ○ △ ▽

Co-Production Rate : 1 - 0.1 - 0.01 - 0.001

12.4.2 Regional Industrial Structure of The Kohoku Region

The following empirical study was designed to grasp the local industrial structure. It was executed in the Kohoku Region in Shiga-Prefecture (780 km^2) near Kyoto, where 1,005 manufacturing workshops employed 22,423 people in 1985. Because JCM data for the present industrial categorization was available for 1983, we use cross sectional data for 1983 and 1988.

a) Existence of Industries and their Vertical Industrial Structure

In 1983 and 1988 there were sixty 3-digit SIC categorized industries in the region; about half of the total number of categories (149). Nine industries withdrew during this five year period and nine additional industries located in the region. Figures 12.1 and 12.2 show the existing industries and vertical linkages between them in 1983 and 1988, respectively. We pick up vertical linkages when either the purchasing rate or the selling rate take a value higher than 0.1. As shown by these figures, the basic vertical structure did not change during the period. It is based on two combinations of components: from fibers to leather and textile products, from wooden and paper materials to machines via the chemical and electric industries. Leather industries (SIC. 242, 247) lost their superiority and withdrew during the 5 year period, although they had strong vertical linkages with other industries. Some of the new industries like surface processed steel (SIC. 265) and electronic measurement equipment (SIC. 307) were connected to many existing industries. The number of significant vertical linkages ($p_{jk}>0.1$ or $s_{jk}>0.1$) was increased from 64 to 71; the vertical structure seems to have become more complicated.

b) Technological Clusters in the Kohoku Region

Industrial clusters derived in the national level analysis are helpful in evaluating regional technical characteristics. When at least one of the products contained in a certain cluster is produced in the region, another product in the same cluster is likely to be easily produced in the region based on technological linkages or market commonality. Therefore the number of clusters means a variety of regional technologies from the point of view of horizontal linkages. Codes of the industries existing in 1988 are shown in bold print, italic industrial codes mean that industry was recently introduced or withdrew. Table 12.4 shows the existence of clusters in the Kohoku Region by columns. Forty-six clusters existed in 1983. Two technologies (No. 12, No. 44) diminished due to withdrawals of industries (*1). The other 44 technologies were still present in the region in 1988. Three new industries using existing technology appeared (*2). In two of the ten industries which withdrew the technology was not diminished because some industries using the same technology still remained (*3). In this type of withdrawal, manpower may easily move to the remaining industries. Seven new technologies were introduced by new industries (*4). As a result, the number of clusters was increased from 46 to 51.

c) Industrial Structure by Compactness Indexes

Here we evaluate the regional industrial structure of the Kohoku Region by the compactness indexes defined in the previous section. Table 12.5 shows their value in 1983 and 1988. Their technological compactness increased from 2,887 to 3,184; an increase of the technological synergy effect. On the contrary, both the horizontal market compactness and the vertical market compactness decreased slightly from 3,053 to 2,961, from 74.92 to 72.24, respectively. Despite the increase in the number of vertical linkages as shown by Figures 12.1 and 12.2, a decrease of vertical compactness means that each vertical linkage lost significance.

In order to assess the industrial structure's influence in industrial location, average linkages are calculated for newly located and withdrawn industries. Table 12.6 summarizes the results. At first glance, only technological relationships influence

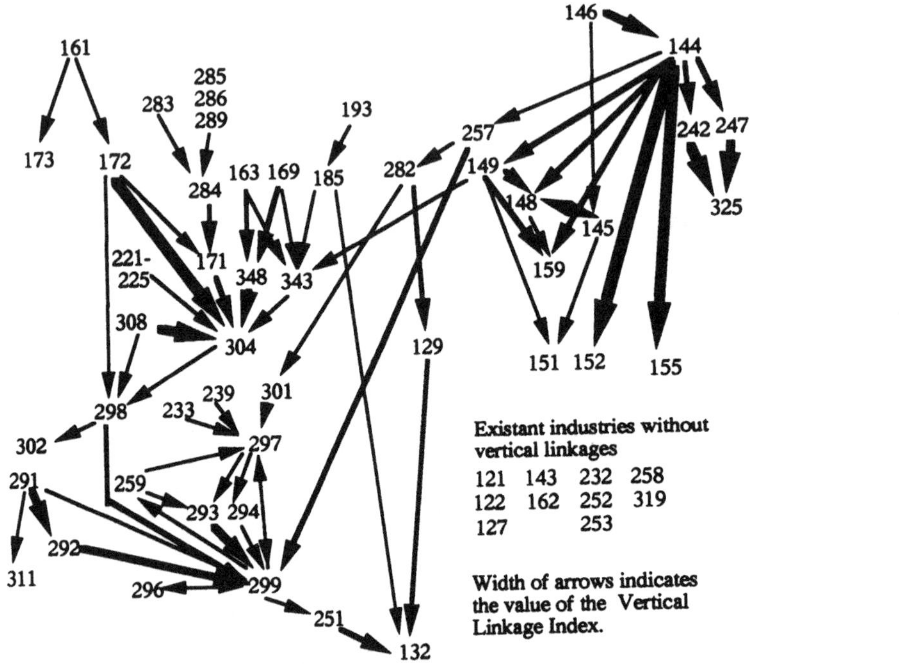

Figure 12.1: Vertical linkages in Kohoku region in 1983

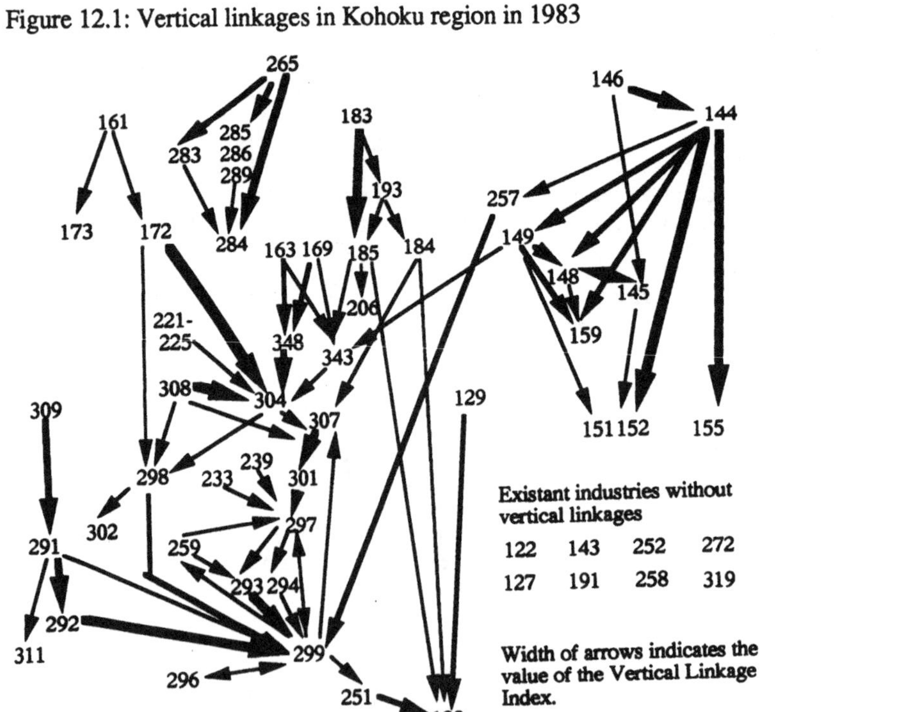

Figure12.2: Vertical linkages in Kohoku Region in 1988

industrial location; the average value of technological linkages for newly located industries (0.817) much exceeds the value for withdrawn industries (0.451). This situation does not hold for horizontal or vertical market linkages; withdrawn industries had a greater value (51.71) of H_j's than the newly located industries (37.77), and the average value of the vertical linkage index for newly located industries is less than half of that for withdrawn industries.

Table 12.4: Products using each technology and their existence in the Kohoku region

Tech No.	Products using technology (SIC code)	Kohoku 83	Kohoku 88
1	**349**,345	y	y
2	***265***(*4),**261**		y
3	212,215		
4	241,248		
5	209,331		
6	261,287		
7	128,135		
8	202,271		
9	154,248		
10	**293**,**302**,331	y	y
11	**151**,*242*(*3)	y	y
12	249,*247*(*1)	y	
13	**144**,**149**,**159**	y	y
14	**223**,229	y	y
15	***191***(*4), 192		y
16	123,131		
17	212,219		
18	**258**,**259**	y	y
19	**185**,*183*(*2),189	y	y
20	201,202,203,205,209,213, 331		
21	**293**,**297**,**299**,**319**, 306,	y	y
	315	y	y
22	**294**,**296**,305,314,321,322	y	y
23	***265***(*4),287		y
24	**233**,**239**	y	y
25	126,128,135		
26	**311**,**312**	y	y
27	**129**,181	y	y
28	**297**,**302**,305,331	y	y
29	**144**,**146**,142,204	y	y
30	**294**,**319**,315,331	y	y
31	**296**,**299**,331	y	y
32	**293**,321,331	y	y
33	**302**,203,205,209,331	y	y
34	321,322		
35	**304**,305,306,331	y	y
36	**173**,*162*(*3)	y	y
37	203,204,213		
38	**143**,147	y	y
39	**151**,**152**	y	y
40	***272***(*4),271,279		y
41	**149**,**159**,142,204	y	y
42	201,213		
43	**284**,312,315	y	y
44	243, *232*(*1)	y	
45	**221**,**222**	y	y
46	**294**,**302**,201,331	y	y
47	124,181,182		
48	**308**,***309***(*2)	y	y
49	**296**,**297**,203,312,331	y	y
50	202,262,273,279		
51	205,209,306,315,331		
52	**302**,**319**,331	y	y
53	**293**,315,331	y	y
54	**185**,***183***(*2)	y	y
55	**185**,189	y	y
56	**294**,**297**,**299**,331	y	y
57	**319**,***206***(*2),201,203,215,219	y	y
	305	y	y
58	***265***(*4),261,287,288		y
59	**149**,**159**	y	y
60	**129**,135	y	y
61	301,**304**,306,312	y	y
62	**302**,306,331	y	y
63	**293**,201,331	y	y
64	205,297,321,331		
65	202,203,213		
66	**299**,205,305,331	y	y
67	***272***(*4),201,202		y
68	201,203,321,331		
69	**293**,**296**,331	y	y
70	***307***(*4),306,315,322		y
71	**145**,**155**,245	y	y
72	**259**,261,262,266,271	y	y
73	203,219		
74	**294**,203,295,331	y	y
75	**233**,**259**	y	y

SIC: Industries which existed in 1983 and 1988
SIC: Industries which located in Kohoku between 1983 and 1988
SIC: Industries which withdrew from Kohoku between 1983 and 1988
SIC: Industries which did not exist in Kohoku in 1983 or 1988

Note: *1-*4 correspond to the explanation in the text.

Table 12.5: Industrial compactness in the Kohoku region

	TC Technological Compactness	HC Horizontal Compactness	VC Vertical Compactness
1983	2.887	3053	74.92
1988	3.184	2961	72.24

Table 12.6 shows that vertical linkages have no power to prevent the decline of industries. Horizontal market linkages have no significant role in local industrial location, partly because our indexes picked up the horizontal linkages at the industry level, not at the level of the firm.

Table 12.6: Average value of indexes for located/withdrawn industries

	Tj Technological Linkage Index	Hj Horizontal Linkage Index	Vj Vertical Linkage Index
New Located	0.817	37.77	0.595
Withdrrawn	0.451	51.72	1.260

As a whole, Tables 12.5 and 12.6 illustrate that technological structure has many impacts upon industrial location, but reject the hypothesis that marketing information is an important factor in location.

12.4.3 Evaluation of Candidate Industries for the Region

As a result of the previous analysis, we must ascertain the possibility of the introduction of industries mainly from the view point of regional technological characteristics. It is beneficial to select candidate industries from such industries that can use the existing technologies shown in Table 12.4. The possibility of introduction will be increased if the new industry can use existing technology and if it can provide a workplace for skilled artisans in the region. The regional technological character is also strengthened by new industries. Table 12.7 shows the values of the Technological Linkage Index, the Vertical Linkage Index and the Horizontal Linkage Index for each such candidate industry.

Based on this technological relationship, Spinning (SIC. 142), Leather Gloves, (SIC. 245), Electronic Equipment (SIC. 306), Aerospace (SIC. 315), Measurement Equipment (SIC. 322) and Buttons and Accessories (SIC. 345) are promising. Leather Gloves, Buttons and Accessories also have advantages for marketing information. The Spinning industry has relatively strong vertical linkages to existing industries; the locations of these industries is promising.

12.5 CONCLUSION

An overview of past industrial policies revealed the fragility of an industrial structure based on vertical linkages. The increasing globalization of industrial market competition and the focus on Research and Development activities has strengthen the importance of horizontal linkages in manufacturing industries. In order to find regional technological characteristics and select industries which can utilize and strengthen these regional characteristics, horizontal linkages are very important. This chapter has discussed these horizontal linkages and their political implications in industrial development. New indexes measuring horizontal linkages were proposed and technologies were investigated using the Overlapping Clustering Method. Candidate industries which can utilize and strengthen the existing technology were also selected on the basis of this analysis.

Table 12:7: Index values for candidate industries

Candidate industries (SIC code)		Technology (Cluster No.)	Technological Linkage Tj	Horizontal Linkage Hj	Vertical Linkage Vj
135	Fodder and organic manure	60	0.021	6.02	0.32
142	Spinning	29,41	0-101	33.29	1.36
147	Net and Rope	38	0.008	53.50	0.33
181	Pulpmills	27	0.000	27.70	0.83
189	Other paper products	19,55	0.035	43.28	0.54
192	Publishing	15	0.014	63.91	1.03
201	Chemical manure	46,57,63,67	0.015	53.93	0.11
202	Inorganic chemical industry	67	0.006	68.18	0.41
203	Organic chemical industry	33,49,57,74	0.016	63.69	0.41
204	Chemical fiber and textile	29,41	0.052	43.12	0.64
205	Petrochemical products	33,66	0.000	65.56	0.50
209	Other chemical products	33	0.007	55.80	0.60
215	Pavement material	57	0.000	7.00	0.00
229	Other plastic products	14	0.023	68.99	0.95
245	Leather glove	71	0.667	53.83	0.62
261	Iron by blast furnace	2,58,72	0.010	43.11	0.57
262	Crude steel	72	0.041	44.35	0.60
266	Tempered steel	72	0.007	68.34	1.17
271	Non-ferrous metal refinement	40,72	0.012	37.48	0.30
279	Other non-ferrous metal	40	0.050	34.88	0.82
287	Metal strings	23,58	0.020	54.30	1.17
288	Bolts, nut and metal screws	58	0.008	54.30	1.17
295	Textile machinery	74	0.019	45.83	0.48
305	Computers and accessories	22,28,35,57,66	0.039	44.05	0.53
306	Semi-conductor device and IC	21,35,61,62,70	0.791	47.41	0.44
312	Railway ore, and parts	26,43,61	0.087	55.49	0.30
314	Ships, parts and accessories	22	0.061	58.84	0.74
315	Aerospace	21,30,43,53,70	0.209	38.20	0.24
321	Measures and analyzer	22,32	0.012	57.31	0.98
322	Land measurement equipment	22,70	0.121	0.00	0.00
331	Ordnance	10,15, etc.	0.415	42.16	0.66
345	Buttons and other accessories	1	0.484	50.63	0.71

It is worth noting that our study is along the lines of evolutionary economics suggested by Nelson and Winter (1982). Our concept of local technology corresponds to that of genes in biology. In the biological world, each individual creature lives and grows using a variety of biochemical processes printed in genes. Some biologists insist that each life lives and grows in order to increase the copies of their genes (Nishiyama, 1985). The performance of genes is tested in a natural selection process. Genes can live for many generations. Furthermore, the adaptation or evolution process makes their life much longer. Now, we construct an analogy as following. Each firm obtains profits and grows by production based on technologies (genes). All products (species) decline and their life is limited. However, technology (genes) itself can survive if new profitable products (species) are developed. Further, the research process improves the performance of technology just as the evolution process does. This chapter suggests a quantitative methodology along the line of this biological analogy.

We would like to conclude the chapter by summarizing the tasks for the future:

(1) Local technological information was treated as an ill-formulated skill which could not be standardized or documented in this discussion. However, the introduction of intelligent machines may compensate for such technologies. The development of soft-engineering may also narrow the area for such technologies or skills. Much market information can also be gathered and analyzed by consulting companies which have no direct transactions. Therefore, this area of ill-formulated skills must be carefully assessed by more detailed case studies.

(2) The co-production rate is based on the diversification behavior of firms. That behavior is, needless to say, much influenced by market profitability. Therefore, the co-production rate is a function of market price or the demand for goods. Micro scope analyses of a firm's diversification behavior are needed to provide a theoretical foundation for this index.

(3) This chapter aims to find a way of applying existing technology along the horizontal linkages. The birth of new technology is out of our scope. However, when we discuss the industrial policies for metropolitan areas, this process becomes more important. The ability to develop new technology is strongly needed for metropolises in order to keep their comparative superiority.

REFERENCES

Blin, G.M. and C. Cohen, 1977, "Technological Similarity and Aggregation in Input-Output Systems: A Cluster-Analytic Approach", *The Review of Economics and Statistics*, 59:82-91.

Campbell, J., 1975, "Application of Graph Theoretic Analysis to Interindustry Relationships", *Regional Science and Urban Economics* 5:91-106.

Carter, J.R., 1977,"In Search of Synergy: A Structure-Performance Test", *The Review of Economics and Statistics*, 59:279-289.

Czamanski, S., 1971, "Some Empirical Evidence on the Strengths of Linkages between Groups of Related Industries in Urban-Regional Complexes", *Papers of the Regional Science Association* 27:137-150.

Czamanski, D.Z. and S. Czamanski, 1977, "Industrial Complexes: Their Typology, Structure and Relation to Economic Development", *Papers of the Regional Science Association* 38:93-111.

Dosi, G., 1988, "Sources, Procedures and Microeconomic Effects of Innovation", *Journal of Economic Literature* 26:1120-1171.

Emerson, J. and A. Ringleb, 1977, "A Comparison of Regional Production Structures", *Papers of the Regional Science Association* 39:85-98.

Florence, P.S., 1944, "The Selection of Industries Suitable for Dispersion into Rural Areas, *Royal Statistical Society Journal* 107:93-116.

Hakoda, H., 1987, *Diversification Strategies and Industrial Structural Change*, Shinzansha, Tokyo. (In Japanese).

Hatumei Kyokai (Invention Association), 1985, "Protection and Encouragement of High Technology", Research Report. (In Japanese).

Hewings, G.J.D., 1982, "Regional and Interregional Interdependencies: Alternative Accounting Systems", *Environment and Planning A*, Vol. 14, 1587-1600.

Isard, W., E. Schooler and T. Viertorisz, 1959, *Industrial Complex Analysis and Regional Development*, Wiley, New York.

Larsson, J., 1989, "Product Life Cycle and Infrastructure: A Regional Perspective, *Japan-Sweden Workshop on Infrastructure of C-rich Regions*.

Leontief, W., 1963, "The Structure of Development", *Scientific American*.

Nelson, R.R. and S.G. Winter, *An Evolutionary Theory of Economic Change*, Harvard Press.

Nishiyama, K., 1985, "An Evolutionary Theoretical Model of Firms in an Industry: The Replicon Model", *IEEE Transactions on Systems, Man, and Cybernetics* 15, No. 5, 662-665.

Richter, C.E., 1969, "The Impact of Industrial Linkages on Geographic Association", *Journal of Regional Science* 9:19-28.

Roepke, H., D. Adams and R. Wiseman, 1974, "A New Approach to the Identification of Industrial Complexes Using Input-Output Data", *Journal of Regional Science* 14:15-29.

Rumelt, R.P., 1974, Strategy, Structure and Economic Performance, Division of Research, Harvard Business School.

Slater, P.B., 1977, "The Determination of Groups of Functionally Integrated Industries in the United States using a 1967 Interindustry Flow Table", *Empirical Economics* 2:1-19.

Streit, M.E., 1969, "Spatial Associations and Economic Linkages between Industries", *Journal of Regional Science* 9.

APPENDIX

Japanese Standard Industrial Classification

12 Food
- 121 Stock-firm food products
- 122 Fishery food products
- 123 Agricultural food products
- 123 Salt and condiments
- 125 Sugar and saccharoid
- 126 Cerial products and milling
- 127 Bread and cakes
- 128 Fats and oil
- 129 Other food products

13 Beverages and tobacco
- 131 Soft drinks
- 132 Liquor
- 133 Tea and coffee
- 134 Ice
- 135 Fodder and organic manure

14 Fiber textiles
- 141 Filature
- 142 Spinning
- 143 Yearning
- 144 Weaving
- 146 Dying
- 147 Net and Rope
- 148 Lase and products
- 149 Other fibers

15 Apparel products
- 151 Outer dresses
- 152 Inner dresses
- 153 Hats and headgear
- 154 Furred dresses
- 155 Other dresses and accessories
- 159 Other apparel products

16 Wood and wooden products
- 161 Logging
- 162 Plywood and architectural materials
- 163 Wooden box and containers
- 164 Wooden shoes
- 169 Other wood products

17 Furniture and fixtures
- 171 Furniture
- 172 Religious ornaments
- 173 Fixtures
- 179 Other interior accessories

18 Pulp and paper products
- 182 Pulpmills
- 182 Paper mills
- 183 Special paper
- 184 Paper products
- 185 Paper boxes and containers
- 189 Other paper products

19 Printing, publishing and allied
- 191 Newspapers
- 192 Publishing
- 193 Printing
- 194 Print engraving

20 Chemical products
- 201 Chemical manure
- 202 Inorganic chemical industry
- 203 Organic chemical industry
- 204 Chemical fibers and textiles
- 205 Petrochemical products
- 206 Pharmacy
- 209 Other chemical products

21 Petroleum and coal products
211 Petroleum refining
212 Lubricants and grease
213 Coke
214 Briquettes and solid fuel
215 Pavement materials
219 Other petro or coal products
22 Plastic products
221 Plastic boards, bars and pipes
222 Plastic film and sheets
223 Industrial plastics
224 Foamed plastics
229 Other plastic products
23 Rubber products
231 Tires and tubes
232 Rubber shoes and footwear
233 Industrial rubber products
234 Other rubber products
24 Leather, leather products and fur
241 Leather
242 Industrial leather products
243 Leather shoes material
244 Leather shoes and footwear
245 Leather gloves
246 Leather bags
247 Non-leather bags
248 Fur
249 Other leather products
25 Earthen and related products
251 Glass and glass products
252 Cement products
253 Earthern construction materials
254 Ceramics and pottery
255 Fireproof clay products
256 Charcoal and carbon
257 Grind stones
258 Stone products
259 Other clay products
26 Iron and steel material and product
261 Iron by blast furnace
262 Crude steel
265 Surface processed steel
266 Tempered steel
267 Pig iron
269 Other iron and steel
27 Non-ferrous metal materials and products
271 Non-ferrous metal-elemental refinement
272 Non-ferrous metal-further refinement
273 Non-ferrous alloys
274 Non-ferrous casting
275 Electric wire and cables
279 Other non-ferrous metals
28 Fabricated metal products
281 Metal cans
282 Metal houseware
283 Heating equipment
284 Metal construction materials
285 Pressed metal products
286 Metallurgie powder
287 Metal strings
288 Bolts, nuts and metal screws
289 Other metallic products
29 Ordinary machinery
291 Boilers and engines
292 Agricultural machinery
293 Construction and mining machinery
294 Fabrication machinery
295 Textile machinery
296 Other industrial machinery
297 General parts for machinery
298 Office and service industrial machinery
299 Other machines and accessories
30 Electrical machinery and equipment
301 Electricity transmission and distribution apparatus
302 Household electric equipment
303 Electric lights and illumination
304 Communication equipment
305 Computer accessories
306 Electronic equipment, semi-conductor devices and IC
307 Electric measuring equipment
308 Weak electrical equipment, parts and accessories
309 Other electric machinery
31 Transport equipment
311 Motor vehicles, parts and accessories
312 Railway ore, parts and accessories
313 Bicycles and other light vehicles
314 Ships, parts and accessories
315 Aerospace
319 Other transport equipment
32 Precision equipment
321 Measures and analyzers
322 Land measurement equipment
323 Medical equipment and tools
324 Equipment for physical chemistry
325 Optical equipment and lens
326 Glasses
327 Clocks and watches
33 Ordnance
331 Ordnance
34 Other manufacturing products
341 Precious metals ornaments
342 Musical instruments and records
343 Toys, sporting and athletic goods
344 Pens and other stationery
345 Buttons and other accessories
346 Lacquer ware
348 Other manufacturing products

PART III: KNOWLEDGE, INNOVATION AND TECHNOLOGY NETWORKS

CHAPTER 13

From Knowledge and Technology Networks to Network Technology

Charlie Karlsson
Jönköping International Business School

13.1 INFRASTRUCTURE AND LOGISTICAL REVOLUTIONS

Since the beginning of the 1970's it has become increasingly obvious that the world economy is undergoing a process of structural change that ultimately will have equally as fundamental effects as the industrial revolution had some 150 years ago. Economic observers have had notorious difficulties in anticipating and understanding this radical change of the economic structure. One important reason for this is that so much of the debate about socio-economic development deals with short-run problems such as inflation, unemployment and the balance of payments. In a short-run perspective these factors have great importance, of course, but in the long run they are not decisive for the development of society. If we want to understand the economic, social and cultural developments of regions and nations we must look to those factors that change more slowly and thus have a decisive role for development.

Hence, economic structural change must be seen as a consequence of slow changes in the fundamental conditions of an economy. These basic conditions can be described as the infrastructure of the economy. In our opinion it is the infrastructural capacity, including accessibility to the surrounding world, that determines the long run development of regions as well as of nations. Usually forms of infrastructure are regarded as artefacts. However, this is an inadequate definition. Infrastructure should rather define all those material and non-material factors that are both

* of great collective importance, and
* very durable, in comparison with other types of capital.

Our point of view implies that the durability and collectivity of infrastructure creates an arena or a "possibility landscape" that at the same time facilitates and circumscribes the actions of households, enterprises and political decision makers. By material and non-material infrastructure we mean:

* Durable - almost constitutional - rules for economic and political decision-making; i.e., ownership rights, rules for setting up in business and for competition in the market, social protection systems and other rights and obligations.
* Networks for the transportation of goods, energy, information, human beings and, thus, services and knowledge.
* The level, quality and distribution of scientific, technological and cultural knowledge.
* Basic values and attitudes to development, creativity and entrepreneurship.

In all of these respects Europe and its regions have been subject to a slow but steady change during the last 1000 years. Such a slow change of the infrastructural arena of the economy will sooner or later generate a rapid and fundamental change in its economic and institutional superstructure. It should be emphasized that such complete structural changes have occurred extraordinarily infrequently during the development of the world economy. Economic historians and structural economists have isolated four such large structural changes that, based upon Pirenne (1949, 1956) have been denoted as logistical revolutions by Andersson (1986):

1. The first logistical revolution started in the twelfth century and was based upon improvements in land and, in particular, sea transportation infrastructure on the one hand and upon the creation of a basic infrastructure for commercial communication on the other hand. During this revolution the world witnessed the development of a new Southern European economic system with the city states in northern Italy as its centres and the new economic system of the Hanseatic League that integrated the coasts of Northern Europe. One decisive feature of this first logistical revolution was the integration of the two economic systems that was initiated by the fairs held in Champagne. As a result Western Europe became partly integrated economically for the first time and its first large economic centres were established in Venice, Bruges and Lübeck.

2. The second logistical revolution started in the sixteenth century, partly as a consequence of important developments within sea transportation. The development of a new type of ship - the Caravelle - that could cross the oceans was particularly significant. It was also due to the development of an efficient banking and credit system that, for example, could finance the long distance trade over the oceans with other continents. Initially Anvers and, somewhat later, Amsterdam were the central nodes in this new economic system that reached its peak during the seventeenth century. However, as time passed by London became the centre of gravitation and rose to become the financial centre of the world.

3. The third logistical revolution is known as the industrial revolution and it started at the end of the eighteenth century. It started as a consequence of the dominant position that England had reached within the credit system, maritime commerce, canal building and, later, railway construction. Those new and relatively large scale and sparse communication systems made it possible to establish vertical integration between the raw materials in Northern America and the large scale manufacturing industries in the port cities of Western Europe. The principle of vertical integration was used in one country after the other around the North Atlantic, reaching Sweden in the middle of the nineteenth century. New large cities grew up. Detroit, Glasgow, Gothenburg, Liverpool, Manchester and New York are all examples of cities of the industrial revolution.

4. The fourth logistical revolution started during the post-war period and has manifested itself at almost the same time in a large number of knowledge intensive, often high-tech, regions or corridors. It is based on a rapid expansion of the networks of motorways and airlines and on a greatly increased capacity for information transmission and information processing. It is a period characterized by a rapid knowledge intensification or knowledge deepening that manifests itself through strongly increased R&D investments and a rapid increase of the share of knowledge handlers in the labour force in the Western countries. The pole-star for the development within the different industries is the ambition to increase the use of just in time systems, flexible manufacturing systems and economies of scope within the framework of increasingly refined network relations between the different production units and the parallel ambition to get away from the hierarchical, vertically integrated production systems that were typical after the industrial revolution.

13.2 THE AIM OF THE CHAPTER

The logistical revolutions have been associated with radical changes within all sectors of society. This is particularly true for the direction, volume and organization of higher education and research. The aim of this chapter is to give an overview of the development of higher education and research during the four logistical revolutions with special attention devoted to the development of business schools. We will show how from the very beginning the universities developed as nodes in international knowledge technology networks and how this network organization has been a characteristic of higher education and research for almost 1000 years. We will also try to show that network technology and the art of networking have become central issues for the education at modern business schools. An overview of our main results can be found in Table 13.1.

Table 13.1: The evolution of universities and business schools during the four logistic revolutions

AREA OF INTEREST	TIME PERIOD First logistic revolution	Second logistic revolution	Third logistic revolution	Fourth logistic revolution
Universities and R&D	The birth of the university	Continued foundation of universities The birth and the expansion of learned societies	The birth of professional schools Founding of modern research universities	Rapid growth of the number of universities and of university education
Network activities	Travelling monks "Wanderjahre"	The birth of scientific conferences and scientific newsletters and journals Travelling scientists	International scientific organizations	Large international research programmes Joint authorships Student exchange programmes
Business schools			The birth of the business school	Rapid expansion of the number of business schools Rapid expansion of graduate business studies
Network activities			Interaction and exchange of information at the faculty level	Education in networks and education in networking

In this chapter we simply define a business school as an institution offering education in business subjects at the university level, meaning that at least part of the curriculum has a scientific base. To qualify as a business school an institution must offer university degrees in the form of either a bachelor's degree or a master's degree or both.

13.3 THE NETWORK PARADIGM

The network paradigm has gained increased interest among social scientists during the last decade. The basic mathematical elements of network theory are simple. These elements are *nodes* which are connected by *links*. Together they form a structure that we name a *network*. When analyzing networks, it is of fundamental importance to make a distinction between *links*, on the one hand, and *flows* on the other. The network paradigm has in particular been used in analyses of physical networks and their pertinent flows by system analysts, regional economists, etc. It has also been used by scientists in organization theory, business administration and political science for analyzing networks that are based upon non-physical links. However, it seems that the network paradigm has seldom been used in general analyses of the development of higher education and science (Karlqvist, 1990). The main exception is the analysis of local creative environments such as Vienna, Paris, Silicon Valley, Manchester and Gnosjö, during particular periods. The literature is full of examples of how specialized information in these creative environments has been or is circulating in tightly knitted networks.

This neglect of the network paradigm is rather surprising as intellectual and scientific networks reflect a special kind of organizational network containing both formal relations built upon contracts and, in particular, informal relations built upon friendship and mutual scientific interests. The interaction between the actors in these networks is of a non-physical character with exchanges of information as the most typical flow. The information that is needed to achieve new combinations of pieces of information that creates those synergies is at the heart of creative scientific processes. The effective exchange of qualified information, i.e. knowledge, presupposes communication and direct face-to-face-contacts between individuals. Science actually builds upon communication (Sörlin, 1990). At the interregional and international scale this means that the development of intellectual and scientific networks will be at least partly determined by the development of the physical networks for transportation and communication and their pertinent flows.

These ideas about knowledge growth in networks have been summarized in Figure 13.1 (see Törnqvist, 1990).This figure is totally hypothetical and shows a curve that has borrowed its shape from a learning curve. The knowledge within a certain field and total scientific knowledge grow slowly during certain periods of time. During other phases, however, one can register how total knowledge grows by leaps.

In the figure we also indicate the formation of new networks that appear in sequential order. The nodes in these networks can be individuals, institutional environments or competence areas. Every local network has been built up through contacts between the nodes of this network. As long as the information transmitted through these contacts can generate synergetic effects, the total knowledge increases. However, as time passes by the probability of new combinations of information decreases within a given network and the knowledge level stagnates.

A sudden growth in knowledge will occur when one network suddenly is joined to another network (or when the frequency of interaction between two networks increases significantly). The number of possible new combinations now increases dramatically. Every sudden growth of knowledge in the figure has its counterpart in the formation of new networks. This description of the forces behind the growth of knowledge corresponds to the description above of the forces behind the logistical revolutions.

This theoretical approach can also help us to understand why creative environments so often come and go. They continuously release each other. A node or network that is a source of creativity and renewal during one period can show very clear signs of stagnation or even decline during later periods. It is quite possible to interpret this stagnation or decline as a result of a lack of creation of new links to other nodes or networks.

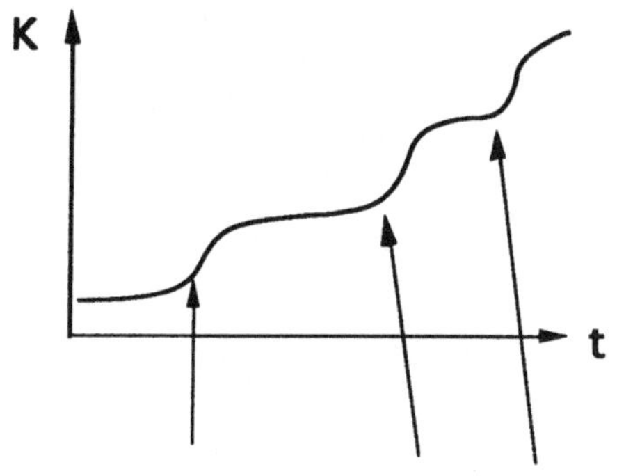

Figure 13.1: The growth of knowledge through networks

13.4 THE BIRTH OF THE UNIVERSITY

Higher education, i.e. education at the university level, can trace its roots back to the first logistical revolution. At that time the modern university grew out of the medieval schools that are known as *studia generalia*. These schools were normally available to students from all parts of Europe. The earliest *studia* arose from efforts to educate priests, monks and civil servants beyond the level that the schools at cathedrals and monasteries could offer. The participation of students from other countries was the most significant difference between *studia* and the schools they grew out of.

The earliest universities were unions of teachers and/or students who later received their charters of foundation from popes and emperors. The University of Bologna, which is considered to be the oldest university, can follow its roots back to the end of the eleventh century. In the 1150's at the law school in Bologna the students created a guild - *universitas scholarium* - that, among other things, paid the teachers their salaries. At the end of the twelfth century there were four *universitas scholarium* in Bologna that corresponded to the four nations: Lombards, Tuscans, Romans and Ultramontanes (a fusion of French, German, English and other nations). In Paris the teachers at the cathedral school at Notre Dame, which became the model of most European universities, formed a masters' guild - *universitas magistrorum* - that was divided in four faculties: theology, law, medicine and philosophy. Later colleges were formed, where the students could live and eat and receive their tuition. Those early *universitas* were copied at a relatively fast rate in a number of cities throughout Europe, so that at the end of the fifteenth century there existed more than 70 universities in Europe.

The background to this rapid university expansion was that the church as well as the medieval cities had a need for knowledgeable and well educated labour. The large and widely spread wealth of the church required people that could write and were skilled in foreign languages, but also people that were well acquainted with the laws of that time. The increased volume of trade during the first logistical revolution also resulted in an increased need for people who could not only write and speak foreign languages, but also people with a knowledge of arithmetic and civil law.

That the universities had their roots in Bologna and Paris was almost totally the result of economic and social factors. A large number of wealthy students had travelled long distances to Bologna for their studies. As the number of students increased the burgers raised the rents. The increased expenses forced the students to organize themselves into guilds in the same way that different kinds of craftsmen had done earlier. The prevailing name for this kind of corporation, whose members had a mutual interest and an independent legal status, was actually *universitas*.

In the year 1158 the university in Bologna was given a jurisdiction of its own through Frederich I Barbarossa's *habita*. Through this act the academic guild got its legal status. However, it was almost 100 years before the universities obtained a legal status.

In Bologna it was dissatisfied students that united and laid the ground for the university, while in Paris it was dissatisfaction among the teachers that resulted in a university forming corporation. In Paris there lived a number of foreign teachers who gave tuition in *artes liberales* at the cathedral school and other schools. It was mainly to strengthen their position against the church and its power over the system of education that the teachers united in the mid-twelfth century. This corporation was given legal status by a bull promulgated by Pope Innocentius III in 1208-09.

At the end of the twelfth century a number of students and teachers moved to Oxford and established a university there and very soon a number of universities were established throughout Europe. The history of the university of Cambridge began in 1209, when a number of disaffected students moved from Oxford. 20 years later Oxford profited by a migration from Paris. The large success of the universities was due to different reasons of course but one extremely important reason was that from the very beginning the universities became nodes in an international university network. One important factor behind this network was the fact that many of the university teachers were monks from the Dominican or the Franciscan order. Those Orders allowed their members to travel from place to place and this made a scientific exchange between different countries possible. Also, other groups travelled from country to country in search for knowledge and so the concept of "Wanderjahre" developed, indicating studies and research breaks in other countries. This was, however, before the principles of modern science had been established and this meant that scientific communication was authoritative and hierarchical. Still, the network played a central role in the diffusion of the established dogma - the scholasticism - as did the establishment of Latin as the common language for scientific communication.

Another important reason behind the success of the universities was specialization. Large coalitions of teachers and students at the same place made more penetrating studies and research possible.

Business education during the first logistic revolution was mainly achieved by sending the sons of merchants abroad when they were quite young. "Dort sollte der Sohn seine praktische Ausbildung erlangen, fremde Sprachen verstehen und Mänschen kennen lernen und so zu einem selbständigen, 'wagenden' Kaufmann herangebildet werden", (Penndorf, 1916, 115). Already at this time the acquiring of communication skills and the building of personal networks were seen as the main objectives for the education of merchants. The merchants of the Hanseatic League sent their sons to Bruges, London, Bergen and Novogorod in particular, so that they could learn French, English, the Nordic languages and Russian. Hence, in those days trade networks and education networks were the same. The Latin schools also played an important role for the training of merchants. "Latein spielte ja für den Kaufmann des Mittelalters eine ganz bedeutende Rolle; es wurde überall verstanden und gewährte daher bei den internationalen handelsbeziehungen bedeutende Erleichterungen" (Penndorf, 1916, 118).

13.5 UNIVERSITIES AND LEARNED SOCIETIES DURING THE SECOND LOGISTICAL REVOLUTION

Above we maintained the view that the first logistical revolution stimulated the birth of the university. Now, turning to the second logistical revolution, we can also register an expansion of the university system that probably was stimulated by developments during this period. This seems to be especially true for its early focus - the Netherlands - where universities were established in Leiden (1575), Groningen (1614), Amsterdam (1632) and Utrecht (1636). During the same period Sweden, then a great power, established universities in Dorpat (1632), Åbo (1640) and Lund (1668). It is natural, at least in the case of the Netherlands, to associate this wave of new universities with the increased demand for knowledge and competence that followed in the wake of the commercial revolution.[1] Germany also witnessed a wave of university founding during this period including Marburg (1527), Jena (1558) and six other universities up to 1672. In Germany, in particular, the founding of new universities was also stimulated by the Reformation and the upheavals that followed.

It must, however, be emphasized that the extent to which universities could serve as centres for the development of new ideas, i.e. as nodes in scientific and intellectual networks, was limited during this period. Their chief goals were to prepare men for the professions of law, technology, and medicine and this precluded the possibility of paying serious attention to intellectual ventures in humanism, natural science, vernacular literature, painting, sculpture, architecture, and music. The flexibility of the universities was further limited by their involvement in the system of estates and in religious controversies. As a result, the intellectual ferment that had started in the universities and had been centered in them until the fifteenth century began to manifest itself both in new types of teaching institutions, such as colleges, Gymnasiums, and academies (to be distinguished from learned societies, which were also called academies) and in organizations that were primarily engaged in research.

The new teaching institutions did no more than the older universities to foster the development of the natural sciences or new developments in literature, painting, sculpture and music. It was left to individuals working on their own to cultivate those two areas, although these individuals were usually connected with some learned society or academy. It is during this period that a new type of scientist developed; one who started to make systematic observations of the surrounding world. This represented the introduction of empirical natural science. Much of the scientific work during this period was concentrated to these academies. The French academy was constituted by cardinal Richelieu in 1635. The first modern academy of science - The Royal Society - was established in 1660.

There are interesting parallels between the emergence of learned societies in the seventeenth century and the early beginning of the universities in the twelfth century. The learned societies started as spontaneous gatherings or networks of people who were interested in scholarship and science and who needed an institutional framework both to facilitate the exchange of ideas and to provide support - psychological and, more rarely, material - for their activities. The exchange of ideas was also stimulated by scientific conferences, a new arena in which scientists could meet. The purpose of the conferences was actually to present innovations and new results. In this respect the scientific conferences were also an important innovation in the field of scientific network building. Scientific communication was also stimulated by newsletters and,

---

[1] With regard to education in business subjects we may note that the first description of double book-keeping was published in 1494: Summa de arithmetica by Pacioli. During the later part of the second logistical revolution in 1675 a book published by Savary - Le Parfait Négociant - strongly influenced what was written during the coming years about business and commerce.

later on, scientific journals for the rapid diffusion of new important results to the international community of scientists.

In the second half of the seventeenth century, the learned societies were granted royal charters and public recognition; princes all over Europe became interested in founding academies. This interest was spurred partly by the then prevailing Mercantilism and the support to the learned societies can be seen as part of a national policy with the main objective of strengthening the economic power of various countries. Where there were few savants, as in Russia or Prussia, they were invited to come from other countries. The outstanding scientists of the seventeenth and the eighteenth centuries - men like Oldenburg, Descartes, Leibnitz, Euler, and Lagrange - spent much of their time as itinerant savants. The academies, scientific conferences, newsletters and itinerant savants together formed the knowledge networks of the second logistical revolution that also eventually become the communication system of modern science. We may also note that even if rather few universities were founded during this period, it was at this time that the first modern university - i.e., the first university that made a clear distinction between basic university education and the training of scientists - was founded in Halle (1694) in Saxony.

The education and training of merchants and businessmen during the second logistical revolution was still mainly carried out by sending young sons abroad with the main objective of learning foreign languages. Of the formal school subjects, it seems as if a knowledge of mathematics and accounting was considered as particularly important during this period. It was also during the second logistical revolution that the need for special schools where merchants could get a proper training and education was first expressed. One early exponent of this view was Thomas Mun who, in the introduction to his book "England's Treasure by Foreign Trade" (1664), stressed the importance of a solid education for merchants. The same was done by Roberts in his book "Map of Commerce" (1700), but it was actually John Locke (1632-1704) who was the first to develop the concept of a reform to create a practical vocational education for merchants. An education that should contain geography, arithmetics, astronomics, geometrics, history, ethics and politics, law and accounting. It has been claimed that it was Jacques Savary (1622-1690) who was the father of the idea of vocational schools for merchants. However, he was mainly an advocate of self-instruction. The first to develop the idea of a full programme for business education at all levels was Paul Jacob Marpenger who presented his ideas in different books including "So notwendig als nützlichen Fragen der Kaufmannschaft" (1714 and 1715) and "Trifolium mercantile aureum" (1723). It seems that he was the first to suggest that faculties or departments of trade and commerce should be included in universities, i.e. that business schools should be founded.

13.6 UNIVERSITIES AND PROFESSIONAL SCHOOLS DURING THE ERA OF THE INDUSTRIAL REVOLUTION

13.6.1 The Birth of the Professional Schools

At the time when the industrial revolution had its breakthrough the universities were being strongly criticized and many intellectuals regarded them as moribund institutions. University instruction was under attack because it almost entirely ignored new developments in science and scholarship. This exclusion of the highest levels of intellectual activity from the universities might not in itself have aroused protest if the universities had not allied themselves, in some of the absolutist countries, with the church and the state in interfering with the freedom of education and publication. In the economically more backward areas, where there were few channels of mobility for talented young people, the universities blocked the only attractive intellectual career opportunities. Many intellectuals, therefore, had the same contempt for the universities that the new entrepreneurs had for the monopolistic guilds, and they envied the privileges of the academies as the middle classes envied the privileges of the aristocracy.

Since the results of university teaching were poor, some critics recommended the replacement of the universities by professional schools. Such a strategy was in line with the thoughts of the Enlightenment movement which put much stress on the importance of practical vocational skills. It would, however, be wrong to look upon the birth of the professional schools as only a supply side phenomena. We must also take the demand side into proper account. The general growth in trade activities and the growth of manufacturing production that was initiated by the industrial revolution increased the demand for people with specialized knowledge and skills. Hence, we can assume that some young people were induced to enter the professional schools as part of their professional career. This is, of course, nothing less than an example of the old dictum that the degree of specialization is limited by the size of the market. The growing labour market led to a development of more specialized occupations, which stimulated some people to search for relevant training. The need to fill these specialized jobs with people with suitable training also stimulated leading merchants, industrialists, and politicians to establish professional schools, We must, however, remember that for a long time most people, even those holding specialized jobs, obtained their occupational training as on-the-job-training.

Professional schools were indeed established, especially in mining and in the new profession of civil engineering, but also in the fields of commerce and business. In Prussia, for example, eight such professional schools were established between 1770-1806 (Shelsky, 1971). In Paris École Polytechnique were established in 1794. These early professional schools formed the pattern for the professional schools that were established during the following century. We can actually say that the establishment of universities with only one faculty is typical for this period. When university education developed in the direction of professional schools there was no longer any reason to keep all faculties under the same roof.

The founding of technical universities started in the early nineteenth century. Sometimes the founding process was carried out in one step, but often the founding process consisted of two or more steps; first a technical school or a technical institute was founded, a number of years later it was given university status. This means that it is somewhat dangerous to compare the founding dates for different technical universities. You may be comparing apples with pears. Anyhow we take the risk and present the following list of founding years: Berlin (1799), Vienna (1815), Stockholm (1827), Dresden (1828), Copenhagen (1829), Gothenburg (1829), Mons (1837), Delft (1842), Zürich (1855) and Munich (1868). Even if science based education started earlier, many of these institutions first obtained the status of technical university in the period 1870-1890.

13.6.2 The Business School - A Product of the Late 19th Century

Universities of economics or business schools are mainly a product of the late nineteenth and the early twentieth centuries. Until then business education had been the province of technical schools and commercial colleges. It was normally an unambitious education, providing little more than basic book-keeping and clerical skills. Commercial education with the ambition of providing education at the university level had started much earlier, however, but none of these earlier schools have survived: Lisbon/"Aulo do Commercio" (1755), Hamburg/Büsch's "Handels-Akademie" (1768), Vienna/"Real-Handlungs-Akademie" (1770), and Württemberg/Karls-Schule's "Handels-Akademie" (1779). In 1782 Karls-Schule obtained the status of a "Hochschule" with five faculties. One of these faculties was a faculty of economics and, according to Gunnarsson (1988), this faculty of economics can be viewed as the very first business school. In 1779 the very first chair in Business administration was introduced at the Karls-Schule (Gunnarsson, 1988). The study programme for the "Handels-Akademie" at Karls-Schule was two years long. A particularly interesting aspect of the study programme is that one-third of it was devoted to the study of modern foreign languages. Thus the link with the schooling of merchants during the first and second logistic revolutions is very clear - communication skills are seen as a major objective in the training of future merchants.

With the exception of Lisbon commercial education with an ambition to provide education at the university level was first established in German speaking countries. These early business schools can, however, be seen as the children of an older tradition which dates back directly to Prussia and the university reforms introduced by Frederick William, the father of Frederick the Great (Spender, 1989). In 1727 he introduced the first chairs in *Cameralism* at the Universities of Halle and Frankfurt-am-Oder. A third chair was introduced in Rinteln in Hessen in 1730. The fourth chair in this area that we know of was introduced at the university of Uppsala in Sweden in 1741[2], which is an indication of the close links between Sweden and Germany in those days. The first professor to hold the chair in Uppsala was Anders Berch, who is viewed as the "father of economics" in Sweden.

Cameralism was the study of administration and the formal application of science to the affairs of the state. Frederick William considered that a properly trained administration was the State's best means of dealing with the changing theological, political, social and legal concepts which were sweeping across Europe. Cameralism's objective was the service of the State through the scientific pursuit of national efficiency, using data collection and analysis, proper accounting and administrative control.[3]

The Cameralist tradition, and the belief that a body of professional public administrators was vital, spread throughout Germany and France. Eventually this concept was brought into the private sector and schools of commerce and business schools were set up with syllabii heavily influenced by the early Cameralist approaches.

Examples of business schools founded during the take off period may be found in Table 13.2. Even here we must be somewhat cautious when using the founding year as an indicator of the year when education at the university level started. Ever since the mid eighteenth-century we have examples of business schools that tried to provide the highest type of commercial education, but remained stuck somewhere between the secondary and university level.

There is, as we can see, an interesting concentration of the establishment of business schools around the turn of the century. While only five surviving business schools had been established before 1895, 31 new business schools were established between 1895 and 1910. Obviously, the schools started to diffuse very rapidly in those years. The rapid diffusion indicates that the commercial and intellectual networks within Europe and the U.S. were now well developed.

From 1908 onwards, however, it seems that business education in Europe lagged behind while it forged ahead in the United States. By 1908 there were 13 business schools in the United States; by 1914 there were 31 schools there; by 1920 there were 65 and by 1951 there were 166 (if departments and colleges teaching business at university level are included) (Barnes, 1989). According to Bowie (1930) the expansion in the U.S. was even more rapid and he reports the following sequence: 1900:7;

[2]Similar chairs were introduced in Åbo in 1747 and in Lund in 1750. Uppsala got a second chair in 1759.

[3]The most important written work from this period is Kaufmannslexikon by Ludovici published between 1752 and 1756. It contains the three main sciences of commerce: "Handlungswissensachaft" (="Die Wissenschafte eine Handlung zu betriben"), "Warenkunde" and "Buchhaltung".

1910:19; 1915:40 and 1930: over 200. During this whole period Britain in particular lagged behind. While more than 80,000 students were enrolled in business schools in the U.S. in 1924, only a few hundred were enrolled in the U.K. (Bowie 1930).

Table 13.2: Founding years for business schools established during the take off period

PERIOD	
- 1890:5	Paris/ESCP (1820), Antwerp (1852), Venice (1868), Paris/HEC (1881), Philadelphia/Wharton School (1881)
1891-1900:15	London/LSE (1895), Vienna (1898), Liège (1898), Chicago (1898), Leipzig (1898), Aachen (1898), California (1898), Mons (1899), St. Gallen (1899), Budapest (1900), Hanover, USA (1900), New York University (1900), Wisconsin (1900), Dartmouth (1900), Vermont (1900)
1901-1910:13	Cologne (1901), Frankfurt am Main (1901), Michigan (1901), Birmingham (1902), Milan (1902), Brussels (1903), Zürich (1903), Manchester (1904), Berlin (1906), Mannheim (1908), Harvard (1908), New York/Columbia (1908), Stockholm (1909), Helsinki (1909)[4], Munich (1910)
1911-[5] :	Helsinki (1911), Königsberg (1915), Copenhagen (1917)[6], Gothenburg (1923)[7], Stanford (1925), Åbo (1927)[8], Bergen (1935), Åbo (1951), Århus (1951)[9].

We can note that the diffusion process was also slow in the Nordic countries. The establishment of business schools in the Nordic countries and their structure were initially strongly influenced by the developments on the continent (in particular, in Germany), but later on developments in the U.S. exerted an increasingly strong influence (ter Vehn, 1956; Olofsson, 1980). A newspaper article from 1903 refers to good examples of business schools in England (LSE) and, in particular, in Germany (Aachen, Leipzig, Frankfurt am Main and planned business schools in Hamburg and Berlin) (Gunnarsson, 1988).

In the U.S. and in Great Britain the business schools were usually established as departments, etc. within existing universities. In continental Europe, on the other hand, many business schools were established as independent, free-standing institutions. As time has passed the situation has changed and today, for example, all the traditional German business schools have become faculties within universities or technical universities. In the Nordic countries, on the other hand, the business schools, with few exceptions, have survived as free-standing institutions. The same has been the case with the business schools in Austria and Switzerland.

---

[4]The Swedish school of business in Helsinki was a senior level commercial school from 1909 until 1927 when it became a school of business.

[5]Some examples only.

[6]It became independent in 1922 after having been part of the merchants' school (Købmandsskolen) from 1917.

[7]Business courses without examinations were offered between 1915 and 1923.

[8]The Swedish business school.

[9]Århus became an independent institution in 1956.

13.6.3 The Mature Industrial Revolution: Demands for a Higher Standard of Management

The fact that technical universities preceded universities of economics reflects the great demand for technical skills for solving the many different technical problems that were associated with the emergence and growth of large scale manufacturing plants during the industrial revolution. Once this large scale production was established, a new demand arose for knowledge and skills associated with management, calculation and accounting for the new large companies in the manufacturing industry and in the banks, insurance companies, etc., that grew in the wake of the growth of the manufacturing industry. The increased sensitivity of the enterprises to the business cycle and the increasingly complicated organization of the enterprises also contributed to the birth of a new type of manager with a more general education in business subjects than had previously been common (Carlsson, 1942). The growth of other sectors of society, mainly retail trade and rail and sea transportation, also led to an increase in the demand for employees with a business education. As a general conclusion we may say that the structural changes within the business sphere led to a demand for managers with theoretical knowledge within economics and the social sciences in the same way as technological developments a century earlier generated a demand for engineers with a solid knowledge of the natural sciences.

The increased demand for management skills was, however, not only the result of the emergence of large business units. Other important changes took place at the same time: for example, the increased reach of markets, the separation of ownership and management, the diffusion of ownership, increased competition - and, as a result, an increased need to rationalize production to bring costs down - applications of science to industry, the rise of organised labour and the decline of *laissez-faire* (Bowie, 1930). All these changes generated a need for a higher standard of management and, hence, a need for management education.

Those interested in founding business schools made frequent reference to the rapid founding of business schools in other countries. Much emphasis was placed upon the need for studies in business and commerce to provide businessmen capable of meeting competition from abroad where such education already existed. Such businessmen needed to be aware of new markets, production methods and methods to organize businesses. It has to be remembered that the early 1900's was a defeatist period and commercial (and technical) education was seen as a panacea for dealing with falling exports and increasing competition from rivals on the world market. Hooper and Graham (1897) wrote, for example, about the need for "securing that the facilities for commercial education in the United Kingdom shall not be inferior to those of any Continental country". In the case of Birmingham, for example, the argument was that "other countries provided such education, so must we" (Smith, 1990). We may note that while most business schools were established according to the Cameralist tradition, the London School of Economics (LSE) deliberately eschewed it. Instead, it was modelled on the *Ecole Libre des Sciences Politiques* established in the 1850's to train men for the Diplomatic Service (Spender, 1989). This model was chosen despite the fact that in 1895 the founder of LSE - Sidney Webb presented and received approval for a report pointing at "the importance of something being done to provide higher and wider instruction in commercial subjects, in order to prevent the English clerk from being ousted by his better educated German rival". (Cited by Caine, 1963, p.41).

The aims of a business education were, however, much broader: "We desire to systematize and develop the special training which is required by men in business and those, who, either as principals or as managers and foremen, will be called upon to conduct the great industrial undertakings in the midst of which our work will be done. In fact it is our aim to do for those engaged in commerce and manufacture what is now done for the professions of law and medicine, and while giving the general education which is calculated to train the mind and the sympathies, we hope also to direct the instruction so as to be of practical advantage to those concerned." (From a letter written by Joseph Chamberlain, who had not been to a University, in 1899, arguing that a Faculty of Commerce should be included in the plans for a Birmingham University; cited by Smith, 1990).

Before the late nineteenth-century business had not advanced far enough to be the object of scientific inquiry or academic research, and the status of the businessman was not high enough to warrant and make possible a truly academic education for business. The novelty of the Wharton School established in 1881 was that it promised a deliberately non-vocational "liberal education in all matters concerning finance and commerce". To put it simply, it sought to create a new kind of general manager, educated to the graduate level in all aspects of running a business.

The possibility of making business techniques and organization the subject of scientific research was, as a matter of fact, probably identified for the first time in 1895 by the German economist Richard Ehrenberg (Barnes, 1989). This means that it was not until the end of the nineteenth century that one can observe a more substantial interest in the systematic study of the economic problems within an individual enterprise, its financing and organization, its income and cost situation, etc. (Carlsson, 1942). Increased competition within industry and retail trade and a growth in the size of individual enterprises to a previously unknown magnitude led at this time to a shift of focus within the enterprises from their technical to their economic problems. The striving to achieve lower production costs stimulated scientific analyses of the production process and its cost elements. The need to increase the sales of mass-produced products demanded careful studies of markets and selling methods.

13.6.4. The Diffusion of the Business School Concept due to Improved Transportation and Communication Networks

Improved transportation and communication networks and increased commercial and scientific contacts can, of course, be seen as the fundamental reasons behind the rapid diffusion of the business school concept during these years. Quite a lot of travelling was also undertaken to study existing business schools. Another important influence came through people studying in foreign countries; when they returned to their home country they not only brought scientific knowledge with them but also information about syllabii, teaching methods, etc. As both travel and study visits occurred in both directions, influences also went in both directions between different countries, including Europe and the United States.

The influences of the Cameralist approach were carried crossed the Atlantic by Americans studying economics in Germany at the turn of the century. One important case was Edmund James, who helped Joseph Wharton to set up his school (Spender, 1989). We can also note that before the creation of the School of Business Administration in Berlin in 1906, people were sent on missions to Paris, London, Antwerp and Vienna. Ignaz Jastrow, Professor of Economics at the University of Berlin was sent to the United States where he studied at close range the existing American academic business schools. On his return, he stressed that the businessman of the twentieth century would no longer be primarily a trader: he would be an administrator. So he suggested that the teaching at the new school begin with business administration and business techniques (Barnes, 1989). Personal networks were probably also important, as is indicated by the fact that the business school in Berlin preceded by two years the creation of the Harvard School of Business Administration in 1908. As Ignaz Jastrow and Frank Taussig of Harvard were personal friends, there is little doubt that one influenced the other. Yet another example was Edwin Gay, the first Dean of the Harvard Business School, who had studied medieval economic history under

Schmoller, the last of the Berlin Cameralists. Gay introduced Schmoller's teaching methods into the newly-created Harvard Business School, aided by Harvard's earlier acceptance of Langdell's similar methods in the Harvard Law School (Spender, 1989).

13.6.5 The Early Management Education Syllabii

Despite some obvious differences, the early management education curriculum was remarkably similar to that in use today with accounting, economics, law, contracts, administration and technology. However, today we see more sociology, psychology and organizational behaviour and the earlier courses on languages, current affairs, and industry specific courses such as banking and publishing are often absent (Redlich, 1957).

One important aspect of the original Cameralist curriculum and its underlying body of knowledge is its practical orientation. Forrester (1989) writes in this connection that "the professors so feared a theory derived from Aristotle, or from the theologians, that they concentrated on practice". Forrester also notes that one important result of the professors' deliberate avoidance of a comprehensive theory on which to base their teaching was that they: "failed to find a methodology adequate to their wide-ranging interests and to the specific functions later to be fulfilled by their students". The Cameralist professors understood that teaching depended on theory. Because the programmes were based on business practice they were forced to change their pedagogical techniques. They developed case-work, seminars and projects. They also sought to develop new theories inductively through detailed observation, recording and analysis.

13.6.6 University Expansion and the Birth of International Scientific Organizations

The era of the industrial revolution is also associated with a wave of university founding. It was during this period, for example, that England got its "red brick" universities, that can be seen as the British answer to the professional schools established on the continent.

It was during the industrial revolution that international meetings between scientists became established. In those days travel was still very time consuming so the number of trips abroad per year was very limited. It is also during this period that international scientific organizations were established, such as The International Geological Congress in 1892 and The International Dental Federation in 1900. The influence of the academies of science decreased, however, as modern research universities were established.

13.7 UNIVERSITIES AND BUSINESS SCHOOLS IN THE KNOWLEDGE ECONOMY

13.7.1 New Business Trends in the Evolving Knowledge-Intensive Network Economy

The knowledge economy develops on an arena that is very different from that of the industrial revolution. The new arena is characterized by networks with fine meshes in multiple layers with a rich supply of information and knowledge and by an increasingly even spatial distribution of information. Together these integrated networks have created the necessary preconditions for a total transformation of the economic systems of the industrialized market economies from what might be called the principle of a vertically integrated resource based economy to the principle of a network economy based upon systems thinking. Several different networks now concentrate their capacity to quite different places than the networks of the industrial revolution. The new arena of multi-layered networks has its points of concentration dispersed more or less at random over the continents in a number of city regions or city corridors where the exploitation of knowledge, modern communications and creative activities within R&D and the arts take place. A new international system of regions - which is characterized by stronger mutual relations amongst themselves than with their often closer national neighbour regions - has developed on the basis of these networks.

This transformation to a knowledge-intensive network economy has meant that several new business trends have started to emerge during the past few decades. The two most important are: the globalisation of markets, communications and resources; and the acceleration of technological change. Globalisation is here used as a catch-all phrase to describe the growing need for companies to treat the world as their stage, if they are to prosper. It is a trend spurred by several factors. First, during the 1980s the world's financial system became more open. Capital, supplies and production are now much easier to locate (in both senses) wherever they are cheapest. Second, improved computer and telecommunications technologies and air travel mean that physical distances between markets are less of a barrier. The third factor is the rise in import penetration in the larger economies. Large domestic markets are no longer only the realm of domestic competitors. The second challenge for business is managing ever-swifter changes in product and process technology. The life spans of product and technology cycles are shrinking at a dizzying speed.

To keep pace, businesses are being forced to toss old assumptions aside. To compete harder, firms are being forced to cut costs. The vertically integrated large enterprises of the industrial revolution are giving way to enterprises which are organized as co-ordinated dynamic networks of business units that behave as independent enterprises. These enterprises are leaner and more flexible than the traditional enterprises. During this reorganization process many enterprises are sweeping away their layers of middle management and demanding managers with new types of skills. These days competitors are as likely to be partners in joint ventures or strategic alliances as they are to be sworn rivals. Suppliers are no longer a simple source of goods, but are being roped in to help design their client's products. Hence, the market relationships between enterprises are being substituted by network relationships. The changes within the business sector are more numerous, however, than those involved in the switch from hierarchies and markets to networks (cf. Paul, 1989). There are changes from a production orientation to a market orientation, from centralization to decentralization, from function to system, from goods-handling to knowledge-handling, from Taylorism to teamwork, from a weak to a hard business culture, from *laissez-faire* to social responsibility, from a national to an international orientation and from management to leadership. Of course, the enterprises are demanding managers with new types of skills, but as we shall see the universities and the business schools are still in the beginning of the transition process to a new type of management education.

13.7.2 A University System with a Fundamental Long-Term Lack of Balance

The development of the knowledge economy got a flying start through the expansion of higher education during the post-war period that culminated during the 1960's. This expansion took the form of both the establishment of new universities and the expansion of capacity at existing universities. An increased number of university graduates was a necessary prerequisite for the knowledge deepening that has characterized the fourth logistical revolution. It may, however, be observed that the changes within the university system have not been qualitative changes. There has generally been a pure quantitative expansion of higher education that has relied on some of the fundamental principles of the industrial revolution: standardization and large scale production. The expansion in many fields of higher education were often made in such a way that it was impossible to preserve the level of quality that previously characterized university education. Education at the graduate level was often not expanded at the same rate as education at the under-graduate level.

The transformation to a knowledge economy has involved strains for a university system that in general has not been adapted to what has been going on. The structure of the undergraduate programmes has not been changed. The undergraduates have not been prepared for employment within the new knowledge intensive systems enterprises. This means that we do not have business economists and engineers who are systems oriented with a good ability in dealing with international relations. Students at the master and the PhD levels have left the universities to take better paid jobs in, for example, the consultancy sector, to work with investigations, development work and applied research. The relative share of external tasks to the universities has increased with the effect of pulling teachers out of basic instruction and researchers out of basic research. The general picture of the university system is a system with a fundamental long-term lack of balance.

13.7.3 Fundamental Changes within Higher Education and R&D

If we start with a summary we can say that history seems to repeat itself. As was the case during the beginning of both the second and the third logistical revolutions, at the beginning of the fourth logistical revolution we can observe how the universities have great difficulties in adapting to the needs of the new economic regime. There are, however, at least some indications that the systems of higher education and R&D are undergoing fundamental changes. We will only give a few examples here.

The main observation we can make is that knowledge production is being performed in international networks. At the macro level this is the era of large mutual international research programmes such as CERN and EUREKA. At the micro level this is the era of joint authorship. It is possible to observe a significant acceleration of co-operation in large scientific networks during the 1980's. While the total number of scientific articles grew by two-thirds between 1975 and 1987, the total number of joint authorships grew two or three times as fast (Table 13.2). This means that the elasticity of joint authorships with respect to the total scientific production is now in the order of two. Such high elasticities are not typical for a state of equilibrium. Rather, the scientific world seems to be in the middle of a spatial structural change. From Table 13.3 it is possible to see that there is a pervading international tendency in all the developed countries.

13.7.4 A Boom in Business Education

It is important to note the tremendous growth in the number of MBA-programmes offered at business schools as illustrated in Figure 13.2. This figure by no means illustrates the growth in the total number of MBA-programmes. It only illustrates the age structure of MBA-programmes, as presented in the Economist's "Which MBA". But disregarding possible biases - in particular, a strong European bias - in the selection process, the diagram still gives us a pretty good picture of the explosive growth of new MBA-programmes during the last decade.

Table 13.3: Growth of authorships and joint authorships between the periods 1974-79 and 1980-1986 in 26 different countries

Country	Authorships		Joint Authorships		Increase in percent	
	1974-79	1980-86	1974-79	1980-86	Authorships	Joint Authorships
China	741	19349	47	3980	2511.20	8369.09
Spain	14410	39465	908	4632	173.87	420.13
Finland	12584	25595	1461	4944	103.39	238.40
Japan	124447	260641	4749	16036	109.44	237.67
Italy	49410	101386	5751	18769	105.19	226.36
West Germany	132354	273129	13059	42496	106.36	225.42
East Germany	20921	43145	1412	4497	106.23	218.48
Netherlands	33638	71578	4883	15111	112.79	209.46
Austria	14545	26218	1909	5653	80.25	196.12
Poland	25204	37266	2362	6960	47.86	194.67
Hungary	16213	24352	1571	4498	50.20	186.31
France	133224	220972	12140	33739	65.87	177.92
Belgium	22292	40832	3737	10128	83.17	171.02
Israel	26361	43810	4048	10943	66.19	170.33
Sweden	37050	66727	5772	15538	80.10	169.20
U.S.A.	1084924	1627707	42974	107901	50.03	151.08
Switzerland	38128	63433	7905	19514	66.37	146.86
Czechoslovakia	22326	31783	1646	4063	42.36	146.84
South Africa	13465	23316	1073	2637	73.16	145.76
England	224409	366010	19230	45906	63.10	138.72
Australia	50454	86988	4608	10824	72.41	134.90
Denmark	19629	34180	3921	9125	74.13	132.72
Soviet Union	129165	190865	2726	6289	47.77	130.70
India	64780	99777	2571	5892	54.02	129.17
Norway	13024	21581	2300	5156	65.70	124.17
Canada	109152	177278	12945	28959	62.41	123.71
Total	2432850	4017383	165708	444190	65.13	168.06

Source: Andersson and Persson (1990)

Another way to illustrate the MBA-boom is shown in Figure 13.3. In the beginning of the 1960's American business schools awarded around 5,000 MBAs a year. In the late 1980's this figure had grown to around 75,000 MBAs, awarded by some 700 business schools (of widely varying quality, of course). This was equivalent to a quarter of all the masters degrees awarded in the U.S. More information about the growth of MBA-programmes in the U.S. is provided in Table 13.4. Elsewhere - mostly in Western Europe - another 10,000 or so students graduated from around 120 business schools and the number of MBAs produced in Europe each year is likely to double during the 1990s (The Economist, March 2nd 1991). However, it would be wrong to look solely upon the MBA-market. We must not forget that a number of business schools in Western Europe produce large numbers of students with a Master of Science in business subjects that without doubt could match the quality of most MBA-programmes. How else can we explain the fact that German managers certainly match the standards of their rivals in other countries? Of course, the number of business-school graduates each year still looks insignificant when set against, say, the combined labour force of America and

Western Europe (300m). Yet their influence far outweighs their number. Around a third of the chief executives of America's biggest 500 companies hold MBAs from America's 20 to schools and Europe is starting to show a similar pattern.

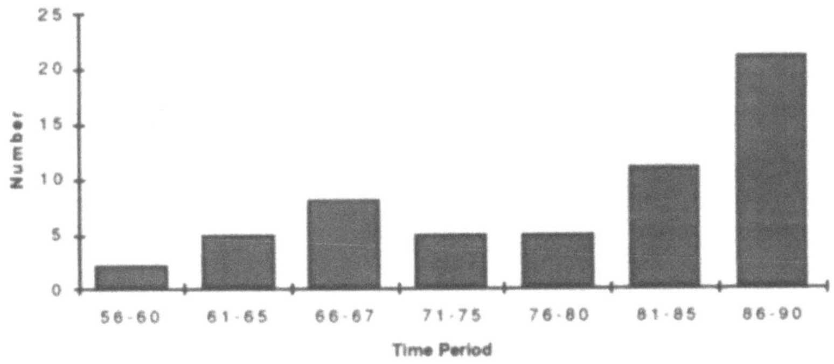

Figure 13.2: The age structure of the MBA-programmes
Source: The Economist's "Which MBA?"

This growth in supply can be seen as a reaction to a growth in demand for new types of knowledge that people need to be able to deal with the new problems that are associated with the emerging network economy. The strength of the demand factors is indicated by the fact that business-school students can expect almost to double their pre-MBA salaries.

In this connection it is also important to mention that there also has been a rapid expansion of undergraduate programs in the field of business studies. For every MBA in the U.S. there are four BBAs.

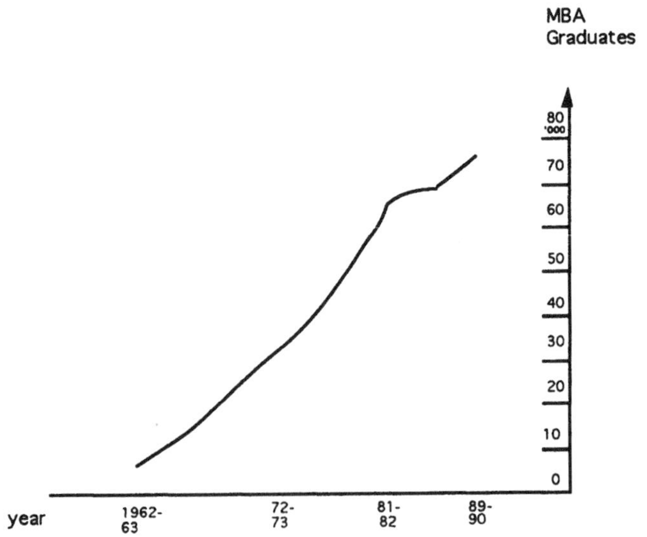

Figure 13.3: The growth of MBA graduates in the U.S.

Table 13.4: A Degree of Growing Popularity

Academic year ending	MBA degrees granted by all U.S. institutions: Number	% increase	MBA degrees granted: % of all masters degrees	% of all degrees (bachelors masters, doctors)
1965	7585		7	1
		181		
1970	21325		10	2
		71		
1975	36364		12	3
		52		
1980	55148		1	4
		21		
1985[1)]	66500		23	23

1) Figures for 1985 are estimates
Source: Cheit (1985)

13.7.5 Changes in Curriculum and Teaching Methods at the Business Schools

If we study the curriculum of business schools and the way in which it has developed over time we may note that the practically oriented curriculum developed around the turn of the century was preserved for almost 60 years. During the late 1950's the relevance of the traditional curriculum was frequently questioned. Two reports published in the U.S. in 1959 confirmed that the academic quality of this rapidly growing university field was uneven and generally too low (Pierson et al, 1959; Gordon and Howell, 1959). Accordingly, they recommended a variety of ways to raise standards, including more academically trained staff, more rigorous course work, more graduate work, and less undergraduate specialization. This programme was to a very large extent and in a very successful manner, adopted by business schools all around the U.S. Certainly, it also influenced business education in Europe despite the argument put forward by Servan-Schreiber (1989) that the American business schools were a key reason for the superiority of American management to that of Western Europe.

However, the success story did not last very long. After ten years the new agenda was criticized by Ashen (1969), who observed that technological and social changes require a managerial approach that the business schools were not structured to provide. Livingstone (1971) criticized the business schools for turning out scholars, scientists, and experts who were eager to act as advisors, but few individuals who were eager to take responsibility for the performance of others. Mintzberg (1975) emphasized that specialist skills, not manager's skills, were being taught. Levitt (1978) claimed that scholasticism was replacing common sense in the business schools.

The main attack against the business schools, their curriculum and their teaching methods was, however, launched by Hayes and Abernathy (1980). They argued that what passes for a sophisticated business curriculum is based on a false, shallow concept of the manager, and, in fact, encourages a preference for (1) analytical detachment rather than the insight that comes from 'hands-on' experience and (2) short-term cost reduction rather than long-term development of technological competitiveness.

American managers, they said, hold the view that there is no need to invent, build or develop anything yourself. Given the capital and good financial management, anything of value can be bought and any problem solved. A sense of commitment - to one's workers, customers, suppliers, even one's fellow managers - is an impediment.

The overall argument was given more authority by Peters and Waterman (1982). The authors, both lecturers at leading business schools, scorned the dominance of quantitative analysis (what they called the numerative, rationalist approach) that teaches the detached analytical justification for all decisions and that has arguably led us astray.

Obviously, something was wrong with the education at business schools. To find remedies a number of projects were launched and a number of reports have been written. Two of the most influential reports are probably Porter and McKibbin (1988) and CAGME (1990). These reports take their starting point in the transformation of the world economy and the new challenges businesses - the ultimate customers of the business schools - have to face during the coming decades.

Taken together we can see that the different reports and the business schools' direct contacts with their clients have resulted in an ongoing transformation of the overall structure and content of business education both in the U.S. and in Europe. We can, for example, see that there is a move away from the "one-shot" MBA approach towards what might be called "continuous learning"; i.e., continuous on-the-job experience combined with continuing, long-term business education. We can also see a change in the focus of many MBA programmes, with more emphasis upon general management skills, leadership, cultural change, corporate vision, creativity, mentoring, entrepreneurship, ethics, environmentalism, the management of change, etc. Taken together these changes in emphasis can be seen as an adaptation of the curriculum and teaching methods to the new realities facing managers in a world where more and more enterprises are organized as dynamic networks. To be able to lead a network a manager must be well trained in the noble art of networking.

One important trend within the business schools is that towards internationalization (and globalization) that, of course, is a result of the fact that global businesses and global markets are here to stay. Internationalization means using international teaching material and international cases, recruiting an international faculty, engaging in joint ventures and alliances with business schools and companies in other countries, recruiting students from other countries, etc. In short, internationalization means that the business schools are engaged in the long, patient creation of a global network of relationships. Those networks also include former students. This network building is perhaps one of the most important characteristics of a business school and for the students the involvement in a good network of overseas contacts may be much more valuable than the education they get. Many business schools put heavy emphasis upon building up strong alumni associations to create what could be called "brand loyalty". The dream is, of course, to educate managers from the corporate cradle to the grave.

Another important trend within business schools in order to meet the demands of their customers is the inclusion of technology management in the curriculum. The objective here is not simply to teach the managers of today and tomorrow how to manage technology, but to show them how they might recast the organizational structure of their firms to gain a competitive advantage from the technology to which all of their rivals have access. To achieve this the schools have to combine disciplines in cross- or multi-disciplinary ventures to show how technology in reality pervades every part of a business.

A third important trend in the field of executive education is the move away from educating individuals to educating whole companies; i.e., the business schools are becoming engaged in the creation of strong international company networks. This type of mutual learning is crucial for four reasons. First the modern, global, fragmented and complex companies need teams of highly capable managers at every level, not least so that they can guide and co-ordinate autonomous subsidiaries. Second, developing new products calls for inputs from every part of a company, which means extensive team-work. Third, multinational companies need multinational teams. Fourth, companies are increasingly marketing their culture or brand image, not simply their product and the appropriate culture must thrive throughout the company.

13.8 CONCLUDING COMMENTS

In this chapter we have given a short overview of the development of higher education and research during the four logistical revolutions, with a special focus upon the development of business schools. In particular, we have tried to show how from the very beginning the universities developed as nodes in international knowledge technology networks. Although the universities seem to have had difficulties in adapting to the new challenges at the beginning of the second, third and fourth logistical revolutions in a long-run perspective they have been able to preserve their role as knowledge nodes and to survive in a changing environment.

Business schools providing education at the university level date back to the second half of the nineteenth century when they developed together with professional schools in other fields. As the business schools have developed they have become major builders of networks including not only other business schools, but also companies and former students. Our casual observations also indicate that the teaching methods in business schools have been changing towards a greater emphasis on team work and the art of networking. The core curriculum seems to be developing in the same direction.

REFERENCES

Andersson, Å.E., 1986, "The Four Logistical Revolutions", *Papers of the Regional Science Association* 69:1-12.

Andersson, Å.E., and O. Persson, 1990, Vetenskapens Noder och Nätverk (The Nodes and Networks of Science), (mimeo).

Ashen, M., 1969, "The Management of Ideas", *Harvard Business Review*, 99-107.

Barnes, W., 1989, *Managerial Catalyst. The Story of London Business School 1964-1989*, Paul Chapman Publishing Ltd, London.

Bowie, J.A., 1930, *Education for Business Management. The Case for the Further Development of Educational Facilities*, Oxford University Press, London.

CAGME, 1990, *Leadership for a Changing World. The Future Role of Graduate Management Education*, Commission on Admission to Graduate Management Education, Graduate Management Admission Council, Los Angeles.

Caine, S., 1963, *The History of the Foundation of the London School of Economics and Political Science*, G. Bell and Sons Ltd, London.

Carlsson, S., 1942, Företagsekonomins Ställning till Övriga Socialvetenskaper (The Relationship between Business Administration and other Social Sciences), *Ekonomisk Tidskrift* 44:200-5.

Cheit, E.F., 1985, "Business Schools and Their Critics", *California Management Review*, 27:43-62.

Engwall, L., 1980, (ed.), *Företagsekonomins Rötter (The Roots of Business Administration)*, Studentlitteratur, Lund.

Forrester, D., 1989, Enlightened Administrators, Paper for the European Accounting Association Congress.

Gordon, R. A., and J.E. Howell, 1959, *Higher Education for Business*, Columbia University Press, New York, NY.

Gunnarsson, E., 1988, Från Hansa till Handelshögskola. Svensk Ekonomundervisning fram till 1909 (From the Hanseatic League to a School of Economics. Swedish Business Education up to 1909), Acta univ. Ups., *Studia Oeconomiae Negotiorum 29*, Uppsala (diss.).

Hayes, R.H., and W.J. Abernathy, 1980, "Managing Our Way to Economic Decline", *Harvard Business Review*, 66-77.

Hooper, F., and J. Graham, 1897, *Commercial Education at Home and Abroad*

Karlqvist, A., 1990 (red) (ed.), *Nätverk. Teorier och Begrepp i Samhällsvetenskapen (Networks. Theories and Concepts in the Social Sciences)*, Gidlunds, Värnamo.

Levitt, T., 1978, "A Heretical View of Management Science", *Fortune*, December 18, 50-52.

Livingstone, J.S., 1971, "Myth of the Well-educated Manager", *Harvard Business Review*, 79-89.

Mintzberg, H., 1975, "The Manager´s Job: Folklore and Fact", *Harvard Business Review*, 49-61.

Olofsson, C., 1980, "Företagsekonomins Tillkomst och Institutionella Utveckling i Sverige 1900-1950" (The Birth and Institutional Development of Business Adminstration in Sweden 1900-1950) in Engwall (ed.), 15-27.

Paul, J.-P., 1989, The Consequences of Mega-Trends on Business Education, *European Management Journal* 7:283-8.

Penndorf, 1916, "Geschichte des Kaufmännischen Unterrichtswesen", in Ziegler, (red.), 115-170.

Peters, T.J., and R.H. Waterman, Jr., 1982, *In Search for Excellence: Lessons from America´s Best-Run Companies*, Harper & Row, New York, N.Y.

Pierson, F.C., et al., 1959, *The Education of American Businessmen*, McGraw-Hill Book Company, New York, NY.

Pirenne, H., 1949, *Economic and Social History of Medevial Europe*, London.

Pirenne, H., 1956), *Medieval Cities. Their Origins and the Revival of Trade*, Garden City, N.Y.

Porter L.W., and L.E. McKibbin, 1988, *Management Education and Development. Drift or Thrust into the 21st Century?*, McGraw-Hill Book Company, New York, NY,

Redlich, F., 1957, "Academic Education for Business", *Business History Review* 31:35-91.

Schelsky, H., 1971, *Einsamkeit und Freiheit*, Düsseldorf.

Servan-Schreiber, J.-J., 1968, *The American Challenge*, 1st English edition, trans. Roland Steel, Athenum, New York, NY.

Smith, B.M.D., 1990, *Education for Management: Its Conception and Implementation in the Faculty of Commerce at Birmingham mainly in the 1900s*, 2nd revision, Centre for Urban and Regional Studies, University of Birmingham.

Spender, J.-C., 1989, "Meeting Mintzberg - and Thinking Again about Management Education", *European Management Journal* 7:254-66.

Sörlin, S., 1990, "Vetenskapens Vägar - Idéhistoriens Nätverk" (The Roads of Science - the Network of the History of Thoughts), in Karlqvist, 1990, (ed), 152-183.

Törnqvist, G., 1990, "Det Upplösta Rummet - Begrepp och Teoretiska Ansatser inom Geografin" (The Disintegrated Space - Concepts and Theoretical Approaches in Geography), in Karlqvist, 1990, (ed), 23-59.

ter Vehn, A., 1956, "Något om Företagsekonomi och Företagsekonomisk Forskning" (About Business Administration and Research in Business Administration), *Ekonomiskt Forum*.

Ziegler, A., 1916, (red.), *Handbuch für das Kaufmännische Unterrichswesen in Deutschland*, Verlag für Handelsliteratur G. A. Gloeckner, Leipzig

CHAPTER 14

Contact Decision Behaviour in a Knowledge Context: A Discrete Choice Modelling Approach Using Stated Preference Data

Manfred M. Fischer — Vienna University of Economics
Rico Maggi — Zurich University
Christian Rammer — Vienna University of Economics

14.1 INTRODUCTION

In several European countries there is currently a great political debate about 1993, and the structural and economic changes which will have to come about in Europe within the coming years. Substantial industrial restructuring is taking place and forward planning in the commercial and financial sectors recognizes the profound changes an open market will bring. All this may force us to ask whether the scientific community is undertaking similar efforts to advance cooperation at a European scale and to attempt to achieve the same lack of barriers to thinking and recognition as in the USA (Fischer, 1989). One step towards this rather ambitious goal involves the Scientific Networks initiated and pushed forward by the European Science Foundation (ESF) located in Strasbourg.

With the ESF Network of European Communication and Transport Activities Research (NECTAR) a successful attempt has been made to mobilize a major part of the scientific potential in Western Europe. Activities of the Network have centred around the following four focal points of joint research: Barriers to Communication (core research area 1), Europe 2020 (core research area 2), Behaviour of Transportation and Communication Users and Suppliers (core research area 3), and Transportation and Communication Policy Development (core research area 4). In core research area 1 the major emphasis is laid

* *first,* on identifying the *major factors and barriers that affect media choice behaviour and contact decision behaviour in the university setting,*
* *second,* on developing a *conceptual framework for analyzing communication behaviour and barriers to communication* in the case of *telephone communication,* and
* *third,* on deepening the *understanding of the role of national borders in economic development.*

For the first field of research, universities have been chosen as the focus as they play an important role in an information society as the creators and disseminators of knowledge. In addition, a university setting provides a test-bed for studying differences in communication behaviour due to organizational, social and cultural factors.

While two earlier papers by the current authors focussed on media choice behaviour (see Fischer et al., 1990a, 1991), in this chapter it is the issue of contact decision behaviour which is being emphasized. The contact decision is conceived as a decision to establish direct or face-to-face contact with a potential contact partner. The main purpose of the present contribution can be summarized as follows: First, to develop a conceptual framework for analyzing contact decision behaviour in the university setting; second, to identify the major factors and barriers influencing contact decision behaviour; and, third, to identify cross-national differences in contact decision behaviour.

The outline of the chapter is as follows. In Section 14.2 the conceptual framework is sketched, while in Section 14.3 an attempt is made to integrate a stated preference experimental design procedure into a discrete choice modelling framework. The choice modelling approach developed emphasizes the influence of the contact decision context, the individual and organizational characteristics of the contact decision maker as well as the existing contact network on the formation of preferences. Section 14.4 is devoted to testing the framework empirically. The analysis relies on face-to-face interviews which were conducted in six major universities in Austria and Switzerland. Empirical results are presented using stated preference models of contact decision behaviour. In the final section the major conclusions of the study are summarized.

14.2 THE CONCEPTUAL FRAMEWORK

The complexities inherent in the process of communication behaviour have led to the development of an integrated framework for communication choice within an university setting, as outlined in Figure 14.1 (see Fischer et al., 1990a). This figure depicts the interaction of a department's supply of communication facilities (such as telephone, facsimile, electronic mail, traditional mail, courier mail, etc.) with the demand for communication in a simplified manner. The demand for communication evolves from the organizational structure of the department, including the department's objectives (especially with respect to research) as well as formal and informal rules governing individual behaviour. Supply and demand result in the need for a certain quantity and type of communication activity. Most of the communication needs are met by communication within the existing contact network, either by using communication media or by travel to face-to-face meetings (conferences, workshops, lectures, etc.), while others may be satisfied only by establishing new direct contacts. An important feature of the conceptual model is the feedback from communication outcomes to both the supply of communication facilities and the demand for communication.

The contact decision segment of the conceptual framework will be expanded in some more detail in the sequel. Contact decisions in the context of this chapter refer to the situation where an individual (termed a contact decision maker) decides on a possible new face-to-face contact which is conceived as a necessary, but not sufficient condition to extend his/her personal (knowledge based) contact network (see Figure 14.2). The choice is conceptualized as including the following stages:

First, the contact decision maker becomes aware of a need to cooperate in a specific context and expects a productivity gain from cooperation with a potential contact partner, where awareness and expectations strongly depend upon his/her own stock of knowledge, research activities and ambitions. The decision maker has individual characteristics (such as profession and status, reputation, scientific ambitions, etc.) and works in a specific institutional environment. Two extreme types of institutional environments may be distinguished: Competitive environments with several incentives in which the quality of academic output is rewarded, and bureaucratic environments where constraints rather than incentives dominate the scene and where the reward system is only loosely related to the quality of academic output. Thus, not only personal characteristics, but also the institutional setting may have strong implications for the formation of contact decision preferences in specific contexts.

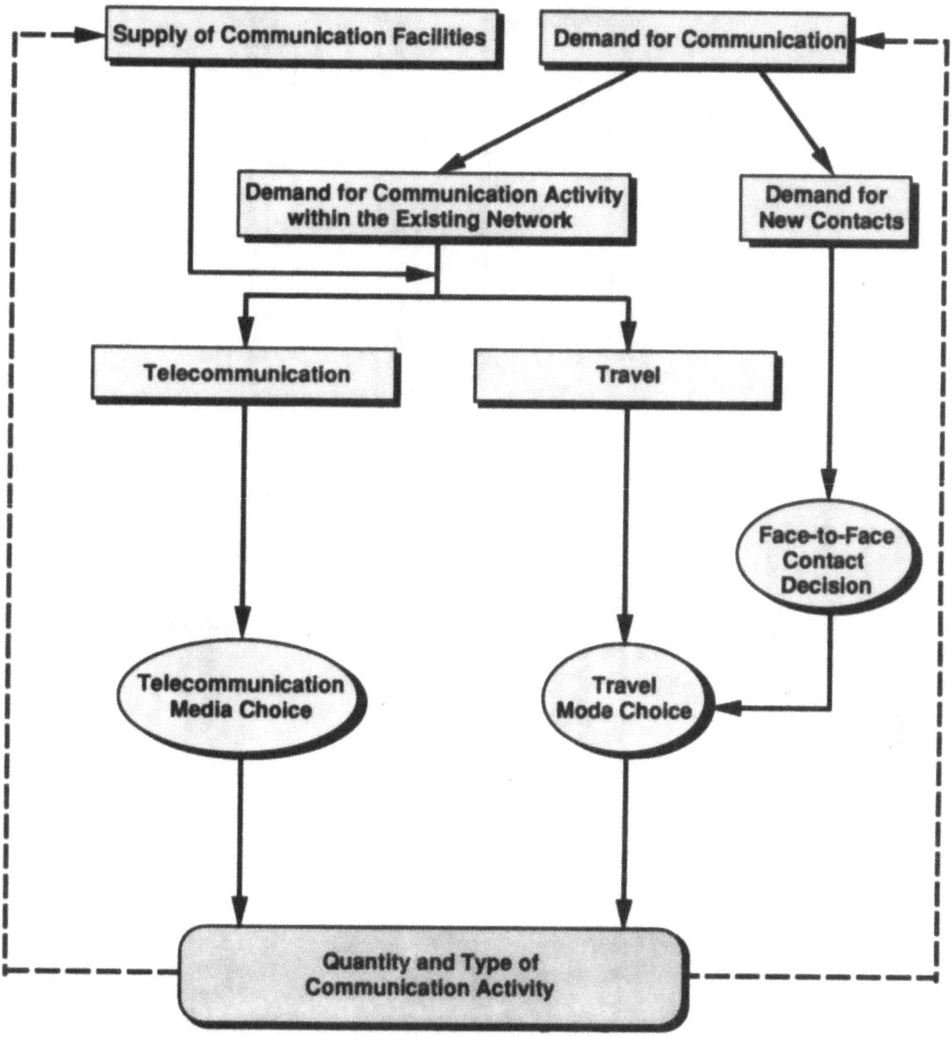

Figure 14.1: Integrated framework for communication behaviour within a university setting

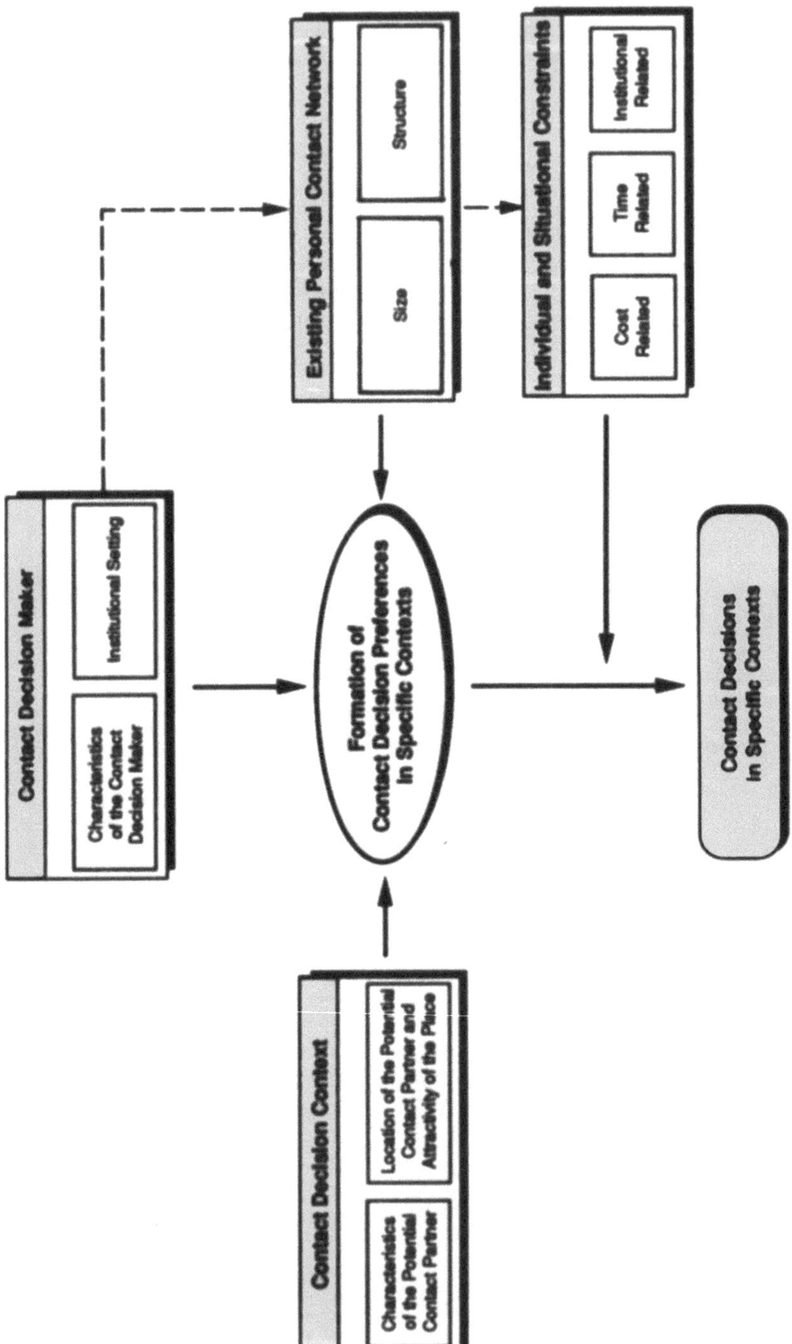

Figure 14.2: A conceptual framework for contact decision behaviour

Second, given the academic's awareness of the contact decision context it is assumed that he/she evaluates the fellow scholar's knowledge potential in relation to his/her own human capital stock. Consequently, individual characteristics of the potential contact person (such as his/her reputation in the academic field, his/her professional status, but also his/her language skills) as well as the reputation of the institution with which he/she is associated, and additionally the attractivity of the city in which the institution is located may be considered as important factors influencing the formation of contact decision preferences.

Third, the contact decision is assumed to depend not only upon the contact decision maker's own knowledge potential, but also upon the knowledge accessible in his/her existing personal contact network. Personal contact networks are conceived as informal immaterial knowledge-based networks where nodes represent academic scholars and links personal relationships.

Finally, the decision maker is subject to restrictions which relate to the rules and norms (culture) of the institution in general, and to the time and cost budgets allocated to travel by the decision maker and the academic's level of mobility in particular.

Table 14.1 presents details of the variables considered to be important for modelling the contact decision behaviour.

Table 14.1: Important variables characterizing the contact decision situation

Contact decision maker	Contact decision context
A. *Personal characteristics*	A. *Characteristics of the potential partner*
* Profession and status	* Professional status
* Age	* Reputation
* Reputation and scientific ambitions	* Language skills
* Language skills	
* Degree of Mobility	
B. *Personal knowledge based contact network*	B. *Location of the potential contact person*
* Size (number of contact persons)	* Location of the university
* Extension of the network	* Attractivity of the city
* Orientation and direction of the network	
* Intensity of network use	
C. *Institutional setting*	
* Rules and norms (culture of organization)	
* Cooperation and mobility incentives	
* Reputation and international competitiveness	
* Travel budget	

14.3 METHODOLOGY

Testing the conceptual framework is based on a micro-based approach which combines the stated preference data and the discrete choice modelling approaches.

14.3.1 The Stated Preference Data Approach

The stated preference data approach widely used in market research offers an attractive empirical setting (see Bates, 1988; Hensher et al. 1988; Wardman, 1988) in which individual contact decision behaviour may be analyzed within the context of discrete choice modelling. The stated preference data approach to data collection provides an extremely useful framework for empirical studies. In particular, it facilitates the analysis of different contact decision situations while allowing the influence of contextual variables to be determined. A key feature of the approach is that individuals are exposed to a set of choice experiments generated by some controlled experimental design procedure so that the independent variables can be made truly independent. The principal drawback of the stated preference data approach is that individuals' stated preferences may not correspond closely to their actual preferences. They may diverge because of a systematic bias in stated preference responses or because of the difficulty in carrying out the stated preference task.

Stated preferences typically predefine the attributes of the choice alternatives (in this case: a binary choice situation) and seek behavioural responses from the characteristics of choice alternatives in the form of either a preference ranking/rating or a choice selection (see Hensher et al., 1988). In this study, the technique of choice selection rather than preference ranking/rating was used. Choice selection designs are easier to complete and the best understood. Choice designs are difficult to control if self-administered; consequently face-to-face interviews have been conducted to ensure that each choice response is an independent assessment.

Survey respondents had to respond to multiple contact decision contexts, each described by carefully chosen independent variables. Behavioural responses were then measured in relation to these experimentally designed contact decision situations. Theoretical reasoning and exploratory analysis revealed that the location of the potential contact partner with a predefined attribute level of five, and his/her reputation, his/her professional status and language skills (each with two predefined attribute levels) were important contextual variables to be used to design the questionnaire contexts (see Table 14.2). The four variables were incorporated into a reduced fractional design with 32 different choice contexts.

Each questionnaire contained two contact decision contexts presented on a card, with a short description of each context variable. An example of one of these contexts is presented below:

> There is an increasing interest in a rather new field of research which attracted your attention recently. You are looking for potential partners to discuss a first draft of a paper in this exciting area. Your attention has been drawn to one of the *leading American full professors* in the field associated with the *University of California at Los Angeles*. Would you take the opportunity to participate at a Conference to be held in Los Angeles next month in order to exchange points of view and to discuss specific research problems of your interest in a face-to-face meeting with the above mentioned scholar?

The second context in this pair was composed of exactly the opposite set of levels on each of the three 2-level attributes (reputation, professional status and language skills) and another location. The pairs of contexts were equally distributed throughout the questionnaire and randomly assigned to the interviewees.

Table 14.2: A reduced fractional design for the contact decision

Block	Professional status	Reputation	Language skills	Location
Block 1	Full professor	high	perfect	Munich
Block 2	Assistant professor	high	perfect	Munich
Block 3	Full professor	low	perfect	Munich
Block 4	Assistant professor	low	perfect	Munich
Block 5	Full professor	high	basic	Prague/Paris
Block 6	Assistant professor	high	basic	Prague/Paris
Block 7	Full professor	low	basic	Prague/Paris
Block 8	Assistant professor	low	basic	Prague/Paris
Block 9	Full professor	high	perfect	Prague/Paris
Block 10	Assistant professor	high	perfect	Prague/Paris
Block 11	Full professor	low	perfect	Prague/Paris
Block 12	Assistant professor	low	perfect	Prague/Paris
Block 13	Full professor	high	basic	Lisbon
Block 14	Assistant professor	high	basic	Lisbon
Block 15	Full professor	low	basic	Lisbon
Block 16	Assistant professor	low	basic	Lisbon
Block 17	Full professor	high	perfect	Lisbon
Block 18	Assistant professor	high	perfect	Lisbon
Block 19	Full professor	low	perfect	Lisbon
Block 20	Assistant professor	low	perfect	Lisbon
Block 21	Full professor	high	perfect	Los Angeles
Block 22	Assistant professor	high	perfect	Los Angeles
Block 23	Full professor	low	perfect	Los Angeles
Block 24	Assistant professor	low	perfect	Los Angeles
Block 25	Full professor	high	basic	Tokyo
Block 26	Assistant professor	high	basic	Tokyo
Block 27	Full professor	low	basic	Tokyo
Block 28	Assistant professor	low	basic	Tokyo
Block 29	Full professor	high	perfect	Tokyo
Block 30	Assistant professor	high	perfect	Tokyo
Block 31	Full professor	low	perfect	Tokyo
Block 32	Assistant professor	low	perfect	Tokyo

14.3.2 The Discrete Choice Modelling Approach

Testing of the contact decision segment of the conceptual framework is based upon the discrete choice modelling approach, with economic random utility theory as the underlying theoretical rationale, using stated preferences. Discrete choice models such as multinominal logit, nested multinominal logit and multinominal probit models are now well established model approaches which are applied in a wide range of fields (see, for example, Ben-Akiva and Lerman, 1985; Fischer and Nijkamp, 1985, and for recent applications in geography and regional science Fischer et al. 1990b). Thus, it is not necessary to review the discrete choice modelling approach in detail, except for some specifics of the application in the empirical section of this chapter.

The contact decision may be characterized as follows. A scholar i (termed a contact decision maker) faces 2 alternatives where alternative k might be the option of realizing a face-to-face contact and alternative l would be not realizing a face-to-face contact. It is assumed that a contact decision maker's preferences among the two choice options may be described by a utility function and that (s)he selects the alternative with the greatest utility. The utility u_{ia} of an alternative $a (a = k, l)$ may be additively separated into a deterministic component v_{ia} and a random component ε_{ia}:

$$u_{ia} = v_{ia} + \varepsilon_{ia} = V(\mathbf{x}_{ia}, \theta) + \varepsilon_{ia} \tag{1}$$

where $\mathbf{x}_{ia}$ is a vector of observed characteristics of individual i and choice option a, θ denotes a vector of parameters. ε_{ia} relates to faulty perceptions of the choice options, idiosyncratic preferences, neglected choice relevant attributes etc.

The probability p_{ia} that contact decision maker i chooses option a is given by

$$\begin{aligned} p_{ia} &= \text{Prob}\,(u_{ia} > u_{ia'}, \text{ for } a \neq a' \varepsilon\{k,l\}) \\ &= \text{Prob}\,(v_{ia} + \varepsilon_{ia} > v_{ia'} + \varepsilon_{ia'}, \text{ for } a \neq a' \varepsilon\{k,l\}). \end{aligned} \tag{2}$$

The functional specification of (2) involves two major steps: First, specifying the probability distribution of the random terms; and second, specifying the functional form of the deterministic component of utility. In the current context the following assumptions are made. It is assumed that $\varepsilon_i = \varepsilon_{ik} - \varepsilon_{il}$ is logistically distributed, i.e.

$$F(\varepsilon_i) = \frac{1}{1 + \exp(-\mu\,\varepsilon_i)} \quad \text{with } \mu > 0, -\infty < \varepsilon_i < \infty \tag{3}$$

where μ is a positive scale parameter.

Moreover, it is assumed that v_{il} and v_{ik} are linear in their parameters, i.e. the choice structures are postulated to be compensatory in nature

$$V(\mathbf{x}_{ia}, \theta) = \theta' \mathbf{x}_{ia} \tag{4}$$

of linear-parameters utilities, the scale parameter μ cannot be distinguished from the overall scale of the θ's. For convenience, μ is assumed to equal one. This corresponds to the assumption that the variances of ε_{ik} and ε_{il} are both $\pi^2/6$, which implies that the variance of $(\varepsilon_{ik} - \varepsilon_{il}) = \pi^2/3$.

Under the above mentioned assumptions, the choice probability for alternative k is given by

$$\begin{aligned} p(k \mid \mathbf{x}_{ia}, \theta) &= \frac{\exp \theta' \mathbf{x}_{ik}}{\exp \theta' \mathbf{x}_{ik} + \exp \theta' \mathbf{x}_{il}} \\ &= \frac{1}{1 + \exp(-\theta' (\mathbf{x}_{ik} - \mathbf{x}_{il}))}\,. \end{aligned} \tag{5}$$

14.4. ANALYSIS AND RESULTS

Scholars associated with the University of Vienna, the Technical University of Vienna, the Vienna University of Economics and Business Administration, the University of Fribourg, the University of Zürich and the Swiss Federal Institute of Technology make up the target population of the study. These classical schools, institutes of technology and the business school may be considered to represent the major types of academic institutions in Austria and Switzerland. So the scholars associated with these universities were targeted for the testing phase of the research. The sample design used relies on exogenous stratification (proportionate stratification). The dimensions for stratification were the type of university, the type of department and the status of the scholar (full professor and assistant professor/docent). The sampling fractions were chosen to be equal to the population shares. Consequently, the sample likelihood of the stratified sample reduces to that of random sampling (see Ben-Akiva and Lerman, 1985, p. 235). The drawing of observations out of each stratum was done randomly and produced a total of 326 questionnaires (188 questionnaires in the Austrian and 138 in the Swiss case).

In order to clarify the effects of context variation on contact decision preferences several context-specific binary choice models were estimated. Börsch-Supan's HLOGIT program was used to estimate the models. HLOGIT estimates maximum likelihood parameters, utilizing a Marquardt-type modified Newton-Raphson procedure.

Three standard goodness of fit measures were used: Rho-squared (at market shares), adjusted rho-squared (at market shares) and the prediction success. Rho-squared is the standard likelihood ratio index which indicates how well the model explains preferences relative to the market shares model where all parameters in the model except the alternative specific constants are set to zero. Rho-squared (at market shares) ρ^2 is defined as

$$\rho^2 = 1 - L^*(\theta)/L(C) \tag{6}$$

where $L^*(\theta)$ denotes the value of the log likelihood function at its maximum and $L(C)$ the value of the log likelihood function when only alternative-specific constants are included. This measure is useful in comparing two specifications. Even if there are no general guidelines for when a ρ^2-value is sufficiently high, McFadden (1979) has suggested that values between 0.2 and 0.4 can be considered to represent a very good fit. A major shortcoming of this measure, however, lies in the fact that it will always increase or at least stay the same whenever new variables are added to the utility function. For this reason we also use the adjusted rho-squared (at market shares)

$$\bar{\rho}^2 = 1 - (L^*(\theta) - K)/L(C) \tag{7}$$

with K denoting the number of parameters. Another informal goodness-of-fit measure refers to the percentage of correct ex-post predictions (the so-called prediction success) which includes those observations for which the model predicted the same contact decision as was actually observed.

Three types of variables are taken into consideration. The *first group* of variables attempts to measure the influence of the personal and institutional characteristics of the contact decision maker. Four alternative-specific socio-economic variables are included: Ages and status (value 1: over 50 years and full professor, value 0: otherwise), institutional setting (value 1: Austria, value 0: Switzerland), the technical orientation of the university (value 1: school of technology, value 0: otherwise), and cooperation incentives measured in terms of the aggregated contact intensity of the institution. Basically these variables reflect the differences in preferences for establishing a new direct contact as a function of age and status, the institutional setting, the technical orientation and cooperation incentives.

The *second group* of variables measures the influence of the existing knowledge based contact network on the contact decision. The orientation of the personal knowledge based contact network (value 1: international orientation, 0: otherwise) of the contact decision maker is used to represent this type of variable.

The *third group* of variables incorporates context specific variables. A first sub-group of these variables relates to the professional status (value 1: full professor, value 0: otherwise), reputation (1: high, 0: low) and language skills (value 1: perfect in English, value 0: otherwise). A second sub-group of context specific variables measures locational characteristics. Travel costs and location specific dummies are used. For the five locations (Munich, Prague/Paris, Lisbon, Los Angeles, Tokyo) four location specific dummies (excluding Munich) have been constructed. These take the value 0 if the perceived costs are prohibitive for realizing a contact with a scholar at the corresponding location, and the value 1 otherwise. The location specific dummies may be viewed as reflecting the perceived attractiveness of the contact place in face of cost considerations.

Finally, the constant is introduced to capture the effects of the unobserved factors and individual idiosyncracies influencing the choice decision.

Two types of stated preference contact decision models were estimated:

* a *base model* estimated on the full sample size of 652 observations (326 questionnaires with two choices each),
* two *national split models* relying only on national segments of the data.

Table 14.3 summarizes the coefficient estimates and the goodness of fit statistics used for the base and the national split models. The adjusted rho-squared (at market shares) values of 0.28 (base model), 0.33 (Austrian model) and 0.27 (Swiss model) show that the models fit reasonably well. The Austrian model fits slightly better than the Swiss one. With the model specifications used 77.5 per cent (base model), 80.33 per cent (Austrian case) and 76.8 percent (Swiss case) of the positive contact decisions are predicted successfully.

Table 14.3 indicates clearly the influence of different institutional environments in Austria and Switzerland for the contact decision through varying levels of significance of the variables characterizing the contact decision maker and his/her institution. The variable of cooperation incentives is the only one significant in both the Austrian and Swiss cases. This variable, however, tends to be much more important in Austrian academia than in Switzerland. This view is also supported by the country specific dummy reflecting country specific differences in the institutional settings, and points to a more favourable institutional academic environment for direct contacts in Austria, a result which calls for further research into the incentives for individual knowledge production in the two countries.

Age interacting with professional status negatively influences the contact decision behaviour. Full professors older than 50 years are less likely to realize a new contact. The international orientation of the personal knowledge based contact network positively affects the contact decision. The association with Institutes of Technology has a negative influence on the contact decision. This may seem strange at first, but it can be explained by the fact that scholars in the engineering field of these institutions tend to be strongly nationally or internationally oriented primarily towards the German speaking countries.

The contact decision context variables have an important influence on preference formation across the two countries considered. The context variables appear to be much more important than the above mentioned characteristics of the contact decision maker. The cost variable is highly significant, has the expected negative sign, and appears to be rather robust across the Austrian and Swiss cases. The same is true for the location

Table 14.3: Parameter estimates of the stated preference contact decision model: The base model and the national split models (t-values in parentheses)

Variables	Base model		Austrian model		Swiss model	
Personal characteristics of the decision maker						
Age and status (1 if older than 50 years and full professor, 0 otherwise)	-0.69	(-2.46)*	-0.95	(-2.46)*	-0.39	(-0.83)
Personal knowledge-based contact network						
Orientation (1 if international, 0 otherwise)	0.83	(3.84)*	1.22	(4.05)*	0.30	(0.88)
Organizational Environment of the Contact Decision Maker						
Institutional setting (1 if Austria, 0 if Switzerland)	0.47	(2.20)*	-.-	(-.-)	-.-	(-.-)
Cooperation incentives (aggregated contact intensity)	0.09	(1.99)*	0.74	(2.24)*	0.21	(2.59)*
Institutes of technology (1 if school of technology, 0 otherwise)	-0.56	(-2.66)*	-2.52	(-2.59)*	-0.07	(-0.18)
Characteristics of the potential contact person						
Professional satus (1 if full professor, 0 otherwise)	0.23	(1.16)	0.16	(0.59)	0.31	(0.98)
Reputation (1 if high, 0 if low)	0.45	(2.26)*	0.54	(1.91)	0.41	(1.32)
Language skills (1 if perfect, 0 otherwise)	0.79	(3.37)*	0.65	(2.03)*	0.92	(2.47)*
Location of the potential contact person and perceived attractiveness of the place						
Prague (Austrian sub-sample) Paris (Swiss sub-sample)	1.83	(5.60)*	0.88	(1.99)*	3.05	(5.09)*
Lisbon	2.93	(7.18)*	5.20	(2.86)*	4.76	(3.21)*
Los Angeles	4.46	(6.45)*	9.99	(2.95)*	6.53	(2.60)*
Tokyo	6.81	(7.14)*	13.73	(2.80)*	11.94	(2.93)*
Travel costs	-0.32	(-5.72)*	-0.69	(-2.63)*	-0.69	(-2.60)*
Alternative-specific constant	-1.72	(-3.60)*	-4.56	(-2.32)*	-1.91	(-2.60)*
Log-likelihood at zero	-459.93		-260.62		-191.31	
Log-likelihood at constant	-442.92		-255.15		-188.68	
Log-likelihood at convergence	-307.55		-158.48		-124.51	
Rho-squared at market shares (adjusted)	0.31	(0.28)	0.38	(0.33)	0.34	(0.27)
Production success	77.5		80.3		76.8	
Observed (predicted) positive contact decisions	57.8	(71.5)	58.5	(70.2)	56.9	(69.2)
Number of observations	652		376		276	

* Significant at the 0.05 level

specific dummies considered in which Tokyo tends to be perceived as a more attractive contact place than Los Angeles, Los Angeles as a more attractive place than Lisbon, and Lisbon as a more attractive one than Prague/Paris. The relatively low parameter value for Prague evidently points to the barrier of the iron curtain, which was still Table present at the time of the survey. Surprisingly, the reputation of the potential contact partner is only weakly significant and professional status does not play a significant role at all. Language skills are found to be important characteristics of the contact person which positively influence the contact decision, especially in Swiss academia.

Finally, it is worth mentioning that the coefficient of the constant is significantly different from zero in the three models, which indicates that some choice-relevant influences have not been taken into account.

14.5 SUMMARY AND CONCLUSIONS

A general framework to analyze contact decision behaviour in an academic environment has been proposed for integrating a stated preference experimental design procedure into a discrete choice modelling framework. The framework has been empirically tested using hypothetical choice experiments. For this purpose face-to-face interviews were conducted in six universities representing classical schools, business schools and technological schools in Austria and Switzerland. The choice modelling approach developed emphasizes the influence of contact decision context specific characteristics, such as the individual and organizational characteristics of the potential contact partner as well as the personal and institutional attributes of the contact decision maker on the formation of preferences.

Empirical results are presented using stated preference models of contact decision behaviour. The results clearly indicate the importance of the contact decision context variables in general and the location specific dummies reflecting the perceived attractiveness of specific contact places in different cultural regions, the cost variable as well as the language skills of the potential contact partner in particular. Several cross-national differences in decision behaviour were identified. First, it has been found that the institutional academic environment in Austria is more favourable for contact making than in Switzerland. Second, cooperation incentives tend to influence the contact decision behaviour in Austria more strongly than in Switzerland. Third, an international orientation of the personal knowledge based contact network has a more positive influence upon contact decisions in Austria than in Switzerland. Finally, the reputation of the potential contact person or - in other words - the expected increase in the knowledge potential associated with a new link in the personal contact network appears to have only a weak influence on the contact decision.

REFERENCES

Bates, J.J., 1988, "Stated Preference Techniques and the Analysis of Consumer Choice", in Wrigley, N. (ed.), *Store Choice, Store Location and Market Analysis*, Routledge, London and New York, 187-202.

Ben-Akiva, M. and S.R. Lerman, 1985, *Discrete Choice Analysis: Theory and Application to Travel Demand*, The Mit Press, Cambridge, Ma., and London.

Fischer, M.M., 1989, "New Scientific Cooperations Across Europe: The ESF Scientific Initiative", *Environment and Planning A* 21:1-2.

Fischer, M.M. and P. Nijkamp, 1985, "Developments in Explanatory Discrete Spatial Data and Choice Analysis", *Progress in Human Geography* 9:515-551.

Fischer, M.M., R. Maggi and C. Rammer, 1990a, "Context Specific Media Choice and Barriers to Communication in Universities", *The Annals of Regional Science* 24:253-269.

Fischer, M.M., P. Nijkamp and Y.Y. Papageorgiou, (eds.), 1990b, *Spatial Choices and Processes*, North-Holland, Amsterdam (= Studies in Regional Science and Urban Economics 21).

Fischer, M.M., R. Maggi and C. Rammer, 1992, Telecommunication Media Choice Behavior in Academia, *Geographical Analysis* 24:1-15.

Hensher, D.A., P.O. Barnard and T.P Troung, 1988, "The Role of Stated Preference Methods in Studies of Travel Choice", *Journal of Transport Economics and Policy* 22:45-58.

McFadden, D., 1979, "Quantitative Methods for Analysing Travel Behaviour of Individuals. Some Recent Developments", in Hensher, D.A. and P.R. Stopher, (eds.), *Behavioural Travel Modelling*, Croom Helm, London, 279-318.

Wardman, M., 1988, "A Comparison of Revealed Preference and Stated Preference Models of Travel Behaviour", *Journal of Transport Economics and Policy* 22:71-91.

CHAPTER 15

A Dynamic Input-Output Model with Endogenous Technical Change

Kiyoshi Kobayashi
Tottori University

Åke E. Andersson
Institute for Futures Studies, Stockholm

15.1 INTRODUCTION

Technical change has played a decisive role in bringing about socio-economical structural changes. It has similarly exerted a strong influence on the history of industrialization, the evolution of political regimes, changes in international relations, and in the acquisition or loss of ability to play a hegemonic role in the world. We all recognize the importance of economic power both domestically and internationally, but technical change which brings about a shift in economic power is not directly dealt with in the framework of economics. We need to focus on this segment of the economy in discussing future growth potentials, employment possibilities, and international relations. The discussion so far, however, has tended to lack a consistent theoretical framework. In order to avoid confusion over the matters arising from this tendency, we have proposed that the discussion be based on a dynamic input-output framework. It seems that it is the most appropriate method for measuring the knowledge orientation of an economy in that it provides a consistent framework encompassing final demand, inter-industrial relationships, and factor inputs. Needless to say, input-output tables form the core of the Systems of National Accounts, and are widely available in many countries, providing both theoretical generality and empirical applicability.

The conventional input-output models developed by Leontief (1966) assume fixed input-output ratios which imply the zero elasticity of substitution between inputs in the production function. Morishima (1958, 1964) introduced variable input-output models in which the input ratios are determined by the optimizing behavior of producers. Though he paved the way for dealing with changes in input-output coefficients over time, the major forces for changing the coefficients are the relative changes in the prices of goods and in wage rates. The existence of ex ante technologies (Johansen, 1972) is presumed, among which ex post technology is chosen given the set of relative prices of goods. There is there no room to envisage technological progress, which may alter ex ante technology itself. Technical change in multisectoral models has also been discussed by Bacharach (1970), Andersson (1981), Andersson and Zhang (1988), Craven (1983) and others. These authors examine the special types of changes in input-output coefficients and the long-term properties of input-output systems that are subject to a specific type of exogenous technical change. No attention is paid to endogenous technological change. Recently, though, Zhang (1990) and Miyata et al. (1990) proposed dynamic I-O models with endogenous technical change, but they did not carry out micro economic analyses on firms' behavior underlying technical change. The main tenet of this chapter is to discuss the properties of Leontief input-output systems in which the technical coefficients change over time due to research and development activities carried out by industries. Technical change, measured as the rate of change of input-output coefficients, is endogenously determined within the economic system as the result of R&D in both the public and private sectors. In this chapter, we refer to technical change as improvement. However, it may also include the effects of substitution between inputs and of the realization of economies of scale as well as the consequences of advances in knowledge.

Public finance economists have recently expanded the classification of economic goods to encompass the degree to which they are rivalrous and the degree to which they are excludable (Cornes and Sandler, 1986). A purely rivalrous good has the property that its use by one firm or person precludes its use by another, whereas a purely non-rivalrous good has the property that its use by one agent in no way limits its use by another. Excludability relates to both technology and legal systems. A good is excludable if the owner can prevent others from using it. While conventional economic goods are both rivalrous and excludable, public goods are both non-rivalrous and non-excludable in character. Basic scientific research is an example of a public good that can be provided in this way via government funding. Technology as a whole may be perceived and even constructed as a non-rivalrous, partially-excludable good (Romer, 1990). Its non-rivalrous character stems from the fact that instructions for combining raw materials are inherently different from those for other economic goods. Once the cost of creating a new set of instructions has been incurred, the recipe may be used over and over again at no additional cost. It is in this sense that technology is non-rivalrous. This chapter treats stocks of technical knowledge as non-rivalrous, partially excludable goods. Rather than being used up in the process of production, our assumption is that knowledge capital can be used many times over at little or no additional cost (Kobayashi, Batten & Andersson, 1991). Such knowledge is made available to other firms or individuals by means of knowledge exchange processes across spatial networks (Batten, Kobayashi & Andersson, 1989). Partial excludability is allowed for by means of an accessibility function, which obviates the fact that an additional dimension of control over knowledge transfers is the geographical one. The inclusion of knowledge stocks among the inputs is hardly new. In addition to the present authors, Evenson and Kislev (1975), Romer (1986), and Adams (1990) have adopted this strategy. However, these studies only dealt with knowledge stocks within a single firm or industry. Our departure from the above studies consists of analyzing interrelations among relative prices, factor demands, outputs and final demands within the framework of a multi-sectoral growth model.

The aim of this chapter is to propose a computable dynamic input-output model to examine the interactions between knowledge accumulation and economic development with endogenous technical change. The model explicitly introduces new sectors called "knowledge sectors" into the traditional dynamic input-output economic system. It is assumed that the growth of knowledge is the main source of technical change. In order to increase knowledge we need inputs and investments just as we do to produce material capital.

The chapter is organized as follows. It commences with a description of the basic structure of our model. Section 15.3 describes firms' long-term decisions on knowledge and physical investments by applying the adjustment-cost approach. Section 15.4 contains the empirical specification of firms' micro-behavior and an input-output description of multi-sectoral interactions between outputs and investment, knowledge, and factor inputs. We also try to investigate the input-output description to a dynamic input-output model with endogenous technical change and discuss the computability of the model.

15.2 BASIC STRUCTURE OF THE MODEL

15.2.1 A Dynamic I-O Model

We begin this section by summarizing a traditional dynamic I-O model without technical change. Let $\mathbf{X}(t)$ be an n-vector whose ith element is $X_i(t)$, the output of the ith good in period t. Let $\mathbf{A} = [a_{ij}]$ be an $n \times n$ matrix, where a_{ij} denotes the current input of the ith good used per unit of the jth good. Let $\mathbf{B} = [b_{ij}]$ be an $n \times n$ matrix, where b_{ij} denotes the quantity of the ith good invested in the jth industry in order to increase the output of that industry by one unit. Let $\bar{\mathbf{D}}(t) = [\bar{D}_i(t)]$ be the vector of final demands, where $\bar{D}_i(t)$ is the final demand of the ith good in period t. Then the total demand for the ith good in period t is

$$X_i(t) = \sum_{j=1}^{n} a_{ij}X_j(t) + \sum_{j=1}^{n} b_{ij}[X_j(t+1) - X_j(t)] + \bar{D}_i(t). \tag{1}$$

The second term of the R.H.S. of this expression can be understood by supposing that, in the production of one unit of the jth good, b_{ij} units of the stock of the jth good are necessary as a capital good. Equation (1) can be rewritten in matrix form:

$$\mathbf{X}(t) = \mathbf{A}\mathbf{X}(t) + \mathbf{B}[\mathbf{X}(t+1) - \mathbf{X}(t)] + \bar{\mathbf{D}}(t). \tag{2}$$

Let $K_{ij}(t)$ be the amount of the stock of the ith good required as capital in the jth industry in period t. Then we have $b_{ij}X_j(t) = K_{ij}(t)$. Let $K_i(t)$ be the total stock of the ith good required as a capital good in the economy in period t. Then we have

$$K_i(t) = \sum_{j=1}^{n} b_{ij}X_j(t). \tag{3}$$

Assume that captal is freely transferable from one industry to another, and assume also full employment of capital so that $K_i(t)$ also denotes the supply of the ith capital as well as its demand ($i=1,\ldots,n$) (Takayama, 1985). Then

$$\mathbf{I}(t) = \mathbf{K}(t+1) - \mathbf{K}(t) = \mathbf{B}[\mathbf{X}(t+1) - \mathbf{X}(t)], \tag{4}$$

where $\mathbf{I}(t)$ is an n-vector whose ith element is the amount of the ith good demanded for investment. It is assumed that $a_{ij} > 0$ and $b_{ij} > 0$ for all $i, j = 1,\ldots,n$, and the matrix $\mathbf{B}$ is non-singular. Thus we can rewrite equation (2) as

$$\mathbf{X}(t+1) = [\mathbf{E} + \mathbf{B}^{-1}(\mathbf{E} - \mathbf{A})]\mathbf{X}(t) - \mathbf{B}^{-1}\bar{\mathbf{D}}(t), \tag{5}$$

where $\mathbf{E}$ is an $n \times n$ identity matrix.

Let us now consider the dual systems of Equation (5). Assume that our economy is equipped with money which can be produced at no cost and which functions as a medium of exchange as well as a unit of account by which the price of each good is measured (Takayama, 1985). Let $\mathbf{p}^{\circ}(t) = (p_1(t),\ldots,p_n(t),p_{n+1}(t))' = (\mathbf{p}(t),p_{n+1}(t))'$ be the price vector in period t, whose ith element $p_i(t)$ denotes the price of the ith good in period t. Assume that the production of all goods takes exactly one period and that prices are constant throughout each period. No individual can affect the interest rate $\rho(t)$ which prevails in the economy throughout period t. Let $\mathbf{a}_{n+1} = (a_{1,n+1},\ldots,a_{n,n+1})'$ be the vector, whose ith element, $a_{i,n+1}$, is the amount of labor necessary to produce one unit of the ith good. Let $p_{n+1}(t)$ be the price of labor (wages) in period t and assume that wages are paid at the end of the period. Denote the ith row of $\mathbf{A}$ by $\mathbf{a}_i$ and the ith row of $\mathbf{B}$ by $\mathbf{b}_i$. Then, the wage cost for the production of one unit of the ith good in period t is given by $W_i(t) = p_{n+1}(t)a_{i,n+1}$. Assume that wages and the costs for materials are paid at the end of the period. Then, the total value of the firm's assets at the beginning of period $(t+1)$ is given by

$$\pi_i(t) + \mathbf{p}(t+1)\mathbf{b}_i, \tag{6}$$

where $\pi_i(t) = p_i(t+1) - p_{n+1}(t+1)a_{i,n+1} - \mathbf{p}(t+1)\mathbf{a}_i$ is the current profit for period t per unit production of the ith goods and $\mathbf{p}(t+1)\mathbf{b}_i$ is the value of the bundle of capital equipment for the production of one unit of the ith goods. On the other hand, by lending at the market interest rate, the firm can obtain

$$[1 + \rho(t)]\mathbf{p}(t)\mathbf{b}_i \tag{7}$$

at the beginning of the $(t+1)$th period. Assuming the competitive arbitrage condition, it should be immaterial in equilibrium whether one lends money or invests it in the production of the ith good. There holds:

$$\pi_i(t) + \mathbf{p}(t+1)\mathbf{b}_i = (1 + \rho(t))\mathbf{p}(t)\mathbf{b}_i. \tag{8}$$

This relation must hold for all $i=1,...,n$. Hence recalling that $\mathbf{a}_i$ is the ith row of $\mathbf{A}$ and $\mathbf{b}_j$ is the ith row of $\mathbf{B}$, we have

$$\mathbf{p}(t+1) = (1 + \rho(t))\mathbf{p}(t)'[\mathbf{E} - (\mathbf{E} - \mathbf{A})\mathbf{B}^{-1}]^{-1} + \mathbf{W}(t), \tag{9}$$

where $\mathbf{E}$ is an $n \times n$ identity matrix and $\mathbf{W} = (W_1(t),...,W_n(t))'$ is the vector of wage costs (value added) (Solow, 1956).

15.2.2 A Dynamic I-O Model with Endogenous Technical Change

Morishima (1958, 1964) provided a significant version of dynamic Leontief systems, whose essential difference from the usual dynamic model is that the technical coefficients, a_{ij} and b_{ij} are no longer constant; rather, they are variable through time in response to changes in the relative prices of goods as well as wage rates. Given the relative prices and wage rate, the equilibrium coefficients are chosen at the beginning of each period through firms' optimizing behavior so that the usual analysis of the dynamic Leontief system with fixed coefficients can be justified. This is an extension of the substitution theorem of Samuelson (Samuelson, 1951) from a static to a dynamic case. Although his introduction of the possibility of factor demand substitution is a very important contribution in coping with technical change, we will argue that Morishima's dynamic substitution theorem holds only for very limited cases. The dynamic substitution in his model can only be made within the set of ex ante technologies. Progress of ex ante technologies are precluded in his discussions on technical substitution among factor demands.

The model in this chapter can be viewed as an equilibrium model of endogenous technological change in which long-term growth is driven primarily by the accumulation of knowledge by forward-looking, profit-maximizing agents. The focus on knowledge as a basic form of capital suggests a natural modification in the formulation of the standard dynamic input-output model. In contrast to physical capital that can be monopolized by firms, knowledge capital can be regarded as a non-rivalrous good. The creation of new knowledge by one firm is assumed to have positive external effects on the production possibilities of other firms because knowledge cannot be perfectly patented or kept secret. Firms may choose to undertake knowledge production to expand their own stocks of knowledge, but eventually this knowledge becomes accessible to other firms. The basic characteristics of knowledge stocks as endogenous public goods are explicitly described in the form of dynamic systems of knowledge accumulation, in Section 15.4.3.

Let us now envisage the basic structure of our dynamic Leontief model with endogenous technical change. Suppose each industry's output can be used for final consumption, factor inputs for production, and investment for either capital or knowledge formation. Denote industries by the numerals $(1,2,...,n)$, which are supposed to fall exclusively into two categories: i) m industries whose outputs are invested for physical capital formation, indexed by $(1,...,m)$, and ii) s industries for knowledge formation, denoted by $(m+1,...,m+s(=n))$. Our dynamic model is operated under the framework of dynamic Leontief models with variable coefficients:

$$\left.\begin{aligned} \mathbf{X}(t+1) &= \{\mathbf{E} + \mathbf{B}(t)^{-1}[\mathbf{E} - \mathbf{A}(t)]\}\mathbf{X}(t) - \mathbf{B}(t)^{-1}\bar{\mathbf{D}}(t), \\ \mathbf{p}(t+1) &= (1+\rho(t))\mathbf{p}(t)'[\mathbf{E} - (\mathbf{E} - \mathbf{A}(t))\mathbf{B}(t)^{-1}]^{-1} + \mathbf{W}(t), \end{aligned}\right\} \tag{10}$$

where $\mathbf{X}(t)$ is an n-vector of outputs in period t; $\mathbf{A}(t)$ is the $n \times n$ matrix of input-output coefficients, $a_{ij}(t)$, in period t; $\mathbf{B}(t)$ is the $n \times n$ matrix of investment coefficients, $b_{ij}(t)$; $\mathbf{p}°(t)$ is the price vector of $n+1$ goods, whose $(n+1)$th element $p_{n+1}(t)$ represents the wage rate. The major difference from the traditional dynamic I-O models is that neither of the coefficient matrices $\mathbf{A}(t)$ and $\mathbf{B}(t)$ are fixed and they vary over time. The main driving force underlying the coefficient's change is the accumulation of physical capital and knowledge stocks. The changes driven in the coefficients' matrices $\mathbf{A}(t)$ and $\mathbf{B}(t)$ due to technical change can be generally expressed by the following dynamic systems:

$$\mathbf{A}(t+1) = \mathbf{F}_A(\mathbf{p}°(t+1), \rho(t+1), \mathbf{B}(t+1)), \tag{11}$$

$$\mathbf{B}(t+1) = \mathbf{F}_B(\mathbf{p}°(t+1), \rho(t), \mathbf{B}(t); \zeta(t), \mathbf{\Omega}(t)), \tag{12}$$

where $\mathbf{\Omega}(t)$ is the vector of knowledge stocks in industries and $\zeta(t)$ is the matrix of parameters representing the stock levels of social infrastructure. Though the levels of knowledge stocks are determined by firms' decisions, their collective outputs eventually play decisive roles in controlling the pace of capital and knowledge formation over the whole industry. The functional forms of $\mathbf{F}_A$ and $\mathbf{F}_B$ are rationalized in Section 15.4.

15.3 MICROECONOMIC ANALYSIS OF KNOWLEDGE ACCUMULATION

15.3.1 A Basic Model

The purpose of this section is to investigate firms' decisions on R&D investment and to develop a dynamic model incorporating inter-industry knowledge spillovers in order to investigate how the pace of capital and knowledge accumulation is regulated by the economic environment and the external economy for knowledge production. Microeconomic analyses of firms' R&D activities were developed by Fellner (1960), Arrow (1962), Kennedy (1964) and others. Though they paved a way to cope with technical change, their theories are essentially static. The starting point of our model is based on the adjustment-cost models for investment and factor demands (Bernstein and Nadiri, 1989). Dynamic duality theory as developed by Rockafeller (1970), Benveniste and Scheinkaman (1979), McLaren and Cooper (1980) and Epstein (1981) is applied to obtain closed-form solutions to the primal problem. A firm is supposed to minimize the present value of its future stream of production costs given the stream of output flows over time, which include adjustment costs associated with capital and knowledge investment. Costs are also regulated by the social availability of knowledge stocks emanating from inter-industrial knowledge spillovers. Indeed, it is this property as an endogenous public good that distinguishes knowledge from capital. The mechanism of knowledge accumulation under knowledge spillovers can be described in the form of dynamic I-O models with endogenous technical change, as will be discussed later in Section 15.4.

Firms' technology with inter-industrial knowledge spillovers can be represented by a production function:

$$Y(t) = H(\mathbf{X}^{\circ}(t),\mathbf{K}(t),\mathbf{G}(t),\mathbf{I}(t),\mathbf{J}(t);\boldsymbol{\zeta}(t),\boldsymbol{\Omega}(t)), \tag{13}$$

where $Y(t)$ is the output flow at time t, $\mathbf{X}^{\circ}(t) = (X_1(t),\ldots, X_n(t), X_{n+1}(t))'$ is the vector of factor inputs, $\mathbf{K}(t) = (K_1(t),\ldots, K_m(t))'$ is the vector of physical capital service flows, $\mathbf{G}(t) = (G_1(t),\ldots,G_s(t))'$ is the knowledge service flow, $\mathbf{I}(t) = (I_1(t),\ldots,I_m(t))'$ is the amount of physical investments, $\mathbf{J}(t) = (J_1(t),\ldots,J_s(t))'$ is the amount of knowledge investments (R&D), $\boldsymbol{\zeta}(t)$ is the technology of knowledge infrastructure and $\boldsymbol{\Omega}(t)$ are the knowledge stocks across the whole industry. It will be assumed that H possesses the standard properties of neo-classical production functions.

The short-term variable costs for factor inputs, which are crucially dependent upon the output $\bar{Y}(t)$, are derived by inverting the production function (13):

$$\begin{aligned} C(t) &= C(\mathbf{p}^{\circ}(t),\bar{\mathbf{K}}(t),\bar{\mathbf{G}}(t),\bar{\mathbf{I}}(t),\bar{\mathbf{J}}(t),\bar{Y}(t);\boldsymbol{\zeta}(t),\boldsymbol{\Omega}(t)) \\ &= \min_{\mathbf{X}^{\circ}(t)} \{\mathbf{p}^{\circ}(t)'\mathbf{X}^{\circ}(t) : \bar{Y}(t) = H(\mathbf{X}^{\circ}(t),\bar{\mathbf{K}}(t),\bar{\mathbf{G}}(t),\bar{\mathbf{I}}(t),\bar{\mathbf{J}}(t);\boldsymbol{\zeta}(t),\boldsymbol{\Omega}(t))\} \end{aligned} \tag{14}$$

where $\mathbf{p}^{\circ}(t) = (p_1(t),\ldots,p_n(t), p_{n+1}(t))'$ is the price vector of factor inputs, whose $(n+1)$th element represents the wage rate. As a vehicle of explanation, let us abbreviate the variables $\boldsymbol{\zeta}$, $\boldsymbol{\Omega}$ in expressing the cost function C.

The basic assumption of the model is that at any point in time, factor prices are exogenous to firms; the plans of outputs, $\bar{Y}(t)$s are exogenously given to them. Given relative prices, the plans of outputs, and social stocks of knowledge resources, each firm tries to minimize the present value of the stream of discounted costs over an infinite horizon by determining the optimal intensity of physical and knowledge investments subject to the process of capital and knowledge accumulation. The firm's problems can be written as

$$\left.\begin{aligned} &\min_{\mathbf{I}(t),\mathbf{J}(t)}\{\int_0^{\infty} [C(\mathbf{p}^{\circ}(t),\mathbf{K}(t),\mathbf{G}(t),\mathbf{I}(t),\mathbf{J}(t),\bar{Y}(t)) \\ &\qquad + \mathbf{p}_I(t)'\mathbf{I}(t) + \mathbf{p}_J(t)'\mathbf{J}(t)]\exp(-\rho t)dt\} \\ &\qquad \dot{\mathbf{K}}(t) = \mathbf{I}(t) - \boldsymbol{\delta}_K\mathbf{K}(t), \\ &\qquad \dot{\mathbf{G}}(t) = \mathbf{J}(t) - \boldsymbol{\delta}_G\mathbf{G}(t), \\ &\qquad \mathbf{K}(0) = \mathbf{K}^0, \\ &\qquad \mathbf{G}(0) = \mathbf{G}^0, \end{aligned}\right\} \tag{15}$$

where ρ is the discount rate, $\mathbf{p}_I(t) = (p_1(t),\ldots,p_m(t))'$ is the relative price vector corresponding to $\mathbf{K}(t)$, $\mathbf{p}_J(t) = (p_{m+1}(t),\ldots,p_n(t))'$ is the relative price vector of knowledge resources $\mathbf{G}(t)$, $\boldsymbol{\delta}_K$ is the diagonal matrix of the depreciation charges of capital goods, and $\boldsymbol{\delta}_G$ is the diagonal matrix of the depreciation rates of knowledge resources.

Let $\boldsymbol{\omega}$ and $\boldsymbol{\theta}$ be the price vectors of service flows of capital goods ($j = 1,\ldots,m$) and knowledge resources ($k = 1,\ldots,s$), respectively. Let $\mathbf{p}_I$ and $\mathbf{p}_J$ be the price vectors of capital goods and knowledge resources, respectively. Then, the depreciation charges of capital goods and knowledge resources are given by $\boldsymbol{\delta}_K\mathbf{p}_I$ and $\boldsymbol{\delta}_G\mathbf{p}_J$, respectively. By deducting depreciation charges from the gross prices of capital and knowledge service flows, $\boldsymbol{\omega}$ and $\boldsymbol{\theta}$, the net prices of capital and knowledge service flows can be given by $\boldsymbol{\omega} - \boldsymbol{\delta}_K \mathbf{p}_I$ and $\boldsymbol{\theta} - \boldsymbol{\delta}_G \mathbf{p}_J$, respectively. Assume that the depreciation rate ρ is equal to the interest rate. Following Walras, the net prices of capital and knowledge service flows are equilibrated with the returns of capital flows if capital markets are perfectly competitive. That is, there hold

$$\left.\begin{aligned} \boldsymbol{\omega} - \boldsymbol{\delta}_K \mathbf{p}_I &= \rho \mathbf{p}_I, \\ \boldsymbol{\theta} - \boldsymbol{\delta}_G \mathbf{p}_J &= \rho \mathbf{p}_J. \end{aligned}\right\} \tag{16}$$

In a perfectly competitive capital market, the prices of capital and knowledge services are given by

$$\left.\begin{aligned} \boldsymbol{\omega} &= (\boldsymbol{\delta}_K + \rho \mathbf{E}_K)\mathbf{p}_I, \\ \boldsymbol{\theta} &= (\boldsymbol{\delta}_G + \rho \mathbf{E}_G)\mathbf{p}_J. \end{aligned}\right\} \tag{17}$$

where $\mathbf{E}_K$ and $\mathbf{E}_G$ are the $m \times m$ and $s \times s$ identity matrices, respectively. By taking account of Equations (17), our problem can be rewritten to

$$\left.\begin{aligned} &\min_{\mathbf{I}(t),\mathbf{J}(t)} \{\textstyle\int_0^\infty [C(\mathbf{p}^\circ(t),\mathbf{K}(t),\mathbf{G}(t),\mathbf{I}(t),\mathbf{J}(t),\bar{Y}(t)) \\ &\qquad + \boldsymbol{\omega}(t)'\mathbf{K}(t) + \boldsymbol{\theta}(t)'\mathbf{G}(t)]\exp(-\rho t)dt\} \\ &\dot{\mathbf{K}}(t) = \mathbf{I}(t) - \boldsymbol{\delta}_K \mathbf{K}(t), \\ &\dot{\mathbf{G}}(t) = \mathbf{J}(t) - \boldsymbol{\delta}_G \mathbf{G}(t), \\ &\mathbf{K}(0) = \mathbf{K}^0, \\ &\mathbf{G}(0) = \mathbf{G}^0. \end{aligned}\right\} \tag{18}$$

Consider now the following inter-temporal optimization problem. At time τ, the firm is endowed with the initial states of quasi-fixed factors, i.e., capital and knowledge stocks, $\mathbf{K}(\tau) = \bar{\mathbf{K}}$ and $\mathbf{G}(\tau) = \bar{\mathbf{G}}$; it decides on the optimal combination of variable factor inputs with a restricted cost function $C(\mathbf{p}^\circ(\tau),\mathbf{K}(\tau),\mathbf{G}(\tau),\bar{\mathbf{I}}(\tau),\bar{\mathbf{J}}(\tau),\bar{Y}(\tau))$; it determines the optimal paces of capital and knowledge investments in order to minimize the discounted sum of future costs. The firm has static expectations on the future states of the prices of capital and knowledge services, $\boldsymbol{\omega}$, $\boldsymbol{\theta}$, the level of outputs $\bar{Y}$, and the discount rate ρ. In other words, current relative prices and output levels are expected to persist indefinitely. As time goes, the firm revises its expectations and production plans with updated information on factor prices. The problem is to choose the optimal time path of quasi-fixed factors $\mathbf{K}(t)$ and $\mathbf{G}(t)$ to:

$$\left.\begin{aligned} &\min_{\mathbf{I}(t),\mathbf{J}(t)} \{\textstyle\int_0^\infty [C(\mathbf{p}^\circ,\mathbf{K}(t),\mathbf{G}(t),\mathbf{I}(t),\mathbf{J}(t),\bar{Y}) \\ &\qquad + \boldsymbol{\omega}'\mathbf{K}(t) + \boldsymbol{\theta}'\mathbf{G}(t)]\exp\{-\rho(t-\tau)\}dt\}, \\ &\dot{\mathbf{K}}(t) = \mathbf{I}(t) - \boldsymbol{\delta}_K \mathbf{K}(t), \quad (t \geq \tau) \\ &\dot{\mathbf{G}}(t) = \mathbf{J}(t) - \boldsymbol{\delta}_G \mathbf{G}(t), \quad (t \geq \tau) \\ &\mathbf{K}(\tau) = \bar{\mathbf{K}}, \\ &\mathbf{G}(\tau) = \bar{\mathbf{G}}. \end{aligned}\right\} \tag{19}$$

It is clear that for this problem the optimal present value of costs does not depend explicitly upon τ. By the same token, the optimal path of investment rates $\dot{\mathbf{K}}(t)$, $\dot{\mathbf{G}}(t)$ depends only upon the givens $(\bar{\mathbf{K}},\bar{\mathbf{G}},\mathbf{p}^\circ,\boldsymbol{\omega},\boldsymbol{\theta},\bar{Y})$. The optimal accumulation path of quasi-fixed factors can be denoted by $\mathbf{K}^*(t)$ and $\mathbf{G}^*(t)$. Then, the optimal value function V depends upon the initial givens $(\bar{\mathbf{K}},\bar{\mathbf{G}},\mathbf{p}^\circ,\boldsymbol{\omega},\boldsymbol{\theta},\bar{Y})$. Through the definition, we have

$$
\begin{aligned}
&V(\bar{\mathbf{K}},\bar{\mathbf{G}},\mathbf{p}^\circ,\boldsymbol{\omega},\boldsymbol{\theta},\bar{Y}) \\
&= \{\int_\tau^\infty [C(\mathbf{p}^\circ,\mathbf{K}^*(t),\mathbf{G}^*(t),\mathbf{I}^*(t),\mathbf{J}^*(t),\bar{Y}) \\
&\quad + \boldsymbol{\omega}'\mathbf{K}^*(t) + \boldsymbol{\theta}'\mathbf{G}^*(t)]\exp\{-\rho(t-\tau)\}dt.
\end{aligned}
\tag{20}
$$

15.3.2 Optimality Conditions

The derivation of the optimality conditions for the problem (19) is based on Pontryagin's Maximum Principle. The current value Hamiltonian is:

$$
\begin{aligned}
&H(\mathbf{K}(t),\mathbf{G}(t):\mathbf{p}^\circ,\boldsymbol{\omega},\boldsymbol{\theta},\bar{Y}) \\
&\quad = C(\mathbf{p}^\circ,\mathbf{K}(t),\mathbf{G}(t),\mathbf{I}(t),\mathbf{J}(t),\bar{Y}) + \boldsymbol{\omega}'\mathbf{K}(t) + \boldsymbol{\theta}'\mathbf{G}(t) \\
&\quad + \boldsymbol{\lambda}_K(t)'[\mathbf{I}(t) - \boldsymbol{\delta}_K\mathbf{K}(t)] + \boldsymbol{\lambda}_G(t)'[\mathbf{J}(t) - \boldsymbol{\delta}_G\mathbf{G}(t)],
\end{aligned}
\tag{21}
$$

where $\boldsymbol{\lambda}_K(t) = (\lambda_{K1}(t),\dots,\lambda_{Km}(t))'$ and $\boldsymbol{\lambda}_G(t) = (\lambda_{G1}(t),\dots,\lambda_{Gs}(t))'$ are the vectors of costate variables. On the optimal trajectory, the state variables $\mathbf{K}(t)$ and $\mathbf{G}(t)$ and the costate variables $\boldsymbol{\lambda}_K(t)$ and $\boldsymbol{\lambda}_G(t)$ satisfy:

$$\dot{\mathbf{K}}(t) = \frac{\partial H(t)}{\partial \boldsymbol{\lambda}_K} \tag{22}$$

$$\dot{\mathbf{G}}(t) = \frac{\partial H(t)}{\partial \boldsymbol{\lambda}_G} \tag{23}$$

$$\dot{\boldsymbol{\lambda}}_K(t) = \rho\boldsymbol{\lambda}_K(t) - \frac{\partial H(t)}{\partial \mathbf{K}(t)}, \tag{24}$$

$$\dot{\boldsymbol{\lambda}}_G(t) = \rho\boldsymbol{\lambda}_G(t) - \frac{\partial H(t)}{\partial \mathbf{G}(t)}. \tag{25}$$

The optimal solution to problem (19) is of the form:

$$
\left.
\begin{aligned}
\mathbf{K}^*(t) &= \mathbf{K}(t-\tau,\bar{\mathbf{K}},\bar{\mathbf{G}};\mathbf{p}^\circ,\boldsymbol{\omega},\boldsymbol{\theta},\bar{Y}), \\
\mathbf{G}^*(t) &= \mathbf{G}(t-\tau,\bar{\mathbf{K}},\bar{\mathbf{G}};\mathbf{p}^\circ,\boldsymbol{\omega},\boldsymbol{\theta},\bar{Y}).
\end{aligned}
\right\}
\tag{26}
$$

Define the optimal investment function by

$$
\left.
\begin{aligned}
\mathbf{I}^*(t) &= \boldsymbol{\gamma}(\mathbf{K}^*(t),\mathbf{G}^*(t);\mathbf{p}^\circ,\boldsymbol{\omega},\boldsymbol{\theta},\bar{Y}), \\
\mathbf{J}^*(t) &= \boldsymbol{\phi}(\mathbf{K}^*(t),\mathbf{G}^*(t);\mathbf{p}^\circ,\boldsymbol{\omega},\boldsymbol{\theta},\bar{Y}),
\end{aligned}
\right\}
\tag{27}
$$

where

$$\gamma(\mathbf{K}^*(t),\mathbf{G}^*(t);\mathbf{p}^\circ,\boldsymbol{\omega},\boldsymbol{\theta},\bar{Y}) = \mathbf{I}(t-\tau,\bar{\mathbf{K}},\bar{\mathbf{G}};\mathbf{p}^\circ,\boldsymbol{\omega},\boldsymbol{\theta},\bar{Y}),$$

$$\phi(\mathbf{K}^*(t),\mathbf{G}^*(t);\mathbf{p}^\circ,\boldsymbol{\omega},\boldsymbol{\theta},\bar{Y}) = \mathbf{J}(t-\tau,\bar{\mathbf{K}},\bar{\mathbf{G}};\mathbf{p}^\circ,\boldsymbol{\omega},\boldsymbol{\theta},\bar{Y}).$$

Note that γ and ϕ do not depend explicitly upon τ. Varying the initial conditions $\bar{\mathbf{K}},\bar{\mathbf{G}}$, define the synthesized optimal investment functions γ and ϕ. Then the optimal value function V can be redefined in terms of γ and ϕ as:

$$\begin{aligned} &V(\bar{\mathbf{K}},\bar{\mathbf{G}},\mathbf{p}^\circ,\boldsymbol{\omega},\boldsymbol{\theta},\bar{Y}) \\ &= \{\int_\tau^\infty [C(\mathbf{p},\mathbf{K}^*(t),\mathbf{G}^*(t),\gamma(\mathbf{K}^*(t),\mathbf{G}^*(t)),\phi(\mathbf{K}^*(t),\mathbf{G}^*(t)),\bar{Y}) \\ &\quad + \boldsymbol{\omega}'\mathbf{K}^*(t) + \boldsymbol{\theta}'\mathbf{G}^*(t)]\exp\{-\rho(t-\tau)\}\}dt, \end{aligned} \tag{28}$$

where the arguments $\bar{\mathbf{K}}$ and $\bar{\mathbf{G}}$ in V now refer to arbitrary initial conditions. V is differentiable almost everywhere in the interior of the effective domain of V (Rockafeller, 1970). Furthermore, Equation (28) may be now differentiable along the optimal path ϕ and γ to yield the Hamilton-Jacobi equation.

$$\begin{aligned} &\left(\frac{\partial V}{\partial \mathbf{K}^*}\right)\dot{\mathbf{K}}^*(t)+\left(\frac{\partial V}{\partial \mathbf{G}^*}\right)\dot{\mathbf{G}}^*(t) \\ &\quad = \rho V - \{C(\mathbf{p}^\circ,\mathbf{K}^*(t),\mathbf{G}^*(t),\gamma(\mathbf{K}^*(t),\mathbf{G}^*(t)),\phi(\mathbf{K}^*(t),\mathbf{G}^*(t)),\bar{Y}) \\ &\quad + \boldsymbol{\omega}'\mathbf{K}^*(t) + \boldsymbol{\theta}'\mathbf{G}^*(t)\}. \end{aligned} \tag{29}$$

Rather than solving the primal problem given by (18), we can approach the dynamic dual problem of Problem (19) by defining the minimized present value of normalized costs as $V(\mathbf{K},\mathbf{G};\mathbf{p}^\circ,\boldsymbol{\omega},\boldsymbol{\theta},\bar{Y})$, which satisfies

$$\begin{aligned} \rho V(\mathbf{K},\mathbf{G},\mathbf{p}^\circ,\boldsymbol{\omega},\boldsymbol{\theta},\bar{Y}) &= C(\mathbf{p}^\circ,\mathbf{K},\mathbf{G},\mathbf{I}^*,\mathbf{J}^*,\bar{Y}) + \boldsymbol{\omega}'\mathbf{K} + \boldsymbol{\theta}'\mathbf{G} \\ &\quad + \mathbf{V}'_K\dot{\mathbf{K}} + \mathbf{V}'_G\dot{\mathbf{G}}, \end{aligned} \tag{30}$$

where $\mathbf{V}_K = (\partial V/\partial K_1,\ldots,\partial V/\partial K_m)'$ and $\mathbf{V}_G = (\partial V/\partial G_1,\ldots,\partial V/\partial G_s)'$. The investment demand functions implied by Equation (29) can be obtained by differentiating Equation (30) with respect to the rental rates $\boldsymbol{\omega}$, $\boldsymbol{\theta}$. Thus

$$\left.\begin{aligned} \rho\mathbf{V}_\omega &= \mathbf{K} + \mathbf{V}_{K\omega}\dot{\mathbf{K}} + \mathbf{V}_{G\omega}\dot{\mathbf{G}}, \\ \rho\mathbf{V}_\theta &= \mathbf{G} + \mathbf{V}_{K\theta}\dot{\mathbf{K}} + \mathbf{V}_{G\theta}\dot{\mathbf{G}}, \end{aligned}\right\} \tag{31}$$

where

$$\mathbf{V}_\omega = \left(\frac{\partial V}{\partial \omega_1},\ldots,\frac{\partial V}{\partial \omega_m}\right)',$$

$$\mathbf{V}_\theta = \left(\frac{\partial V}{\partial \theta_1},\ldots,\frac{\partial V}{\partial \theta_s}\right)',$$

$$\mathbf{V}_{K\omega} = \begin{bmatrix} \frac{\partial^2 V}{\partial\omega_1\partial K_1} & \cdots & \frac{\partial^2 V}{\partial\omega_1\partial K_m} \\ \vdots & \ddots & \vdots \\ \frac{\partial^2 V}{\partial\omega_m\partial K_1} & \cdots & \frac{\partial^2 V}{\partial\omega_m\partial K_m} \end{bmatrix},$$

$$\mathbf{V}_{G\omega} = \begin{bmatrix} \frac{\partial^2 V}{\partial\omega_1\partial G_1} & \cdots & \frac{\partial^2 V}{\partial\omega_1\partial G_s} \\ \vdots & \ddots & \vdots \\ \frac{\partial^2 V}{\partial\omega_m\partial G_1} & \cdots & \frac{\partial^2 V}{\partial\omega_m\partial G_s} \end{bmatrix},$$

$$\mathbf{V}_{K\theta} = \begin{bmatrix} \frac{\partial^2 V}{\partial\theta_1\partial K_1} & \cdots & \frac{\partial^2 V}{\partial\theta_1\partial K_m} \\ \vdots & \ddots & \vdots \\ \frac{\partial^2 V}{\partial\theta_s\partial K_1} & \cdots & \frac{\partial^2 V}{\partial\theta_s\partial K_m} \end{bmatrix},$$

and

$$\mathbf{V}_{G\theta} = \begin{bmatrix} \frac{\partial^2 V}{\partial\theta_1\partial G_1} & \cdots & \frac{\partial^2 V}{\partial\theta_1\partial G_s} \\ \vdots & \ddots & \vdots \\ \frac{\partial^2 V}{\partial\theta_s\partial G_1} & \cdots & \frac{\partial^2 V}{\partial\theta_s\partial G_s} \end{bmatrix},$$

For the simplicity of expression, let us introduce $\mathbf{\Phi} = [\mathbf{V}'_\omega, \mathbf{V}'_\theta]$, $\mathbf{v} = [\mathbf{K}', \mathbf{G}']'$, $\dot{\mathbf{v}} = [\dot{\mathbf{K}}', \dot{\mathbf{G}}']'$, $\boldsymbol{\mu} = [\boldsymbol{\omega}', \boldsymbol{\theta}']'$, $\mathbf{R} = [\mathbf{V}'_K, \mathbf{V}'_G]'$, and

$$\mathbf{\Psi} = \begin{bmatrix} \mathbf{V}_{K\omega} & \mathbf{V}_{G\omega} \\ \mathbf{V}_{K\theta} & \mathbf{V}_{G\theta} \end{bmatrix}.$$

Then Equation (31) can be rewritten in a simple form:

$$\rho\mathbf{R} = \mathbf{v} + \mathbf{\Psi}\dot{\mathbf{v}}. \tag{32}$$

By solving Equation (32) in terms of $\dot{\mathbf{v}}$, we have

$$\dot{\mathbf{v}} = \mathbf{\Psi}^{-1}(\rho\mathbf{R} - \mathbf{v}). \tag{33}$$

From Equation (33), we get the short-run cost function $C(\mathbf{p}, \mathbf{v}, \boldsymbol{\mu}, \mathbf{I}^*, \mathbf{J}^*, \bar{Y})$:

$$\begin{aligned} &C(\mathbf{p}^\circ, \mathbf{v}, \boldsymbol{\mu}, \mathbf{I}^*, \mathbf{J}^*, \bar{Y}) \\ &\quad = \rho V(\mathbf{v}; \mathbf{p}^\circ, \boldsymbol{\mu}, \bar{Y}) - \boldsymbol{\omega}'\mathbf{K} - \boldsymbol{\theta}'\mathbf{G} - \mathbf{V}'_K\dot{\mathbf{K}} - \mathbf{V}'_G\dot{\mathbf{G}}' \end{aligned}$$

$$= \rho V(\mathbf{v};\mathbf{p}^\circ,\boldsymbol{\mu},\bar{Y}) - \boldsymbol{\mu}'\mathbf{v} - \mathbf{R}'\dot{\mathbf{v}}. \tag{34}$$

By substituting Equation (33) into Equation (34), we have

$$C(\mathbf{p}^\circ,\mathbf{v},\boldsymbol{\mu},\mathbf{I}^*,\mathbf{J}^*, \bar{Y}) = \rho V - \boldsymbol{\mu}'\mathbf{v} - \mathbf{R}'\boldsymbol{\Psi}^{-1}(\rho\mathbf{R} - \mathbf{v}). \tag{35}$$

Applying Shephard's lemma, the inputs of intermediate good $\mathbf{X}^\circ = (X_1,...,X_n,X_{n+1})'$ are given by

$$\begin{aligned}\mathbf{X}^\circ &= \frac{\partial C(\mathbf{p}^\circ,\mathbf{v},\boldsymbol{\mu},\mathbf{I},\mathbf{J},\bar{Y})}{\partial \mathbf{p}^\circ} \\ &= \rho\frac{\partial V}{\partial \mathbf{p}^\circ} - \frac{\partial\{\mathbf{R}'\boldsymbol{\Psi}^{-1}(\rho\mathbf{R} - \mathbf{v})\}}{\partial \mathbf{p}^\circ}.\end{aligned} \tag{36}$$

From the definition of the cost function (14), it is clear that the minimized present value of normalized costs, V, is regulated by those variables representing inter-industrial knowledge spillovers, $\boldsymbol{\Omega}$, and the quality of knowledge infrastructure, $\boldsymbol{\zeta}$. Thus, both variable factors and investment demand functions derived from V are also conditional to the accessibility to social knowledge stocks.

15.4 MODEL SPECIFICATION

15.4.1 Specification of Value Functions

Given a specification of value functions that correspond to Problem (19), Equations (33) and (36) provide straightforward procedures for generating functional forms of dynamic factor demands which can serve as the basis for estimation. Moreover, the characterizing properties of V, which are reflected as restrictions on the parameters of the demand functions, constitute the exhaustive testable implications of the model.

Let us first characterize the forms of cost functions permissible for the adjustment cost approach. It is assumed that cost function, C, satisfies the following regularity conditions:

A.1 $C \geq 0$.
A.2 C is non-decreasing with respect to Y, **I**, and **J**, and non-increasing in terms of **K** and **G**.
A.3 C is convex with respect to **I** and **J**.
A.4 The value of the integral (28) is bounded for arbitrary bounded values of $\mathbf{K},\mathbf{G},Y,\mathbf{p}^\circ,\boldsymbol{\omega}$ and $\boldsymbol{\theta}$.
A.5 Problem (19) has a unique optimal trajectory for arbitrary bounded values of $\bar{\mathbf{K}},\bar{\mathbf{G}},Y,\mathbf{p}^\circ,\boldsymbol{\omega}$, and $\boldsymbol{\theta}$.
A.6 For arbitrary $\mathbf{K},\mathbf{G},Y$, there exist $\mathbf{p}^\circ,\boldsymbol{\omega}$, and $\boldsymbol{\theta}$, such that Problem (19) possesses the optimal investment demands $\mathbf{I}^*$ and $\mathbf{J}^*$ at arbitrary base period $t = \tau$.

A.1-A.3 are self-explanatory and standard. **A.4** asserts that there are well-defined factor demand functions associated with Problem (19). **A.5** asserts the uniqueness and stability of a steady state. Initial points which cannot sustain feasible paths would never be observed. The irregular initial conditions are ruled out from discussions below.

Hence, optimal value functions corresponding to Problem (19), where C satisfies **A.1-A.5** may be shown with one qualification noted below to be characterized by the following set of properties:

B.1 $V \geq 0$.

B.2 V satisfies the properties:

1) $\rho V_K - \boldsymbol{\omega} - V_{KK}\dot{K} - V_K\ddot{K} - V_{GK}\dot{G} < 0$,
2) $\rho V_G - \boldsymbol{\theta} - V_{KG}\dot{K} - V_{GG}\dot{G} - V_G\ddot{G} < 0$,
3) $V_K < 0$,
4) $V_G < 0$,
5) $\rho V_Y - V_{KY}\dot{K} - V_{GY}\dot{G} > 0$.

B.3 V is strictly concave with respect to prices of factor inputs, $\mathbf{p}^{\circ}, \boldsymbol{\omega}$, and $\boldsymbol{\theta}$.

B.4 $\mathbf{I}(=\dot{K} + \delta_K K)$, $\mathbf{J}(=\dot{G} + \delta_G G)$, and $\mathbf{X}$ must satisfy $\mathbf{I} \in \Omega^{+m}$, $\mathbf{J} \in \Omega^{+s}$, and $\mathbf{X} \in \Omega^{+n}$, respectively, where Ω^{+k} is an open set in the non-negative orthant of a k-dimensional Euclidean space.

B.5 For arbitrary $\bar{K}$,and $\bar{G}$, the capital and knowledge accumulation trajectories, $\dot{\mathbf{v}} = \Psi^{-1}(\rho\mathbf{R} - \mathbf{v})$, converge to a unique stable stationary point.

15.4.2 Investment and Variable Factor Demand Functions

Functional forms of dynamic factor demand functions may be obtained by hypothesizing a functional form for V that is consistent with properties (**B**) and by applying Equations (33) and (36). We hypothesize the following functional form for V:

$$pV = \frac{1}{2}\alpha_0 Y + \frac{1}{2}\boldsymbol{\upsilon}'\boldsymbol{\Xi}\boldsymbol{\upsilon}Y + \rho(\boldsymbol{\upsilon}'\boldsymbol{\Theta} + \boldsymbol{\iota}')\mathbf{v} + (\boldsymbol{\upsilon}'\boldsymbol{\Theta}\boldsymbol{\zeta} + \boldsymbol{\beta}'_0)\boldsymbol{\Omega} + \boldsymbol{\upsilon}'\boldsymbol{\Theta}\boldsymbol{\pi}Y. \tag{37}$$

The parameters in the R.H.S. of (37) are given by the scalars α_0 and the vectors $\boldsymbol{\upsilon} = (\boldsymbol{\mu}, \mathbf{p}, p_{n+1})'$, $\boldsymbol{\iota} = (\iota_1,\ldots,\iota_n)'$, $\boldsymbol{\beta}_0 = (\beta_{01},\ldots\beta_{0s})'$, $\boldsymbol{\pi} = (\pi_1,\ldots,\pi_n)'$. The vector $\boldsymbol{\Omega} = (\Omega_1,\ldots,\Omega_s)$ is a s-vector whose ith element, Ω_i, represents the social knowledge stocks produced by knowledge industry i. $\boldsymbol{\zeta} = \{\zeta_{ij} : i = 1,\ldots s; j = 1,\ldots,n\}$ is a $s \times n$ matrix, whose ζ_{ij} implies inter-industrial knowledge transferability of the jth type of social knowledge into the ith production division of the concerned industry. The parameter matrices $\boldsymbol{\Theta}$ and $\boldsymbol{\Xi}$ can be decomposed into submatrices:

$$\boldsymbol{\Theta} = \begin{bmatrix} \sigma_1 \\ \sigma_2 \end{bmatrix},$$

$$\boldsymbol{\Xi} = \begin{bmatrix} \xi_{11} & \xi_{12} \\ \xi_{21} & \xi_{22} \end{bmatrix},$$

where σ_1 is an $n \times n$ symmetric matrix, σ_2 is an $(n+1) \times n$ matrix, ξ_{11} is $n \times (n+1)$, ξ_{12} is $n \times (n+1)$, $\xi_{21}(=\xi_{12})$ is $(n+1) \times n$, and ξ_{22} is $(n+1) \times (n+1)$.

The form of the value function is generally quadratic in the factor prices and linear in the outputs, capital stocks, and social knowledge stocks. There are two attractive features of this functional form (Bernstein and Nadiri, 1989). First, it is self-dual in the sense of

Lau (1976), since the form of the function is preserved over the production, cost and value functions. Second, the form is consistent with aggregation across production technologies (Epstein, 1981). The corresponding investment functions implied by the specification of the value function V, can be given by

$$\dot{\mathbf{v}} = \sigma_1^{-1}[\xi_{11},\xi_{12}]\boldsymbol{\upsilon}Y + (\boldsymbol{\pi} + \boldsymbol{\zeta\Omega})Y + (\rho\mathbf{E} - \sigma_1^{-1})\mathbf{v}, \tag{38}$$

where $\mathbf{E}$ is an $n \times n$ identity matrix. The investment demand functions are linear in the predetermined variables. Notice that from Equation (38) the net investment equations are of the flexible accelerator variety (Epstein and Denny, 1983) with

$$\dot{\mathbf{v}} = -\mathbf{M}(\mathbf{v} - \mathbf{v}^*) + \boldsymbol{\zeta\Omega}Y, \tag{39}$$

where $\mathbf{M} = \sigma_1^{-1} - \rho\mathbf{E}$ is the adjustment matrix which governs the pace of the knowledge and capital accumulation processes and $\mathbf{v}^* = (\sigma_1^{-1} - \rho\mathbf{E})^{-1}\{\sigma_1^{-1}[\xi_{11},\xi_{12}]\,\boldsymbol{\upsilon} + \boldsymbol{\pi}\}Y$. The vector $\mathbf{v}^*$ denotes the long-run steady states of the capital and knowledge stocks when no knowledge spillovers among industries exist. In other words, there is no external economy in knowledge production (so that $\boldsymbol{\zeta} = \mathbf{o}$). In the case without inter-industrial knowledge spillovers, the accumulation processes of capital and knowledge are fully governed by a traditional multi-sectoral accelerator model. The term, $\boldsymbol{\zeta\Omega}Y$, in Equation (39) represents the external effects by knowledge spillovers. The existence of an external economy may shift the long-run steady states from $\mathbf{v}^*$ to $\mathbf{v}^* + \mathbf{M}^{-1}\boldsymbol{\zeta\Omega}Y$.

Let us take a discrete approximation with

$$\dot{\mathbf{v}} = \mathbf{v}(t+1) - \mathbf{v}(t), \tag{40}$$

and

$$\mathbf{v} = \mathbf{v}(t) \tag{41}$$

so that Equation (38) becomes

$$\begin{aligned}\mathbf{v}(t+1) = {} & \mathbf{b}_0 Y^e(t+1) + \boldsymbol{\Lambda}_1\boldsymbol{\upsilon}^e(t+1)Y^e(t+1) \\ & + \{(1+\rho(t))\mathbf{E} - \boldsymbol{\Lambda}_2\}\mathbf{v}(t) + \boldsymbol{\zeta}(t)\boldsymbol{\Omega}(t)Y^e(t+1),\end{aligned} \tag{42}$$

where $\mathbf{b}_0 = \boldsymbol{\pi}$, $\boldsymbol{\Lambda}_1 = \sigma_1^{-1}[\xi_{11},\xi_{12}]$, and $\boldsymbol{\Lambda}_2 = \sigma_1^{-1}$. $\boldsymbol{\upsilon}^e(t+1)$ and $Y^e(t+1)$ are firms' expectations about the factor prices and outputs to be produced in period $t+1$, respectively. Assume that firms have static expectations about these variables, i.e., $Y^e(t+1) = Y(t)$ and $\boldsymbol{\upsilon}^e(t+1) = \boldsymbol{\upsilon}(t)$. Hence, by dividing both sides of Equation (42) by $Y(t)$, we see that the capital coefficients of dynamic I-O systems, $\mathbf{b}^e(t+1)$ can be modified by

$$\mathbf{b}^e(t+1) = \mathbf{b}_0 + \boldsymbol{\Lambda}_1\boldsymbol{\upsilon}(t) + \{(1+\rho(t))\mathbf{E} - \boldsymbol{\Lambda}_2\}\mathbf{b}(t) + \boldsymbol{\zeta}(t)\boldsymbol{\Omega}(t), \tag{43}$$

where $\mathbf{b}^e(t+1) = \mathbf{v}(t-1)/Y^e(t+1)$ is the vector of planned capital coefficients which will be capitalized at the beginning of period $t+1$. Equation (43) describes how the capital coefficients are endogenously modified over time in response to changes in prices, knowledge stocks, and the quality of infrastructure.

By applying Equation (36), we can also get the variable factor demand function:

$$\mathbf{X}^\circ = \mathbf{a}_0 Y + \boldsymbol{\Gamma}_1\boldsymbol{\upsilon}Y + \boldsymbol{\Gamma}_2\mathbf{v}, \tag{44}$$

where $\mathbf{a}_0 = -\xi_{21}\sigma_1^{-1}\iota$, $\Gamma_1 = [\xi_{21},\xi_{22}] - \sigma_2\sigma_1^{-1}[\xi_{11},\xi_{12}] - \xi_{21}\sigma_1^{-1}[\sigma_1,\sigma_2]'$, $\Gamma_2 = \sigma_2\sigma_1^{-1}$. The demands for intermediate goods are determined by firms' short-term interests, given the predetermined long-term variables, **G** and **K**. Hence, Equation (44) can be interpreted as firms' short-term equilibrium conditions. The discrete approximation of Equation (44) can be given by

$$\mathbf{X}^{\circ}(t+1) = \mathbf{a}_0 Y(t+1) + \Gamma_1 \boldsymbol{\upsilon}(t+1)Y(t+1) + \Gamma_2\boldsymbol{\upsilon}(t+1). \tag{45}$$

By dividing both sides of Equation (45) by $Y(t+1)$, we have

$$\mathbf{a}^{\circ}(t+1) = \mathbf{a}_0 + \Gamma_1\boldsymbol{\upsilon}(t+1) + \Gamma_2\mathbf{b}(t+1), \tag{46}$$

where $\mathbf{a}^{\circ}(t) = (a_1(t),...,a_n(t),a_{n+1}(t))'$. In Equation (46), the vector $\mathbf{a}_0$ and the matrices Γ_1 Γ_2 are constant over time. Thus, the input-output coefficients are endogenously modified based on both the prices of the factor inputs $\boldsymbol{\upsilon}(t+1)$ and the capital coefficients $\mathbf{b}(t+1)$. The input-output coefficients are indirectly controlled by the social knowledge stocks and the quality of infrastructure through the adjustment of capital coefficients.

15.4.3 A Dynamic I-O Model

So far we have investigated the accumulation processes of physical and knowledge capitals within a single industry. In deriving the factor and variable demand functions (42) and (45), it is assumed that firms have static expectations about the prices of physical and knowledge capitals of the next period, $\boldsymbol{\upsilon}(t+1)$, which are exogenous to them. In dynamic input-output models, the equilibrium prices of factor demand inputs are to be endogenously determined given the wage rate and the interest rate. Hence, given the current input and investment coefficients $\mathbf{a}(t)$ and $\mathbf{b}(t)$ in period t, and the current relative prices of goods, $\mathbf{p}(t)$, the relative prices of goods in period $(t+1)$, $\mathbf{p}(t+1)$ can be estimated according to Equation (9). Let us now summarize our dynamic systems:

$$\left.\begin{aligned} \mathbf{p}(t+1) &= (1+\rho(t))\mathbf{p}(t)'[\mathbf{E} - (\mathbf{E} - \mathbf{A}(t))\mathbf{B}(t)^{-1}]^{-1} + \mathbf{W}(t), \\ \mathbf{a}^{\circ}_j(t+1) &= \mathbf{a}_{j0} + \Gamma_{j1}\boldsymbol{\upsilon}(t+1) + \Gamma_{j2}\mathbf{b}_j(t+1), \\ \mathbf{b}_j(t+1) &= \mathbf{b}_{j0} + \Lambda_{j1}\boldsymbol{\upsilon}(t) + \{(1+\rho(t))\mathbf{E} - \Lambda_{j2}\}\mathbf{b}_j(t), \\ &\qquad (j = 1,...,n), \end{aligned}\right\} \tag{47}$$

where $\boldsymbol{\upsilon}(t) = (\boldsymbol{\mu}(t),\mathbf{p}^{\circ}(t))'$, $\mathbf{a}_j(t+1)$ is the input-output coefficient vector of industry j ($j = 1,...,n$), in period $(t+1)$ and $\mathbf{b}_j(t+1)$ is the investment coefficient of industry j ($j = 1,...,n$) in period $(t+1)$. The dynamic system (47) describes the process of economic growth with technical change. As shown in Equation (47), if there exist no knowledge spillovers among industries, the relative prices play decisive roles in regulating the pace of technical change.

Let us now close our dynamic system (47) by specifying the modes of knowledge spillovers across industries. Let $G_{ij}(t)$ be the amount of stock of the ith type of knowledge required as capital in the jth industry in period t. Then we have $b_{ij}(t)X_j(t) = G_{ij}(t)$, ($i = m,...,m+s$, $j = 1,...,n$). Let $G_i(t) = \sum_{j=1}^{n} b_{ij}(t)X_j(t)$ be the knowledge produced by the ith industry and stored as capital goods in the economy in period t. Then, we have

$$\boldsymbol{\Omega}(t) = \mathbf{B}^{\circ}(t)\mathbf{X}(t), \tag{48}$$

where $\mathbf{B}^\circ(t) = \{b_{ij}(t); i = m + 1,...,n; j = 1;...,n\}$. Assume that knowledge capital is partially-excludable. Let the parameters $\zeta_j(t) = \{\zeta j_{ki}(t); k = 1,...,n; i = m + 1,...,n\}$ in Equation (43) be the accessibility to social knowledge stocks in the economy, of which each element represents the psychological frictions for inter-industrial knowledge diffusion. Thus, the external effects caused by knowledge spillovers in the production of the ith industry can be expressed in the following equation:

$$\zeta_j(t)\mathbf{\Omega}(t) = \zeta_j(t)\mathbf{B}^\circ(t)\mathbf{X}(t), \tag{49}$$

where $\mathbf{B}^\circ(t) = (\mathbf{b}_{m+1}(t)', ...,\mathbf{b}_n(t)')'$, where $\mathbf{b}_i(t)$ is the *i*th column vector of the investment coefficients' matrix $\mathbf{B}(t)$. The matrix of accessibility coefficients, $\zeta(t)$, may change in the long run due to institutional innovation and infrastructural improvement. The changes in these parameters are rather slow, but their impact on economic development is very decisive. The simplest expression of accessibility change might be expressed by a linear model:

$$\zeta_j^i(t + 1) = \zeta_j^i(t) + \boldsymbol{\eta}_{ij1}\mathbf{X}(t) + \boldsymbol{\eta}_{ij2}\boldsymbol{\kappa}(t), \tag{50}$$

where $\zeta_j^i(t) = [\zeta_{1i}^i(t),...,\zeta_{ni}^i(t)]'(i = m + 1,...,n)$, $\boldsymbol{\eta}_{ij1} = \{\eta^{ij}_{1kh} : k,h = 1,...,n\}$, $\boldsymbol{\eta}_{ij2} = \{\eta^{ij}_{2kg} : k = 1,...,n; g = 1,...,q\}$ are the parameter matrices, and $\boldsymbol{\kappa}(t) = (\kappa_1(t),...\kappa_q(t))'$ is the vector of governmental expenditures. Our dynamic input-output model with endogenous technical change is now fully described by adding Equations (49) and (50) to the system (47):

$$\left.\begin{aligned}
\mathbf{X}(t + 1) &= \{\mathbf{E} + \mathbf{B}^{-1}(\mathbf{E} - \mathbf{A}(t))\}\mathbf{X}(t) - \mathbf{B}^{-1}\bar{\mathbf{D}}(t),\\
\mathbf{p}(t + 1) &= (1 + \rho(t))\mathbf{p}(t)'[\mathbf{E} -(\mathbf{E} - \mathbf{A}(t))\mathbf{B}(t)^{-1}]^{-1} + \mathbf{W}(t),\\
\mathbf{a}^\circ_j(t + 1) &= \mathbf{a}_{j0} + \boldsymbol{\Gamma}_{j1}\boldsymbol{\upsilon}(t + 1) + \boldsymbol{\Gamma}_{j2}\mathbf{b}_j(t + 1),\\
\mathbf{b}_j(t + 1) &= \mathbf{b}_{j0} + \boldsymbol{\Lambda}_{j1}\boldsymbol{\upsilon}(t) + \{(1 + \rho(t))\mathbf{E} - \boldsymbol{\Lambda}_{j2}\}\mathbf{b}_j(t) + \zeta_j(t)\mathbf{\Omega}(t),\\
\zeta_j^i(t + 1) &= \zeta_j^i(t) + \boldsymbol{\eta}_{ij1}\mathbf{X}(t) + \boldsymbol{\eta}_{ij2}\boldsymbol{\kappa}(t),\\
\mathbf{\Omega}(t) &= \mathbf{B}^\circ(t)\mathbf{X}(t), \qquad (i = m + 1,...,n; j = 1,...,n),
\end{aligned}\right\} \tag{51}$$

where $\mathbf{B}^\circ(t) = (\mathbf{b}_{m+1}(t)',...,\mathbf{b}_n(t)')'$.

15.5 CONCLUSION

This chapter has proposed a computable dynamic input-output model to examine interactions between knowledge accumulation and economic development with endogenous technical change. The model explicitly introduced new sectors called knowledge sectors into the traditional dynamic input-output economic system. It has been assumed that the growth of knowledge is the main source of technical change. In order to increase knowledge we need investments just as we do capital formation. Our framework should help to unravel endogenous technical change in which long-run growth is driven primarily by the accumulation of knowledge by forward looking and profit-maximizing agents.

The current chapter focuses on technical change as improvements. In order to develop a more comprehensive dynamic input-output model with endogenous technical change, the following research is still needed: 1) analyses of interactions between quality development and price dynamics, 2) an investigation of the long-term dynamic behavior of the proposed model, 3) incorporation of a spatial dimension, i.e., the development of a multi-regional dynamic input-output model with endogenous technical change, 4) empirical estimations of dynamic input-output models with endogenous technical change. Although more development and sophistication might be expected, our model is encouraging in that it seems to capture the essential mechanism controlling the accumulation of physical and knowledge stocks, technical change and economic development.

REFERENCES

Adams, J.D., 1990, "Fundamental Stocks of Knowledge and Productivity Growth", *Journal of Political Economy* 98:673-702.

Andersson, Å.E., 1981, Structural Change and Technological Development, *Regional Science and Urban Economics*, 11:351-361.

Andersson, Å.E. and W.B. Zhang, 1988, Decision Centralization and Decentralization in a Dynamic Economic System, *Journal of Computational and Applied Mathematics*, 22:317-337.

Arrow, K.J., 1962, "Economic Welfare and the Allocation of Resources of Invention", in Nelsen, R.R., (ed.), *The Rate and Direction of Invention Activity*, Princeton University Press.

Bacharach, M., 1970, *Biproportionate Matrices and Input-Output Change*, Cambridge University Press, Cambridge.

Batten, D.F., K. Kobayashi and Å.E. Andersson, 1989, "Knowledge, Nodes and Networks: An Analytical Perspective", in Andersson, Å.E., D.F. Batten and C. Karlsson (eds.), *Knowledge and Industrial Organization*", pp. 31-46, Springer Verlag, Heidelberg.

Benveniste, L.M. and J.A. Scheinkaman, 1979, "Differentiable Value Functions in Concave Dynamic Optimization Problems, *Econometrica* 47:727-732.

Bernstein, J.I. and M.I. Nadiri, 1989, "Research and Development and Intra-industry Spillovers: An Empirical Application of Dynamic Duality", *Review of Economic Studies* 56:249-269.

Cornes, R. and T. Sandler, 1986, *The Theory of Externalities, Public Goods and Club Goods*, Cambridge University Press, Cambridge.

Craven, J., 1983, "Input-Output Analysis and Technical Change", *Econometrica* 51:3, 585-598.

Epstein, L.G., 1981, "Duality Theory and Functional Forms for Dynamic Factor Demands", *Review of Economic Studies* 48:81-95.

Epstein, L.G. and M., Denny, 1983, "The Multivariate Flexible Accelerator Model: Its Empirical Restriction and Application to U.S. Manufacturing", *Econometrica* 51:647-674.

Evenson, R.E. and Y. Kislev, 1975, *Agricultural Research and Productivity*, Yale University Press, New Haver, Conn.

Fellner, W., 1960, "Two Propositions in the Theory of Induced Innovations", *The Economic Journal* LXXI, 282:305-308.

Johansen, L., 1972, *Production Functions*, North-Holland.

Kennedy, C., 1964, "Induced Bias in Innovation and the Theory of Distribution", *The Economic Journal* LXXIV:541-547.

Kobayashi, K., D.F. Batten and Å.E. Andersson, 1991, "The Sequential Location of Knowledge Oriented Firms over Time and Space", *Papers of Regional Science* 70:4, pp. 381-397.

Lau, L., 1976, "A Characterization of the Normalized Restricted Profit Function", *Journal of Economic Theory* 12:131-163.

Leontief, W.W., 1966, *Input-Output Economics*, Oxford University Press.
McLaren, K.R. and R.J. Cooper, 1980, "Intertemporal Duality: Application to the Theory of the Firm", *Econometrica* 48:7, 1755-1762.
Miyata, Y. and E. Yamamura, 1990, "The Model Reference Adaptive System in the Dynamic Input-Output Model", *Papers of the Regional Science Association* 68:57-70.
Morishima, M., 1958, "Prices, Interest and Profits in a Dynamic Leontief System", *Econometrica* 26:358-380.
Morishima, M., 1964, *Equilibrium Stability and Growth - A Multi-Sectorial Analysis*, Oxford Clarendon Press.
Rockafeller, R.T., 1970, "Conjugate Convex Functions in Optimal Control and the Calculus of Variations", *Journal of Mathematical Analysis and Applications* 32:411-427.
Romer, P.M., 1986, "Increasing Returns and Long-Run Growth", *Journal of Political Economy* 94:5, 1002-1037.
Romer, P.M., 1990, "Engogenous Technological Change", *Journal of Political Economy* 98:71-102.
Samuelson, P.A., 1951, "Abstract of a Model Concerning Substitutability in Open Leontief Models", in Koopmans, T.C., (ed.), *Activity Analysis of Production and Allocation*, Wiley, New York.
Solow, R.M., 1956, "A Competitive Valuation in a Dynamic Input-Output System", *Econometrica* 27:65-94.
Takayama, A., 1985, *Mathematical Economics*, Cambridge University Press.
Zhang, W.-B., 1990, *Theory of Economic Development*, Springer-Verlag.

CHAPTER 16

Networks for Process Innovation by Firms: Conjectures from Observations in Three Countries

Manfred M. Fischer
Vienna University of Economics

Börje Johansson
Royal Institute of Technology, Stockholm

16.1 INTRODUCTION

This chapter attempts to assemble and systematise empirical observations of the diffusion of IT-applications to various sectors and locations in several countries. The notion IT refers to information technology, which comprises microprocessors and their electronic equivalents in the form of integrated and interlinked circuit devices.

Potential applications of IT are innumerable in manufacturing. IT can be integrated in new products, which corresponds to product innovations, and combined with production equipment leading to process innovations. The IT-based automation of manufacturing processes involves a family of technologies which are situated at the intersection of computer science and manufacturing engineering. Manufacturing automation depends in particular on the congruency between the evolution of software and hardware.

16.1.1 Background

Over the past decade empirical evidence has been collected to demonstrate that the introduction of IT-solutions is spatially differentiated within regions and nations. Many of these studies compare the speed of the introduction of IT in industrial establishments, according to (i) location, (ii) sector, and (iii) the internal resources of the unit. This chapter tries to understand these processes by means of a network paradigm, which is used to describe the contact and communication possibilities of every individual establishment. Our focus is on how ordinary economic links are used as parts of innovation networks.

The presentation is based on two individual publications: (1) Fischer and Alderman (1991), and (2) Johansson and Karlsson (1991). The first report presents an econometric assessment of information collected from studies in Britain (at the Centre for Urban and Regional Development Studies at the University of Newcastle upon Tyne) and in Austria (at the Department of Economic and Social Geography of the Vienna University of Economics and Business Administration). The book by Johansson and Karlsson evaluates a series of Swedish studies, in particular, Karlsson (1988).

16.1.2 Outline of the Chapter

In Section 16.2 the nature of information technology is described, and different types of IT-applications are classified. The notion of process innovation is defined and related to a technique concept formulated by Nelson and Winter (1982). Innovations are also distinguished with respect to firms producing standardised and differentiated products.

Section 16.3 introduces the concepts of economic and innovation networks. Both refer to intangible, capital-like links between economic actors. The key actors in the innovation networks we study are sellers and buyers of IT-applications. The analysis shows how traditional determinants of technology adoption and innovation behaviour can be modelled in this framework.

Section 16.4 presents empirical observations about Swedish, British and Austrian regions. By combining these different results, the importance of various economic networks is assessed. The chapter ends with a set of conclusion-like conjectures. The authors also argue that their presentation should be followed by new and penetrating studies of innovation networks.

16.2 INFORMATION TECHNOLOGY AND PROCESS INNOVATIONS

One of the key features of the technological wave in the manufacturing industry is the application of computer control to traditional production processes. In this section information technology and process innovations are characterised and analysed in relation to standardised and non-standardised products.

16.2.1 Characteristics of the Technology

Here, information technology is assumed to comprise hardware (computers, microprocessors, switchboards, etc.) and software (systems, programmes, etc.). IT-applications in industry are used to speed up various operations and to increase their precision. In particular, IT-systems facilitate the coordination of interdependent activities. Three types of activities or operations are focussed upon in this chapter: (i) numerically controlled computer operation of machinery, (ii) design and construction activities aimed at varying product attributes and using the computerised representation of each product, and (iii) commercial and administrative activities. Category (i) is represented by NC and CNC machines, and (ii) by CAD-technology.

The diffusion of IT-applications is often claimed to have a pervasive influence on economic systems, affecting all spheres of manufacturing operations, from design to the production and management of the firm as a whole. The diffusion of information technologies and their applications is also often considered to have far-reaching implications for the competition between, and the globalisation of, industrial enterprises - affecting all spheres of manufacturing engineering, from the control of material flows, fabrication, finishing, quality control and assembly activities, all the way to marketing management. However, a dominant feature of the IT-evolution during the 1970s and 1980s has been its cost-reducing applications in manufacturing (Davies, 1979; Hagedoorn, 1989; Rees, Briggs and Oakey, 1984; Thomas and Le Heron, 1975).

16.2.2 Process Innovations and Types of IT-Applications

The term "process innovation" is conventionally used to describe a change of production technique in a production unit or an industrial establishment. In our presentation, the term process comprises all of the operations or activities of an establishment. This corresponds to the approach suggested by Nelson and Winter (1982). They argue convincingly that an establishment (or a firm) has a set of interdependent process routines or regularities which combine to form a compound technique or process.

The adopted definition means that the technique of an establishment includes the following processes: (i) distribution, (ii) production, (iii) "routinised" design and construction and (iv) management, administration and commercial activities. Improvements in the above systems are all recognized as process innovations. By improvements we refer to reductions of costs and improvements in quality and accuracy.

Process improvements thus refer to all innovations which are not classified as product development. The latter term here signifies resources which are allocated to investments in new products. A tool-producing firm which constructs customer-specific versions for each new order may of course be said to develop new products. The criterion we rely on distinguishes between (i) "routinised" design and construction, and (ii) product innovations (Fischer, 1991; Oakey, Thwaites and Nash, 1980).

In the report by Fischer and Alderman (1991) three categories of IT-applications are considered, while the study by Johansson and Karlsson (1991) identifies four categories. Table 16.1 describes the distinctions.

Table 16.1: Categories of IT-applications

Fischer-Alderman report	Johansson-Karlsson report
(1) Automation of the production process: (a) tools, machinery (b) manufacturing systems	(1) Automation of production and distribution processes
(2) Automation and coordination of of commercial and administrative activities	(2) Automation and coordination of management and administration
(3) Design support	(3) Design and construction engineering support
	(4) Logistical systems for coordination of interdependent flows in the firm

For categories (1)-(3) in Table 16.1 we refer to the specification given in the Fischer-Alderman report. As a consequence the following IT-applications are considered:

(1a) NC and CNC machines
(1b) CAM and CAD-CAM systems
(2) Office machines and systems
(3) CAD and CAE computers

16.2.3 Standardised and Non-Standardised Products

The degree of standardisation of a product group has a major influence on the market structure and innovation behaviour of firms. One can distinguish between two levels of standardisation:

(i) A product group or, in other words, a market consists of product variants. The product group is standardised if all of the variants are equivalent to the extent that price is the only criterion for the decision-making of buyers.

(ii) A firm may supply a standardised product, which at the same time competes in a market with either (a) standardised products or (b) differentiated products.

In neither of the above cases does the individual firm adjust its deliveries and product attributes to customers who specify individual requirements or preferences with regard to attributes. In case (ii, b) the firm has selected a particular market segment, i.e., a specific customer group. The first case refers to mass markets and mature products, including staple goods. In such markets price competition prevails.

A firm which supplies non-standardised products adjusts its product attributes to each specific customer or group of customers. Every delivery may be unique. For firms of this type the capability to adapt product characteristics to specific requirements becomes an essential aspect of their everyday production activity. In these markets price is just one of several criteria which influence the decision-making of a buyer.

The degree of product standardisation is assumed to be strongly correlated with (i) industrial sector, (ii) spatial location, and (iii) the demand for different types of IT-applications (Malecki, 1980; Nasbeth and Ray, 1974). Table 16.2 summarises these hypotheses.

Table 16.2: Product standardisation and innovation characteristics

	STANDARDISED PRODUCTS	NON-STANDARDISED PRODUCTS
CHARACTER OF INDUSTRIAL SECTORS	(1) Raw material and staple goods (2) Weak demand for variation of product attributes (3) Price dominates buyers' criteria	(1) Many types of customers (2) High bid-prices for diffetiation of product attributes (3) Large innovation opportunities
LOCATION ATTRIBUTES	(1) Regions with low input costs (2) Sites with low land prices and rent levels (3) Regions with low knowledge intensity	Accessibility to (1) customers with strong purchasing power, (2) suppliers of technical solutions pertinent to product development and design, (3) a labour force with a university education, (4) rich information network within the regions
TYPES OF IT-APPLICATIONS	Production process automation - by means of , e.g., NC, CNC and CAM systems	(1) Support systems for design and construction such as CAD and CAE (2) Systems which combine design and construction with process automation

16.3 ACTORS, INTERACTORS AND NETWORKS

According to the model we apply, a firm interacts with the suppliers of its current inputs and equipment by means of economic interaction links. These are channels for material flows, communications and control. They have capital characteristics, since they are established through investments made by the individual supplier and the receiving firm. On links of this kind new product ideas, new equipment and new systems are supplied and diffused. From this perspective they constitute innovation networks. Such links have the nature of intangible capital. The formation and maintenance of such networks requires tangible networks in the form of interaction infrastructure.

16.3.1 Innovation Networks and Spatial Infrastructure

The adoption of IT-applications of the kind described in Tables 16.1 and 16.2 is the outcome of the interaction between a seller (technology deliverer) and a buyer (technology adopter). The performance of a firm in selling its products depends on (i) the characteristics of its equipment, (ii) the technological competence of its staff, and (iii) its capability to acquire information about each customer's demands for technical solutions. The latter includes information about every customer's willingness to pay for various combinations of attributes as regards the pertinent IT-applications. Moreover, such information is usually the outcome of intense communication between the supplier and the adopter, in a process in which both are learning.

The capacity of a potential adopter to buy appropriate IT-equipment depends on (i) its internal resources, including the finances, education and competence of the staff, (ii) its economic networks and the regional milieu which influence the possibilities to establish new links to suppliers of technology.

Obviously, one can identify two search processes. First, every seller by definition tries to find customers. Second, in addition, each customer may search on its own for potential deliverers (Foxall, 1987; von Hippel, 1988). In this context we wish to pose the following four questions:

(i) How can an IT-deliverer utilise existing economic networks to come into close contact with potential customers?

(ii) Which existing economic networks can a potential IT-adopter (customer) use in its search process?

(iii) Which tangible infrastructure networks are especially important as facilitators of and vehicles for the formation of links between pairs of IT-deliverers and adopters?

(iv) Is it possible to identify other features of the economic milieu of a region which affect the formation of innovation networks?

Empirical evidence suggests that both deliverers and customers use their established economic networks to find information and to make new contacts. In this process a firm contacts selected actors in its own network, and then attempts to make use of each actor's network. An establishment that belongs to a multi-establishment corporation (or multi-firm concern) will, in the same way, make use of its intra-corporation network.

The formation processes addressed in (iii)-(iv) above are affected by locational features. Such a regional influence has been emphasised by e.g. Thwaites et al. (1981, 1982), Ewers et al. (1980), Anderstig and Karlsson (1989) and Fischer and Menschik (1990). Our basic hypothesis as regards infrastructure focusses on the conditions for close contacts between people. This includes interregional as well as intraregional network properties. A second regional feature is the richness and density of all firms in a region (Johansson, 1987, 1990; Fischer, 1989). These conditions tend to weigh against peripheral regions and to provide advantages for central urban regions. As a general rule, purely locational or regional differences in relation to technology adoption are explained by the difference between each location's accessibility to active technology suppliers. It is in the spirit of this conclusion that we investigate network characteristics.

16.3.2 Determinants of IT-Adoption Behaviour

In Fischer and Alderman (1991) four categories are introduced as determinants of innovation behaviour in general. These are (i) the internal properties of the establishment, (ii) factors related to the establishment's interactions with its technical and economic environment, (iii) location properties of the regional environment, and (iv) the political and institutional framework within which the establishment has to operate.

In the subsequent presentation the fourth factor is put aside, since our three studies do not provide us with any particular insights in this area. At the same time, in the following the first three factors will be narrowed down to reflect the phenomenon of IT-adoption alone.

The first factor refers to the resources of an establishment and its particular characteristics. This includes the financial resources available to buy new equipment as well as the extent to which the establishment has the capacity to make independent decisions about its investments. This latter capacity also depends on the nature of the labour force - in particular, its competence profile and the education level of its various employment categories (Karlsson, 1988). In Mansfield (1971), the education level and experiences of the management are emphasised. Often these factors are correlated with the size of an establishment (Rothwell, 1986; Rothwell and Zegveld, 1982; Pavitt et al, 1985).

What other characteristics of an establishment are important? The nature of IT-applications that a firm will contemplate depends on its product mix and markets. A production unit with a few, standardised products is attracted by solutions which help it to automate the production process. In a firm which produces non-standardised products there is considerable demand for equipment that can facilitate design and construction so that deliveries can be adjusted to satisfy the specific requirements of each customer. Such differences between two establishments are usually reflected by differences in their market and delivery networks.

In recent decades some researchers have focussed on the interdependence between (i) the manufacturer-active paradigm (MAP), and (ii) the customer-active paradigm (CAP), as described by e.g., Foxall (1987). For the types of IT-applications considered here, there is usually a deliverer or supplier who combines various pieces of equipment into custom-made systems. This means that the supplier has a strong incentive to be active in the search for customers and pertinent technical solutions. This is the MAP case. However, the larger and the more complex an organisation is, the more likely that there will be a combination of MAP and CAP.

Suppliers of IT may use networks within a corporation in order to reach its different establishments. Another option for a supplier is to try to trace the networks of its own well-known and/or most recent customers. It is in this context that the local and regional environment exhibits its pros and cons. A regional milieu promotes the diffusion of new technical solutions by facilitating the formation and maintenance of technological delivery links and other economic couplings that indirectly can support the associated communication process (Johansson and Larsson, 1986).

Table 16.3 provides an overview of the analysis in Section 16.3. The table has four main categories: (i) the resources of individual establishments, (ii) intra-firm networks, (iii) external economic networks, and (iv) the regional infrastructure and local milieu.

The focus is on the role of economic networks as channels for information flows, communication, negotiations and transactions.

Table 16.3: Factors determining the introduction of IT-applications

RESOURCES OF THE INDIVIDUAL ESTABLISHMENT	RESOURCES WHICH INFLUENCE THE CAPACITY TO DIGEST INFORMATION AND TO LEARN: - Education level of employment categories - The staff's earlier experience with IT - The vintage of the establishment
INTRA-FIRM NETWORKS	(1) POSITION OF THE ESTABLISHMENT IN THE CORPORATION'S INTERNAL NETWORK - Single vs multiplant organisations - The establishment is a head office, divisional headquarter, regional headquarter or a branch unit (2) DISTANCE TO HEAD OFFICE AND MAJOR OWNER(S) - Head office is located in the region, in the country or outside the country - Major owner is located in the region, in the country, or outside the country
EXTERNAL NETWORKS OF THE ESTABLISHMENT	(1) ORDINARY ECONOMIC LINKS: - Delivery of equipment - Delivery of software - Consultants etc. (2) PUBLIC ECONOMIC NETWORKS AND INFORMATION SOURCES: - Conferences, seminars, courses - Fairs and exhibitions - Journals
REGIONAL INFRASTRUCTURE AND LOCAL MILIEU	AN ARENA FOR ESTABLISHING AND RENEWING INNOVATIONS AND OTHER ECONOMIC LINKS: - Interregional passenger transportation - Passenger transportation within the region - Facilities for public economic networks - Knowledge and competence profile of the region's labour force - Quality and richness of the region's information networks for local and global communication

16.4 EMPIRICAL OBSERVATIONS

In this section we assemble information from the Fischer-Alderman and Johansson-Karlsson studies. Their econometric results are arranged in such a way that we can assess the usefulness of the network model introduced in the previous section.

The basic question asked in this section is: how fast can an establishment introduce a particular application? Two approaches are used. In the first approach, units are classified into (i) non-adopters and (ii) adopters. The latter group may be further classified into early and late. In the second approach a continuous scale of introduction tempo is used. Non-adopters are then assigned a hypothetical future adoption date.

16.4.1 Results from Swedish Studies

The presentation in this section will focus on engineering industries as a whole. This group can also be decomposed into (i) electrical engineering and instruments, (ii) transport equipment, and (iii) machinery equipment. Other sectors considered are (a) iron and steel, (b) chemicals, and (c) food processing industries. We present results in response to the following questions:

- What are the important network properties?
- What are the important location attributes?
- How do standardised and non-standardised production differ?

There are three external networks which influence the rate of the introduction of IT-solutions for both (i) administration and (ii) process automation. The rate of introduction is faster when

- the establishment receives information from deliverers of IT-equipment,
- the establishment receives information from fairs and exhibitions,
- consultants contact the establishment. This factor is especially influential when the production unit does not employ any university engineers.

Next, in Table 16.4, we describe the properties of intra-corporation networks that stimulate the technological adoption of process automation (and administration automation). Pronounced stimulation is observed when the establishment is located in a peripheral region, while (a) the head office and (b) the major owner are located in another region within Sweden. The ownership condition is the stronger of the two factors. A positive effect also exists for establisments which have their head office abroad.

Table 16.4: Major factors which enhance the speed of IT-introduction in Sweden. (The engineering industry)

EXTERNAL NETWORKS	(1) Important information sources are (a) fairs and exhibitions and (b) deliverers of equipment (2) Many customers and a large export share
INTERNAL NETWORK	For establishments in peripheral regions, (1) the major owner is located outside the region but within the country (2) the head office is located outside the region - also outside the country
REGIONAL CHARAC-TERISTICS	(1) The establishment is located in a metropolitan region for complex and reasonably complex applications. (2) Supply of fairs, exhibitions and consultancy services
ESTABLISH-MENT ATTRIBUTES	(1) Old establishment vintages for complex applications (2) Number of persons with university education (and engineering qualifications)

Several of the conclusions in Table 16.4 are related to the complexity of the applications. In terms of the applications that we have selected in Table 16.2, the complexity order is increasing as follows: (1) NC and CNC, (2) CAM, CAD and CAE and (3) CAD/CAM-systems..

The demand for education and competence increases along the described sequence. A parallel sequence is the sectoral order (i) iron and steel, (ii) metal and machinery equipment, and (iii) instruments and electrical equipment. In the latter sector there is a faster rate of introduction of all types of applications.

16.4.2 IT-Applications in British Regions

The first of the surveys in Britain referred to here took place in 1981, and was followed by postal questionnaires and telephone interviews during the period 1986-1988. In general, the adoption of IT-equipment started earlier in Britain than in Austria. There is a lag of approximatively 5-8 years. Swedish regions have a larger share of earlier adopters of NC, CNC as well as CAD and CAE than Austria. At the same time, on the average British regions seem to be ahead of comparable Swedish regions. Factors which influence the speed of IT-introduction in Britain are presented in Table 16.5.

The peripheral Northern region in Britain has similar adoption levels to those of the metropolitan core, for various applications. Moreover, the industrial heartland of iron-based areas together with the metropolitan hinterland appear to have the highest degree of IT-adoption. In addition, Alderman et al. (1988) have demonstrated that for process-automation technologies (NC, CNC, etc.), inter-industry diffusion rates vary more than interregional ones. In particular, CAD-adoption started earlier and has been much faster in the electro-technical sector than elsewhere. A similarly strong tendency exists in Sweden for electrical and telecommunications products. In Britain it is establishments that belong to multi-plant organisations which are most likely to adopt all of the categories of IT-equipment considered. This observation supports the assumption that intra-corporation networks are important channels for the diffusion of technical solutions.

As regards CAD and CAE-applications, the British observations support the hypothesis that product differentiation and R&D-activities are correlated with the introduction of design equipment. As in the Swedish studies, the size of the R&D staff seems to be more relevant than its share of total employment.

16.4.3 IT-Applications in Austrian Regions

The process of IT-introduction displays much more pronounced regional discrepancies in Austria than in Britain. In general, this form of regional influence has the same strength in Austria and Sweden. In particular, the Austrian periphery seems to be lagging behind with regard to NC and CNC-adoption. Table 16.6 provides an overview of factors which influence the speed of IT-introduction in Austria.

The Austrian data set exhibits a distinct difference vis-a-vis that for Britain and Sweden. In Austria, the propensity to introduce new IT-equipment is higher among the small and independent firms. In particular, we can note that the branch plant group performs poorly. In Austria, as in Sweden, the electrotechnical sector reveals an outstanding IT-intensity as regards automation, compared to other industries.

IT-applications for administration and management are strongly size-dependent in Austria. For these applications the metropolitan locations have significantly higher levels of adoption than other regional environments.

Table 16.5: Factors which influence the level and speed of IT-introduction in Britain

EXTERNAL NETWORKS	(1) The most important information sources are demonstrations and visits by deliverers of equipment. (2) Differentiated products (multiple customer links) stimulate CAD and CAE-introduction. (3) Standardised products stimulate process automation applications.
INTERNAL NETWORKS	(1) Multi-plant organisations have higher adoption rates than single-plant firms. (2) Introduction is faster in headquarters than in branch plants.
REGIONAL CHARAC-TERISTICS	The metropolitan hinterland is the location with highest introduction frequency. This correlation is somewhat stronger for CAD-applications.
ESTABLISH-MENT CHAR-ACTERISTICS	(1) Introduction of equipment for administration is correlated with young establishments. (2) CAD-applications as well as equipment for process automation are instead associated with older establishment vintages. (3) All types of applications are introduced more rapidly in large than in small establishments.

With regard to CAD-applications, the metropolitan hinterland is the most favourable location. In Britain the peripheral areas have the lowest probability of introduction, while in Austria it is the traditional iron-based region.

16.5 CONCLUSIONS AND CONJECTURES

This chapter deliberates on studies about the diffusion of IT-applications in three countries, with a particular focus on the differences between peripheral and metropolitan regions. Two aspects are highlighted: (i) the degree of product standardisation, and (ii) intangible economic/communication networks and tangible infrastructure networks determining the regional environment.

As regards the degree of standardisation/differentiation of products and market networks, the results are conclusive:

(i) Establishments with standardised products have a clear bias towards NC and CNC-systems, and away from CAD and CAE-systems.

(ii) Establishments operating in markets with differentiated products and numerous customer groups have a larger frequency of design-oriented IT-applications.

Table 16.6: Factors which influence the level and speed of IT-introduction in Austria

EXTERNAL NETWORKS	(1) The most important sources of information are journals and exhibitions. (2) Product differentiation influences CAD-adoption positively, while product standardisation is associated with process automation techniques.
INTERNAL NETWORKS	Multi-plant organisations perform poorly in the sense that branch plants are slow introducers of equipment for automation of manufacturing and administration processes.
REGIONAL CHARAC-TERISTICS	(1) The levels of CNC-adoption are low in peripheral regions and high in the traditional iron-based region. (2) The metropolitan hinterland provides a favourable environment for the introduction of both process and administration automation. To some extent the same is true also for CAD-applications.
ESTABLISH-MENT CHARAC-TERISTICS	(1) Establishment vintage has little significance, with a small positive bias for old vintages. (2) Size has no positive effect on CAD-applications, a small positive influence on CNC-applications, and a positive effect on the introduction of equipment for management and administration.

16.5.1 Comparing the Influence of Network Properties

Economic networks and infrastructure networks for transportation and communication are complementary. The better the conditions for close person contacts are, the higher the probability that lasting links will be established between economic actors. However, if the regional infrastructure properties are weak, then it becomes indeed important for a firm to maintain and strengthen its established economic network, and to make the best possible use of it.

The regional environment determines the likelihood that innovation networks shall grow rich and become intensively utilised. Fundamental environment-related features are the number, size and diversity of located suppliers of new technical solutions in a region. Here we can observe a basic interdependency in the form of a cumulative equilibrium condition - the regional supply of IT-deliverers grows large in response to the expectation that numerous, interesting customers will be found in the region.

The regional environment is linked with the external networks of a firm. To a certain extent external and internal networks are substitutes. In a poor environment an establishment may still receive information, stimuli and services via the internal networks of the multi-plant corporation to which it belongs.

The analysis has revealed a set of network conditions which influence IT-adoption in the same way in both Britain and Sweden:

- o The most important sources of information in Britain and Sweden are visits and demonstrations by deliverers of equipment. In Austria the decisive information comes from journals and exhibitions.
- o Equipment for the automation of production and administration processes diffuses faster to units belonging to multi-establishment organisations in Britain and Sweden.
- o The more complex an IT-application is, the more the speed of technology diffusion /adoption tends to be dependent on a metropolitan environment. Again, this conclusion can only be drawn from British and Swedish observations.

In all three countries we can observe that peripheral regions provide an environment in which the adoption of IT-applications is delayed compared to other locations.

From the Swedish studies we can note two further observations which emphasise the importance of a firm's network properties. The first observation refers to IT-systems designed for logistical control of a firm, where information, financial and resource flows inside and outside the firm are coordinated according to systems performance criteria. In order to introduce systems of this kind, firms need supporting networks to a much larger degree than to introduce equipment for process automation (Johansson, 1987; Johansson and Karlsson, 1991).

The second observation refers to firms using IT-equipment as components in their development of new products. For this kind of product development the location in a metropolitan region seems to be an important precondition. The reason for this is that such firms have definite network requirements. They frequently have several R&D-links to university departments and researchers, as well as to other R&D active organisations and firms (Karlsson, 1988).

16.5.2 Why Networks?

Our emphasis on networks has a strategic motivation. The concept of economic (and innovation) networks refers to structure in two ways. First, a network consists of durable capital-like (intangible) links between economic subjects. Capital and durability represent rigidities, which are "structural phenomena". Second, an economic network refers to a pattern of links between economic actors. By relating adoption and diffusion to structural phenomena of this type, it becomes possible to formulate models and hypotheses which refer to slowly changing structures. This brings meaning to the concept of science in the term "social science". The described strategy has great prospects for regional science, since in regional analyses a main focus is on the interdependence between tangible infrastructure networks, and intangible interaction networks.

REFERENCES

Alderman, N., S. Davies and A.T. Thwaites, 1988, "Patterns of Innovation Diffusion", Technical Report, CURDS, University of Newcastle upon Tyne.

Anderstig, C. and C.Karlsson, 1989, "Spatial Diffusion of Information Technology in Sweden", in Andersson, Å., D. Batten and C. Karlsson, (eds.), *Knowledge and Industrial Organization,* Springer-Verlag, Berlin, 157-176.

Davies, S., 1979, *The Diffusion of Process Innovations,* Cambridge University Press, Cambridge.

Ewers, H.J., R.W. Wettmann, H. Krist, J. Kleine, and F-J. Bade, 1980, *Innovations-orientierte Regionalpolitik,* Schriftenreihe "Raumordnung" des Bundesministers für Raumordnung. Bauwesen und Städtebau, Nr. 06042, Bonn-Bad Godesberg.

Fischer, M.M., 1989, "Innovation, Diffusion and Regions", in Andersson, Å., D. Batten, and C. Karlsson, (eds.), *Knowledge and Industrial Organization*, Springer-Verlag, Berlin, 47-62.

Fischer, M.M., 1991, "Technological Change and Innovative Behaviour in Industry: A Conceptual and Methodological Framework", WSG-Discussion Paper 10, Department of Economic and Social Geography, Vienna University of Economics and Business Administration.

Fischer, M.M. and N. Alderman, 1991, "Process Innovations and Innovation Behaviour at the Firm Level: Some Empirical Findings from a Cross-National Comparison", Paper presented at the workshop "Europe and Innovation Networks, in Vienna March 1991.

Fischer, M.M. and G. Menschik, 1990, *Innovationsaktivitäten in der Österreichischen Industri. Eine Empirische Untersuchung des Betrieblichen Innovationsverhaltens in ausgewählten Branchen und Raumtypen,* Hirt Verlag, Vienna.

Foxall, G.R., 1987, "Strategic Implications of User-Initiated Innovation", in R. Rothwell and J. Bessant, (eds), *Innovation: Adaptation and Growth,* Elsevier, Amsterdam, 25-36.

Hagedoorn, J., 1989, *The Dynamic Analysis of Innovation and Diffusion: A Study in Process Control,* Pinter, London and New York.

Hippel, von, E., 1988, *The Sources of Innovation*, Oxford University Press, Oxford.

Johansson, B., 1987, "Information Technology and the Viability of Spatial Networks", *Papers of the Regional Science Association* 61:51-64.

Johansson, B., 1990, "Innovation Process in the Urban Network of Export and Import Nodes: a Swedish Example", in Nijkamp, P., (ed.), *Sustainability of Urban Systems*, Avebury, Aldershot, 219-246.

Johansson, B. and C. Karlsson, 1991, *Från Brukssamhällets Exportnät till Kunskapssamhällets Innovationsnät*, Länsstyrelsen i Värmland, Karlstad.

Johansson, B. and J. Larsson, 1986, "Characteristics of the Firm and the Speed of Technology Adoption", *Ricerche Economiche*, 4:675-695.

Karlsson, C., 1988, Innovation Adoption and the Product Life Cycle, Umeå Economic Studies No. 185, University of Umeå, Sweden.

Malecki, E.J.,1980, "Corporate Organisation of R&D and the Location of Technological Activities", *Regional Studies* 14:219-234.

Mansfield, E., et al., 1971, *Research and Innovation in the Modern Corporation*, Macmillan, London.

Nasbeth, L. and F.D. Ray, 1974, *The Diffusion of New Industrial Processes*. Cambridge University Press, Cambridge.

Nelson, R.R. and S.G. Winter, 1982, *An Evolutionary Theory of Economic Change,* Belknap, Harvard University Press, Cambridge, Mass.

Oakey, R.P., A.T. Thwaites and P.A. Nash, 1980, "The Regional Distribution of Innovative Manufacturing Establishments in Britain", *Regional Studies* 14:235-253.

Pavitt, K., M. Robson, J. and Townsend, 1985, The Size Distribution of Innovating Firms in the UK: 1945-1983, SPRU paper.

Rees, J., R. Briggs R.P. and Oakey, 1984, "The Adoption of New Technology in the American Machinery Industry", *Regional Studies* 18:489-504.

Rothwell, R., 1986, "The Role of Small Firms in the Emergence of New Technologies" in Freeman, C., (ed.), *Design, Innovation and Long Cycles in Economic Development*, Frances Pinter, London, 231-248.

Rothwell, R. and W. Zegveld, 1982, *Innovation and the Small and Medium Sized Firms: Their Role in Employment and in Economic Change*, Frances Pinter, London.

Thomas, M.D. and R.B. Le Heron, 1975, "Perspectives on Technological Change and the Process of Diffusion in the Manufacturing Sector", *Economic Geography* 51:231-251.

Thwaites, A.T., A. Edwards and D.C. Gibbs, 1982, "Interregional Diffusion of Production Innovations in Great Britain", Final report to the Department of Trade and Industry and the Commission of the European Communities. *CURDS*, University of Newcastle upon Tyne.

Thwaites, A.T., R.P. Oakey and P.A. Nash, 1981, "Industrial Innovation and Regional Development", Final report to the Department of the Environment, *CURDS*, University of Newcastle upon Tyne.

CHAPTER 17

New Technology and Regional Development in the European Snowbelt Towards a New Emerging Network?

Eddy Blaas
Free University, Amsterdam

Peter Nijkamp
Free University, Amsterdam

17.1 PROLOGUE

Economic dynamics has always been accompanied by spatial dynamics: economic heartlands come and go. The economic history of mankind shows many fluctuations in the functioning of major economic power blocks. For example, the dominant position of Western Europe is in fact fairly recent (see Olson, 1982). But within Western Europe remarkable shifts can also be observed during recent centuries, for instance, from Italy to the Iberian Peninsula and then to North-West Europe. Major changes have taken place in Europe, especially since the Industrial Revolution in which – besides geographical shifts of economic core areas – drastic (inter)sectoral shifts have also been observed (cf. Suarez-Villa, 1989). In recent years there has been an increasing recognition of the rise in economic importance of the regions around the Alps: the 'European Snowbelt'.. This snowbelt comprises the following regions: Baden-Württemberg, Lombardia, Rhône-Alpes, Provence-Alpes Côte d'Azur, and Switzerland and Austria. The rapid growth of the European Snowbelt is not only reflected in the rise in service employment in these areas, but also in the emergence of innovative new technology firms. This development seems to confirm the popular view that industries of tomorrow are not being established in the industrial regions of the past.

Our chapter aims to analyze whether technological innovations form the background of the economic development of the European Snowbelt and whether this area offers favourable incubator conditions for new technology in the European network system. After a concise survey of some important general issues in the field of space, development and technology, the various regions in the European Snowbelt will briefly be reviewed, in order to test the above assumptions on the competitive position of this area.

17.2 INNOVATION AND NEW TECHNOLOGY

Innovation and new technology have been focal points of economic research in the past decade (see for a survey Davelaar, 1990). Neo-Schumpeterian paradigms have played a major role in the economic analysis of new technological regimes (Freeman, 1986; Kleinknecht, 1987). A stage of economic upswing is usually induced by the simultaneous development (and successful application) of several major technological breakthroughs in a limited number of sectors, which – through process and product innovations – lead to a rapid rise in productivity, efficiency and the market coverage of firms. Such key sectors penetrate the whole economy, so that the intensity of diffusion of new technological findings is an especially critical variable (see Ewers and Wettmann, 1980 and Keeble and Wever, 1986).

The 'new frontier' sectors are not randomly distributed over all nations or regions of an economic system, but are clustered in space and time. They are in particular concentrated in those regions which offer favourable seedbed conditions for new and successful initiatives (see Cappellin and Nijkamp, 1990). Regions with an open attitude towards new developments, with an economic multifunctionality, with favourable communication networks and with sufficient flexibility in politics and organizational networks tend to become the winners in this competitive game (see Ewers and Gornig, 1990). In general, advanced regions appear to have better access to all such incubator conditions, and as a consequence regional dynamics and economic dynamics are often two sides of the same coin (cf. Hall, 1985).

Clearly, there may be significant variation in the development of a regional system compared to that of the nation as a whole, as the seedbed conditions for new technologies show many differences. Consequently, different countries may exhibit large differences in regional transformation processes. Some evidence on this observation can also be found in a recent cross – comparative study on the evolution of urban agglomerations in various countries (see Nijkamp, 1990).

Thus, Schumpeterian waves of economic restructuring appear to discriminate among various regions or cities (Galbraith, 1985; Porter, 1990). In the past decade especially the information and communication sector has often been regarded as the key sector in the so-called fifth Kondratieff wave (see Hepworth, 1989). This sector comprises inter alia computers, electronic capital goods, telecommunications equipment, optical fibres, robotics, ceramics, data banks, information services, microelectronics and bio-technology. The knowledge and information components appear to be extremely important in this new technology sector, which has led some authors to the conviction that so-called 3C-regions (regions with creativity, competence and communication) are the most promising areas for spatial - economic dynamics (cf. Andersson, 1985).

The new technology sector does not only use advanced equipment: it is also geared toward tailor-made production or service delivery and is at the same time flexible with respect to new market requirements (cf. Malecki, 1985). Furthermore, the competition in the new technology sector is often strong.

It is noteworthy that the precise demarcation of the new technology sector is usually fraught with many difficulties (Malecki, 1987). The following indicators are often used in empirical research: intensity of R&D, share of highly skilled labour force, extent of risk investments, share in public financing for research, degree of patents and licenses obtained, speed of outdating of new products and production processes etc.

Very often the degree of innovation of a sector (or a region) is measured by means of employment growth, although in various cases R&D expenditures, patents, productivity (or profit) rises or rises in market shares are also used. Nevertheless, a sharp definition of the new technology sector appears hard to obtain (see Pottier, 1985), although there is a common opinion that in most cases information technology, biotechnology and medical technology, new materials technology, and energy technology belong to the new technology frontier. Therefore, in our chapter we will follow this common notion on the coverage of the new technology regime.

17.3 INNOVATION AND SPACE

The creation of a 'new technology' niche in a region is often regarded as a guarantee for regional revitalization. However, the regional innovation potential is a multi-faceted phenomenon which shows much variation, as is also witnessed by Silicon Valley, the Greater Boston area, the London-Bristol corridor, the Dutch Randstad, or the greater

Barcelona area. Thus it is not possible to estimate a priori a survival rate or a success rate of innovative regions. Furthermore, such areas are very sensitive to economic business cycles, especially as far as the routine (often mass) production that is related to basic innovations is concerned. 'Farming out' of standardized products associated with innovative activities has become common practice.

In this context, Johansson (1987) makes a distinction between knowledge-intensive, service-oriented and production activities. The innovative potential of a region is mainly determined by knowledge intensity, to a lesser degree by service delivery and to a minor degree by executive production and goods handling (cf. also Weiss, 1985).

Empirical evidence has shown that in a national system of regions specific areas with an apparently high economic potential continue to exist and that such regions are often also able to generate spin-offs to other areas (Markusen et al., 1986). Traditional location theories fail to provide a satisfactory explanation for spatial dynamics related to technological innovation. Not only has the (mainly footloose) service sector become much more important, but qualitative aspects of the labour market, new forms of industrial organization and complex network configurations in which many firms operate have also exerted a far reaching impact on entrepreneurial decision making.

The innovative potential of a region is thus a complicated inter-play of the regional production environment and the regional production structure (cf. Coffey and Polèse, 1984). In almost all cases, a minimum set of critical success factors appears to play a decisive role for innovative potential. These factors are: an efficient physical network structure and communication infrastructure (Markusen, 1985; Oakey et al., 1985); a good quality of life and a high quality amenity environment (Hall, 1985; Molle, 1985; Rees and Stafford, 1986); an academic research climate (Breheny et al., 1985; Markusen, 1985; Saxenian, 1985); the presence of an urban agglomeration (Davelaar, 1990; Nijkamp et al., 1990; Storper and Scott, 1989; Thomas, 1986; Whetten, 1987). For some years there has been a growing interest in the role of network structures in the explanation of spatial dynamics related to technological innovation. In the next section we will discuss some aspects related to this issue.

17.4 NETWORK STRUCTURES AND TECHNOLOGICAL INNOVATION

For the development of innovative new technology firms the presence of network structures also functions as a seedbed condition. Technological development (or innovation) is nowadays increasingly regarded as a collective phenomenon of interaction between a multitude of actors. Innovations become the product of a network of actors (Lecoq, 1990). In this context, we can distinguish between physical networks and social-economic networks. The first refers to infrastructure and telecommunication networks and the latter to networks based on interrelations between different actors for different purposes (for example, because of the interchange of knowledge, goods and/or information; see Ewers and Wettmann, 1980). Of course these two kinds of networks cannot be regarded as totally separate, because they are interdependent configurations of economic actors.

As for any kind of economic activity, it is also important for innovative technological activities to have access to high quality infrastructure and telecommunication networks. These networks provide contact and communication possibilities between the various actors and are an important factor in their competitive position (Kamann and Boekema, 1989). In the near future all networks will play a critical role in the interchange of information, because the increasing importance of the service sector and the knowledge component in industrial products requires integrated network configurations. The latter is a specific characteristic for innovative technological activities: products in this field have a very high knowledge component. This knowledge component concerns not only scientific knowledge but also informal knowledge concerning product technology, materials used, successful product/market combinations etc. It goes without saying that learning-by-doing and learning-by-using obtained by means of relationships with suppliers, consumers and even competitors are also important in this respect.

Differences in relationships between economic actors in social-economic networks are an important factor in explaining the differences in the development of regions concerning specific sectors (Dunford, 1989). The actors not only pertain to various entrepreneurs in the production domain, but also to actors in an institutional setting (government institutions, unions etc.) and supporting services (financial institutions, marketing offices etc.). Specifically, for innovative high technological activities it is important to have (more or less formal) relationships that lead to technological transformation and the diffusion of technological know-how. Knowledge networks play a crucial role in the success of technological innovation (Gelsing, 1989). Therefore, it is evident that high technological firms have to be located in an infrastructure network of a high quality.

Up-to-date information – by using network relations – is also important for the sectors active in the high-tech area because of drastic fast changes in the market and the demand for customized products. In practice, this leads to network formations like combined research projects, combined knowledge enlistment etc. Innovative and technological activities can no longer only be explained by technology-push or demand-pull theories. These kinds of activities are increasingly the result of an active cooperation between various actors in the high-tech sector. The costly research and development activities are in many cases only possible when the different actors make use of synergetic effects. Besides, nowadays a technological innovation is usually a result of different small innovations.

Each actor in a social-economic network has his own set of other – more or less relevant – actors. The more important actors he has in his enviroment the greater the chance that he will be successful in his activities (Kamann, 1988). When one is dealing with innovative high technological activities – with a high knowledge component – the sensitivity of distance to other actors is high. Therefore, for these kinds of activities it is important to be settled within the nodal point of a specific network. As a rule urban and regional centres function as nodal points for both the physical and social-economic networks in a hierarchical structure. High-tech activities may be expected to be located in these nodal points of a spatial network (Nijkamp et al., 1990). However, for some years it has seemed that different urban centres can perform at the same level. This might lead to a multi-polar structure of urban and regional centres. Also it seems that the development of a network for new activities (for example, innovative activities) is much easier in an environment not dominated by an old structural network (Kamann, 1989). Besides, in Europe, traditional patterns of competition – within national borders – are increasingly being replaced by vigorous competition on a multi-national scale, since traditional boundaries are disappearing. Regions of different countries are becoming part of an economic network. These developments could lead to a tendency in which large metropolises lose part of their innovative potential in favour of medium-sized cities (Malecki, 1983). The network economies in the French regions of Provence-Alpes-Côte-d'Azur and Languedoc Rousillon, based on innovative small and medium-sized companies maintaining linkages among themselves and with large enterprises, could function as an example. Here various forms of expertise in collaborative networks transcend the older types of industrial strategies based on internal concentration (Hansen, 1990). Besides these French regions the "Third Italy" is an example of a territorial network of small businesses maintaining more or less formal relations (Lecoq, 1990).

Having now presented in a nutshell various considerations regarding the economic geographical driving forces of regional innovative behaviour, in the next section we will provide some empirical material on the development of the European Snowbelt as a frame of reference for judging the validity of the above considerations.

17.5 THE EUROPEAN SNOWBELT: SOME FACTS

The geographic delineation of the European Snowbelt and its constituent regions is given in Figure 17.1. This subdivision is based on the so-called NUTS classification used by the Eurostat Bureau of the European Community.

Country	Region	
Germany	Baden-Württemberg	Karlsruhe
		Stuttgart
		Freiburg
		Tübingen
	Bayern	Unterfranken
		Oberfranken
		Mittelfranken
		Oberpfalz
		Schwaben
		Niederbayern
		Oberbayern
Italy	Nord-Ovest	Piemonte
		Valle d'Aosta
		Liguria
	Lombardia	Trentino-Alto Adige
	North East	Veneto
		Friuli-Venezia Giulia
France		Alsace
		Franche-Comté
		Rhône-Alpes
		Provence-Alpes-Côte d'Azur
Switzerland	Total	
Austria	Total	

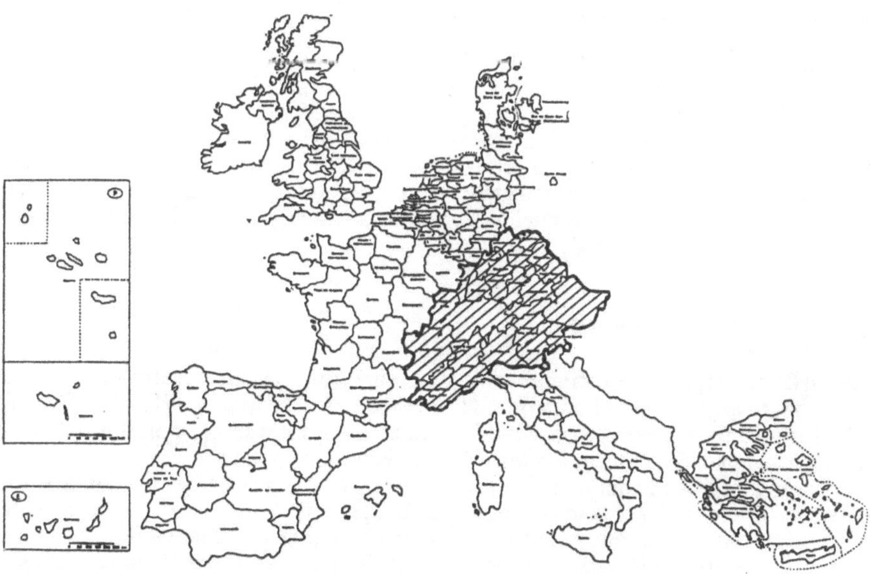

Figure 17.1 The European Snowbelt and its constituent regions.

Due to a lack of statistical data from Eurostat Austria and Switzerland could not be further subdivided. In this section some general regional indicators will be presented, viz: population, gross national and regional product and employment.

17.5.1 Population

Figure 17.2 presents the demographic developments in the regions under consideration.

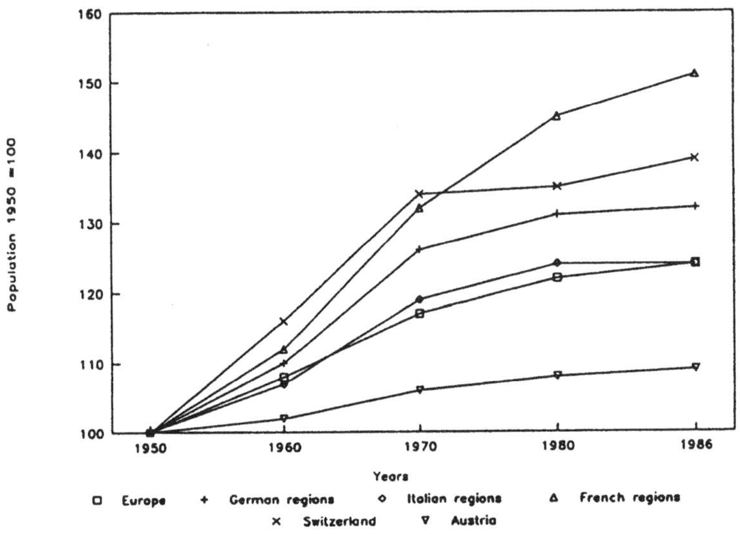

Figure 17.2 Population growth for the period 1950-1986

It appears that all regions in the European Snowbelt – with the exception of Austria – have exhibited a relatively strong population growth compared to Europe as a whole. The four French regions in particular, have grown quickly, followed by Switzerland and the German regions. A more detailed investigation of the available data leads to the conclusion that in France the Rhône-Alpes and Provence-Alpes-Côte d'Azur regions and in Germany the Baden-Württemberg region have the highest growth rates. The Italian regions run more or less parallel to Europe as a whole, albeit that Lombardia has the highest regional growth in Italy.

17.5.2 Gross National Product

Data on GNP per region are difficult to obtain. Therefore, in our description of data results we have only presented data on GNP for the years 1950 and 1985 (see Figure 17.3). Unfortunately a comparison with Europe as a whole was not possible, so that only the regional data for France, Germany and Italy are given here.

Both German regions under consideration appear to have a better performance than Germany as a whole; this holds in particular for Baden-Württemberg, while the growth of Bayern (Bavaria) is apparently characterized by a 'catching-up' manoeuvre.

The Italian regions showed a remarkable achievement, especially Nord-Ovest, although Nord-Est appeared to have a higher growth rate. Both regions now exhibit a converging process at a higher growth rate than Italy as a whole.

The four French regions show a pattern which is more or less equal to the French average with the exception of the relatively better performance of Alsace.

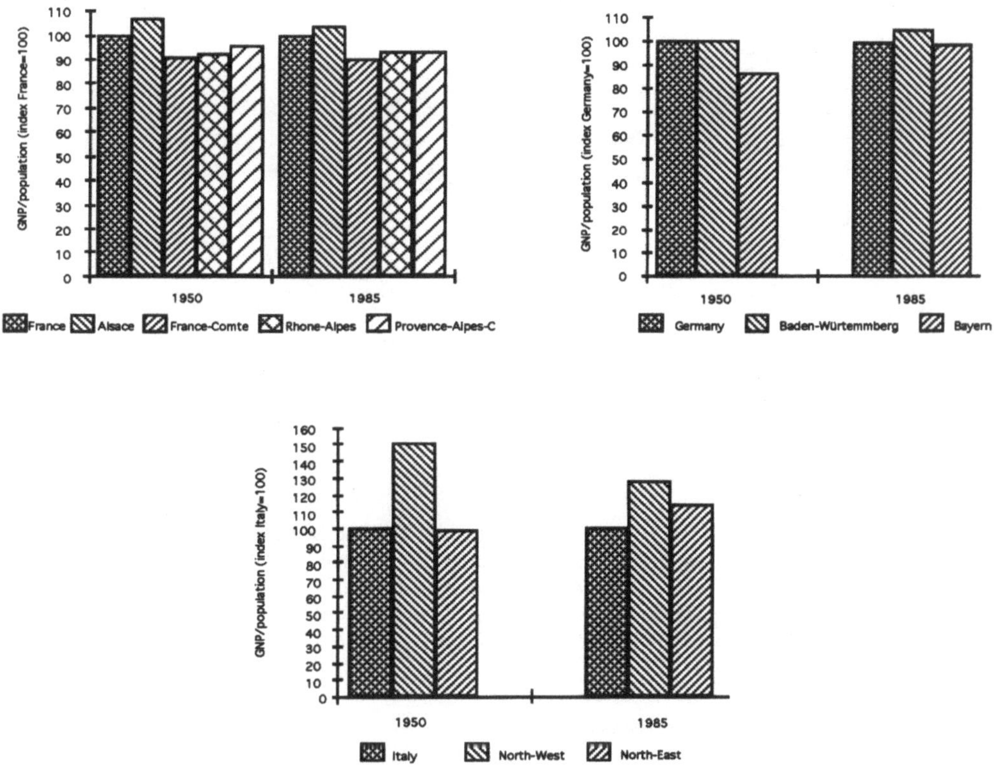

Figure 17.3 GNP per capita for the years 1950 and 1985; index GNP for each country = 100

17.5.3 Employment in Three Main Sectors

The employment data are subdivided according to three main sectors: agriculture, industry and services. The data for the period 1970-1984 are given in Figures 17.4-17.6, respectively.

The agricultural sector appears to follow largely the European trend, with the clean exception of Switzerland, where agriculture is much less a declining sector (which is related to the strong Swiss food industry).

The industrial sector in almost all of these regions is also marked by fairly uniform development processes, i.e., a decline in industrial employment. In this framework Austria offers a clear exception, which may be due to the strong public support for Austrian industry, a situation which is gradually changing and which will also make Austria follow the European trend. It is worth mentioning, however, that thus far all regions considered still have an industrial employment level which is above the European average.

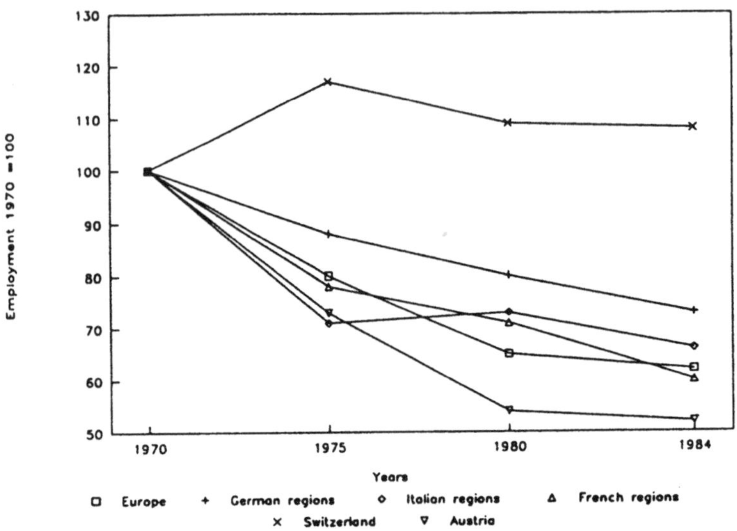

Figure 17.4 Employment development in the agricultural sector for the period 1970-1984

The service sector has become the leading sector in Europe (measured by the number of jobs): Europe as a whole has shown a remarkable growth in the employment in the service sector from 1970 onward. All regions concerned show an employment growth that is above the European average (with Austria as a latecomer). The Italian regions in particular demonstrate a high performance in this respect.

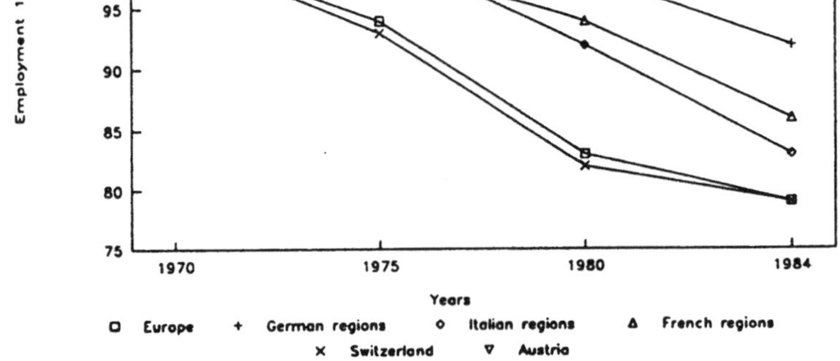

Figure 17.5 Employment development in the industrial sector for the period 1970-1984

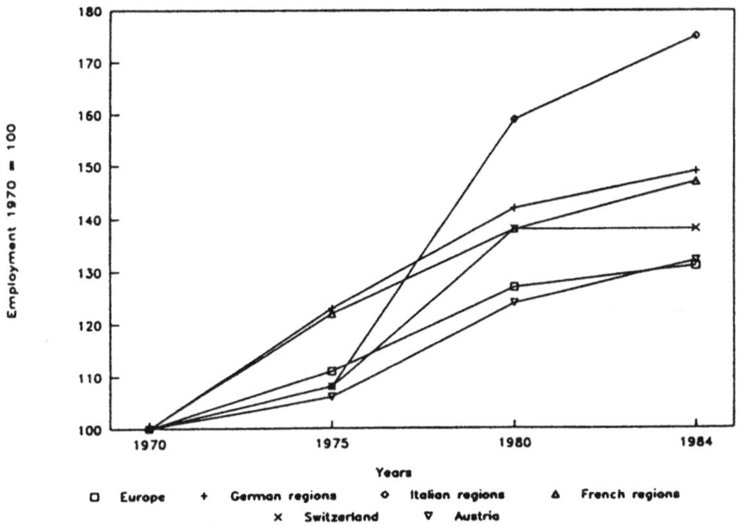

Figure 17.6 Employment development in the service sector for the period 1970-1984

The overall qualitative results for the three indicators used and for all regions under consideration are presented in Table 17.1.

The qualitative information given in Table 17.1 forms an indication of the economic potential of the regions. The employment in all regions is here related to the service sector, as this sector is usually regarded as the new leading sector in Europe (in subsequent sections the position of the new technology will be considered).

The results from Table 17.1 are interesting. For the past decades we may draw the conclusion that only the Baden-Württemberg region has a positive score in all respects. However, the Italian regions show a strong growth in the service sector. Finally, in general the French regions considered in our analysis score better than their own country and run more or less parallel to Europe. Switzerland and Austria can only be compared vis-à-vis Europe; it turns out that Switzerland is performing fairly well.

Having now presented an overall picture of the socio-economic development of the European Snowbelt, the question arises as to what the position and contribution of the new technology sector have been. Unfortunately the data on this sector are incomplete, often inconsistent and not comparable at a European scale. Since Eurostat was unable to provide a comprehensive data set, primary and secondary data sources had to be used. Much support was provided by the various Chambers of Commerce in the regions concerned, who were very helpful in collecting and providing the data requested (for more details see Blaas, 1990). Nevertheless, the current data situation precludes a systematic cross-comparative analysis of new technology developments in all regions concerned. In this respect the present study is exploratory in nature. Nevertheless, in various cases interesting data could be gathered and they have been used in subsequent sections to give a more detailed account of new technology developments in these regions and in the European Snowbelt as a whole.

In order to arrive at comparable data, the standard sectoral classification of the European Community (NACE) is used. This classification is also employed by Eurostat. It is a four-digit subdivision, but at the regional level only data on a two-digit classification are available. This leads of course to a less refined analysis of regional

Table 17.1 Development of the various indicators

	With regard to Europe		With regard to G, F or I		
	pop.	ess.	pop.	GNP	ess.
Germany	o	++			
Baden-Württemberg	+++	+++	+++	+	++
Bayern	-	++	o	++	o
Italy	o	++++			
Nord-Ovest	o	+++++	o	---	+
Lombardia	++	+++++	++		++
Nord-est	--	++++	--	++	o
France	+	+			
Alsace	++	+++	o	o	+++
Franche-Comté	+	+++	o	o	+++
Rhône-Alpes	+++	+++	++	o	++
Prov. Alpes-Côte d.A.	++++		++++	o	o
Switzerland	++	+			
Austria	--	o			

GNP = Gross national product per capita, ess = Employment in the service sector, pop = Population, G = Germany, F = France, I = Italy

sectoral developments. Given this background, the following 6 new technology sectors have been distinguished:

a. nuclear fuels industry
b. chemical industry; man-made fibres industry
c. manufacture of office machinery and data processing machinery
d. electronics - electronical engineering
e. manufacture of other means of transport
f. instrument engineering

17.6 NEW TECHNOLOGY IN THE FRENCH SNOWBELT

The French Snowbelt comprises the regions of Alsace, Franch-Comté, Rhône-Alpes and Provence-Alpes-Côte d'Azur. In the presentation of the above mentioned three indicators it turned out that these regions scored above both the French and the European average. Hence it is relevant to seek an explanation based on the new technology paradigm discussed in Sections 17.1-17.3.

For French Snowbelt regions the location quotient based on the regional employment share of each of the new technology sectors vis-à-vis the nation as a whole was calculated. The results are presented in Figure 17.7.

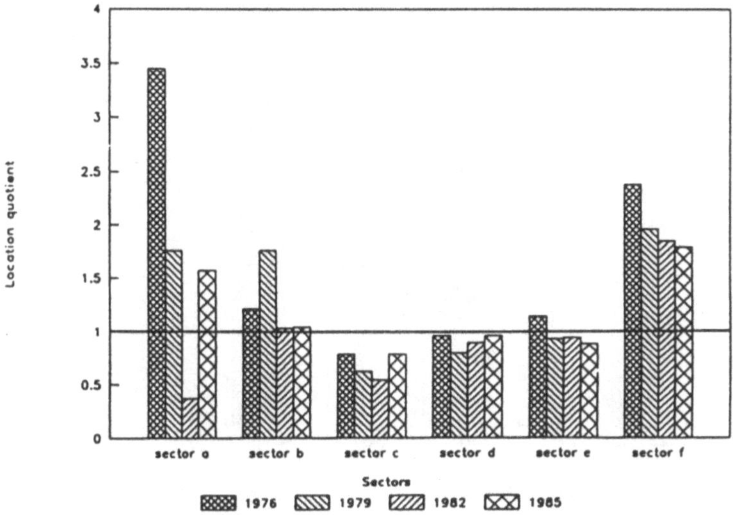

Figure 17.7 Location quotients for the French Snowbelt

It turns out that during the period 1976-1985 sectors a, b and f were intensely concentrated in the French Snowbelt, while sectors d and e were more or less equal to the French average. Only sector c (computers and information technology) is under-represented.

A further analysis of Figure 17.7 brings to light that the above sectors have not grown disproportionately; on the contrary, the location quotients tend to decline during the period considered. This implies that for the French Snowbelt as a whole an increased concentration of the new technology sector has not taken place.

A more detailed investigation of location quotients at the level of the individual regions (see Blaas, 1990) indicates more variety among the four French Snowbelt regions. Provence-Alpes-Côte d'Azur is clearly attracting more activities in sectors b, d and f, while France-Comté scores relatively higher for sector d. On the other hand, Alsace is not shown to be a very strong region.

17.7 NEW TECHNOLOGY IN THE GERMAN SNOWBELT

The booming development of the German Snowbelt regions has also been analyzed in the same way as the French regions (see Figure 17.8). Unfortunately not all relevant data could be found.

The picture shown by Figure 17.8 leads to a straightforward conclusion: the German Snowbelt regions have a concentration of sectors c, d, and f. But analogously to the French regions – there is no increasing concentration of new technology sectors in these German regions.

A more detailed analysis at the level of individual regions (i.e., Bayern and Baden-Württemberg) shows that Bayern is exhibiting a slight rise in the location quotient for most new technology sectors. This confirms the above mentioned 'catching up' development.

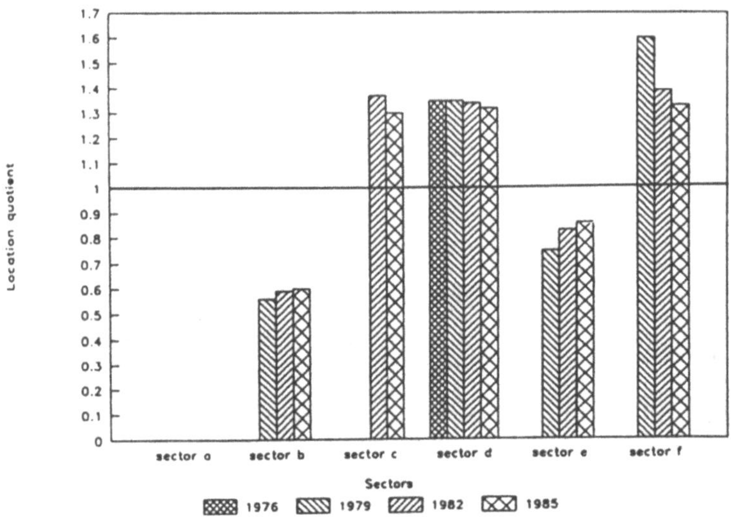

Figure 17.8 Location quotients for the German Snowbelt

The conclusion is that the strong growth of the German Snowbelt regions is certainly not reflected in a strong growth of the location quotients for new technology sectors. The strong points of these regions are the close links between fundamental and applied research, a sound mix of multinational and specialized small and medium sized enterprises, and a well developed network system among firms.

17.8 NEW TECHNOLOGY IN THE ITALIAN SNOWBELT

The regions belonging to the Italian Snowbelt exhibit a relatively strong growth in employment in the service sector. Some information on the location quotients for the six new technology sectors for the Italian Snowbelt is given in Figure 17.9.

It appears that the Italian Snowbelt has a location quotient above 1 for almost all new technology sectors. This conforms with the North-South division of the Italian economy. Figure 17.9 indicates, however, that the location quotients tend to decline, so that this suggests a slight convergence in the concentration of new technology firms over the country as a whole.

A more detailed analysis of some further available data for the sub-regions leads to the conclusion that Lombardia has by far the highest share of employment in the new technology sectors, except for office machines and fine-mechanical and optical instruments (which are more concentrated in the Nord-Ovest). Data at the level of urban agglomerations obtained from Camagni and Pompili (1990) indicate that the urban centers of Torino/Genova, Milano and Venice play a dominant role in the Italian Snowbelt (with the strongest position held by the Milano metropolitan area). However, a strong position of new technology sectors does not necessarily imply the highest concentration of supporting advanced services (in this respect Venice scores better), nor the highest concentration of R&D (in this respect Torino/Genova score better). Thus, in the Italian Snowbelt the pattern is also less straightforward than that suggested by the existing literature, although various basic seedbed conditions (accessibility, adequate network structure, quality of life, trained personnel and academic research climate) are certainly satisfied.

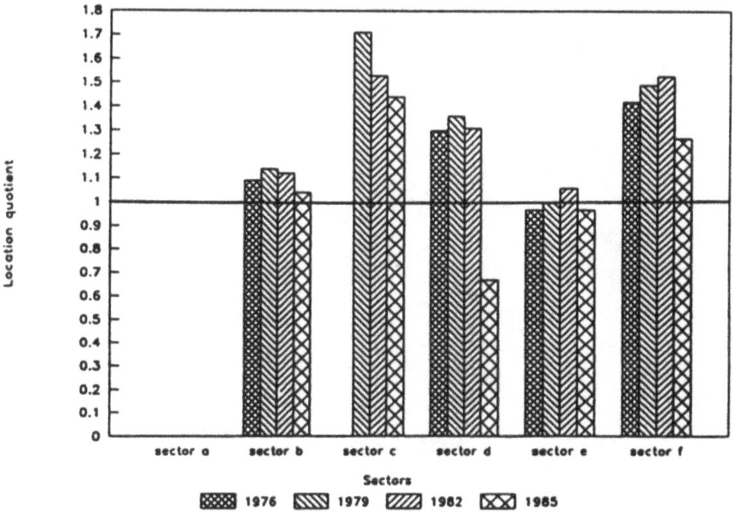

Figure 17.9 Location quotients for the Italian Snowbelt

17.9 NEW TECHNOLOGY IN THE AUSTRIAN/SWISS SNOWBELT

Unfortunately for Austria and Switzerland insufficient data were available at both the sectoral and the regional level, so the calculation of location quotients could not be carried out. Other data (see Blaas, 1990) indicate however that R&D in Austria has stayed behind the European average, although there is at present a steady increase. Switzerland, on the contrary, has a high share of R&D, comparable to the levels of Sweden and the Netherlands. This explains why Austria - in contrast to Switzerland - is performing less well regarding the development of new technology sectors (see also OECD, 1989). Also, the role of the Austrian government in providing incentives for high-tech research is less pronounced than in other countries.

A major problem for both Austria and (to a lesser extent) Switzerland is the relatively strong presence of old-line industries (iron and steel, the petrochemical industry, the wood and paper industry), for which the innovative potential is relatively low. Furthermore, Austria is hampered by the strong state intervention in large-scale traditional industries. Most of the new technology sectors in Austria are found in the Vienna agglomeration. Both Austria and Switzerland show a less well developed industrial network structure in comparison to France, Italy and Germany. In conclusion, the incubation conditions for the new technology sector have not been extremely favourable for Austria, but they have been relatively more favourable for Switzerland. This may explain the differences between these two countries.

17.10 CONCLUDING REMARKS

In this exploratory chapter on the regional incubator conditions of the European Snowbelt it turned out that the new technology sector has played an important role in the performance of these regions (see Figures 17.10 and 17.11)

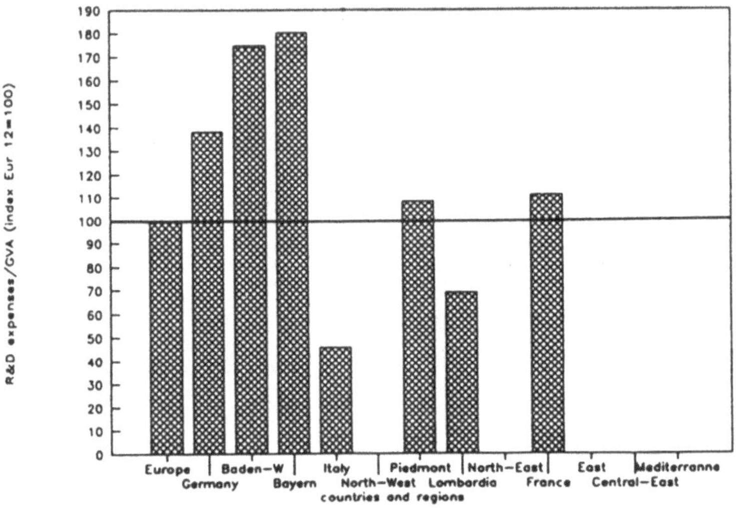

Figure 17.10 Private sector expenditure on R and D/Gross value added in 1983
Source: Boekhout and Romkema (1989)

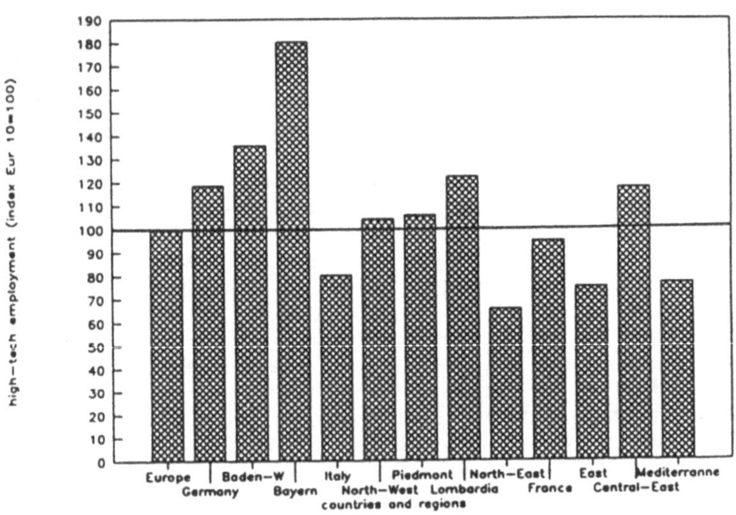

Figure 17.11 High-tech employment in 1981; Eur. 10 = 100;
Source: Boekhout and Romkema (1989)

Due to the lack of sufficient comparative empirical data a satisfactory test on the validity of the incubator hypothesis for the European Snowbelt could not be undertaken. However, various seedbed conditions which have proven to be valid in many other empirical studies were also fulfilled in this area. It is noteworthy that the European Snowbelt is not exhibiting a uniform economic and technological development structure. There is a great diversity among the sectoral compositions, production environment and entrepreneurial spirit in these regions. Apart from being located around the Alps, there is no cohesive network structure between the regions in the different countries involved, while the reasons for having attracted new technology sectors differ drastically among these regions (witness also the difference between Bayern and Lombardia). Thus the innovative potential of the regions in the European Snowbelt seems to depend more on their geographical position in the heartland of Europe than on Neo-Schumpeterian incubation conditions. Despite the relatively higher rise in employment in the service sector, the average Gross Regional Product per capita in the Snowbelt regions does not appear to be rising much faster than the national average. Most Snowbelt regions have, however, developed a relatively favourable pattern of new technological activities, without too much cohesion between these activities in the different countries. In this context it is noteworthy that most of these regions have an above average share of R&D expenditure. Participation in industrial organizational network structures at a regional/national scale also appeared to be an important factor in most regions.

In conclusion, the European Snowbelt regions have a somewhat better performance than most other European regions, but their growth and new technology orientation is – with the exception of a few regions – certainly not spectacular. In this sense there is no reason to regard them as Neo-Schumpeterian regions par excellence.

REFERENCES

Andersson, Å., 1985, "Creativity and Regional Development", *Papers of the Regional Science Association* 56:5-20.

Blaas, E.W., 1990, High-Tech en Regionale Ontwikkeling, Study paper, Department of Regional Economics, Free University, Amsterdam.

Boekhout, I.J. and S.A. Romkema, 1989, *Verschuiving van Economische Zwaartepunten in Noordwest-Europa: Fictie of Realiteit ?*, NEA, Rotterdam.

Breheny, M., P. Cheshire and R. Langridge, 1985, "The Anatomy of Job Creation? Industrial Change in Britain's M4 Corridor", in P. Hall and A. Markusen, (eds.), *Silicon Landscapes*, Allen and Unwin, Boston.

Camagni, R. and T. Pompili, 1990, "Competence, Power and Waves of Urban Development: an Italian Example", in P. Nijkamp, (ed.), *Sustainability of Urban Systems; a Cross-National Evolutionary Analysis of Urban Innovation*, Avebury Aldershot, England.

Cappellin, R. and P. Nijkamp, (eds.), 1990, *Technological Change in a Spatial Context*, Avebury, Aldershot, England.

Coffey, W.J. and M. Polise, 1984, "The Concept of Local Development: a Stages Model of Endogenous Regional Growth", *Papers of the Regional Science Association* 55:1-12.

Davelaar, E.J., 1990, *Regional Economic Analysis of Innovation and Incubation*, Avebury, Gower Aldershot, England.

Dunford, M., 1989, "State-industry Relations, Inter-firm Relations and Regional Development", *Geography*, pp. 20-28.

Eurostat, *diverse publications*.

Ewers, H.J. and R.W. Wettmann, 1980, "Innovation-oriented Regional Policy", *Regional Studies* 14:161-179.

Ewers, H.J. and M. Gornig, 1990, "Openness of Urban Systems as Prerequisite for Urban Sustainability: a German Example", Forthcoming.

Freeman, C., 1986, "The Role of Technical Change in National Economic Development" in A. Amin and J.B, Goddard, (eds.) *Technological Change, Industrial Restructuring and Regional Development*, Allen and Unwin, London.

Galbraith, C.S., 1985, "High-technology Location and Development: the Case of Orange County", *California Management Review*, 28:98-109.
Gelsing, L., 1989, "Knowledge Networks, Industrial Flexibility and Innovation" in F. Borum and P.H. Kristensen, (eds.), *Technological Innovation and Organizational Change: Danish Patterns of Knowledge, Networks and Culture*, New Social Science Monographs, Copenhagen.
Hall, P., 1985, "The Geography of the Fifth Kondratieff" in P. Hall and A. Markusen, (eds.), *Silicon Landscapes*, Allen and Unwin, Boston.
Hansen, N., 1990, Innovative Regional Milieux, Small Firms, and Regional Devlopment: Evidence from the Mediterranean France, *The Annals of Regional Science*, 24:107-123.
Hepworth, M.E., 1989, "The Electronic Comeback of Information Space", *Tijdschrift voor Economisch en Sociale Geografie* 80:30-43.
Johansson, B., 1987, "Information Technology and the Viability of Spatial Networks", *Papers of the Regional Science Association* 61:51-64.
Kamann, D.J.F., 1988, Externe Organisatie: een Inleiding, Universiteitsdrukkerij, Groningen.
Kamann, D.J.F., 1989, "Actoren binnen netwerken" in F.W.M. Boekema and D.J.F. Kamann, (eds.), *Sociaal-Economische Netwerken*, Wolters-Noordhoff, Groningen.
Kamann, D.J.F. and F. Boekema, 1989, "Netwerken: een introductie" in *Sociaal-Economische Netwerken*, Wolters-Noordhoff, Groningen.
Keeble, D. and E. Wever, 1986), "Introduction" in D. Keeble and E.Wever , (eds.), *New Firms and Regional Development in Europe*, Croom Helm, London.
Kleinknecht, A., 1987, *Innovation Patterns in Crisis and Prosperity*, MacMillan, London.
Lecoq, B., 1990, Industrial Organization, Technological Change and Regional Developments: A Network Approach, Paper presented at the 30th European Congress of the Regional Science Association, Istanbul.
Malecki, E., 1983, Technology and Regional Development, *International Regional Science Review*, Vol. 8, No. 2:89-125.
Malecki, E., 1985, "Industrial Location and Corporate Organization in High-technology Industries", *Economic Geography* 61:345-369.
Malecki, E., 1987, "Research and Development and the Geography of High-technology Complexes" in J. Rees and A. Stafford (eds.), *Technology, Regions and Policy*, Rowman and Littlefield, New Jersey.
Markusen, A.R., 1985, "High-Tech Jobs, Markets and Economic Development" in P. Hall and A. Markusen, (eds.), *Silicon Landscapes*, Allen and Unwin, Boston.
Markusen, A., P. Hall and A. Glasmeier, 1986, *High-Tech America*, Allen and Unwin, Boston.
Molle, W.T.M., 1985, "Begrips- en plaatsbepaling" in W.T.M. Molle, (ed.), *Innovatie en Regio: Onderzoekprogramma Technologie en Economie*, Staatsuitgeverij, Den Haag.
Nijkamp, P., (ed.), 1990, *Sustainability of Urban Systems; a Cross-National Evolutionary Analysis of Urban Innovation*, Avebury, Aldershot, England.
Nijkamp, P., H. Bouman and B. Verhoef, 1990, "High-Tech Employment - Place and Competence", *International Association of Applied Psychology* 39:207-222.
Oakey, R.P., A.T. Thwaites and P.A. Nash, 1985, "The Regional Distribution of Innovative Manufacturing Establishments in Britain", *Regional Studies* 14:235-253.
OECD, 1989, *OECD Science and Technology Indicators: Report no. 3, R and D, Production and Diffusion of Technology*, OECD, Paris.
Olson, M., 1982, *The Rise and Decline of Nations: Economic Growth, Stagflation and Social Rigidities*, Yale University Press, New Haven and London.
Porter, M.E., 1990, *The Competitive Advantage of Nations*, The Free Press, New York.
Pottier, C., 1985, "The Adaption of Regional Industrial Structures to Technical Changes", *Papers of the Regional Science Association* 58:59-72.

Rees, J. and H.A. Stafford, 1986, "Theories of Regional Growth and Industrial Location" in J.Rees and A. Stafford, (eds.), *Technology, Regions and Policy*, Rowman and Littlefield, New Jersey.

Saxenian, A., 1985, "The Genesis of Silicon Valley" in P.Hall and A. Markusen, (eds.), *Silicon Landscapes*, Allen and Unwin, Boston.

Storper, M. and Scott, A.J., 1989, "The Geographical Foundations and Social Regulation of Flexible Production Complexes" in J. Wolch and M. Dear, (eds.), *The Power of Geography*, Urwin Hyman.

Suarez-Villa, L., 1989, *The Evolution of Regional Economics: Entrepeneurship and Macro-economic Change*, Praeger, New York.

Thomas, M.D., 1986, "Growth and Structural Change: the Role of Technical Innovation", in A. Amin and J.B. Goddard, (eds.), *Technological Change, Industrial Restructuring and Regional Development*, Allen and Unwin, London.

Weiss, M.A., 1985, "High-technology Industries and the Future of Employment" in P.Hall and A. Markusen, (eds.), *Silicon Landscapes*, Allen and Unwin, Boston.

Whetten, D.A., 1987, "Interorganizational Relations" in J.W. Lorsch, (ed.), *Handbook of Organizational Behaviour*, Englewood Cliffs: Prentice Hall.

CHAPTER 18

Non-Linear Dynamic Economic Structure: Infrastructure and Knowledge in a Two-Sector Growth Model

Wei-Bin Zhang
Institute for Future Studies, Stockholm

18.1 INTRODUCTION

In the future, as in the past, the great ideas must be simplifying ideas.
André Weil

In this chapter we propose a nonlinear dynamic model to analyze interactions between knowledge, infrastructure and economic development. Our approach is developed within the framework of Uzawa's two-sector model (Uzawa, 1961, 1962). It is well known that Uzawa's two-sector model has one advantage over the Solow-Swan one-sector growth model, in that the Uzawa model can be used to analyze economic structure. But as knowledge and infrastructure are omitted, it is impossible to discuss how economic structure is affected by knowledge accumulation and governmental intervention in infrastructure construction. We introduce knowledge and infrastructure into the system as "public goods" in order to ascertain the conditions under which it is economically effective for a government to increase infrastructure with network and diffusion properties and to investigate how economic structural changes may take place as a result of accumulated knowledge.

The issues raised in this study have been examined by, e.g., Andersson, Anderstig and Hårsman (1990), Andersson, Lakshmanan and Zhang (1992), Beckmann and Puu (1985), Batten, Kobayashi and Andersson (1989), Johansson and Karlsson (1990), Lakshmanan (1989), Snickars (1987), Westin (1990) and Zhang (1990, 1991, 1992). Here we hope to provide some new insights into the complexity of interactions between knowledge, infrastructure and economic development within the framework of neoclassical economic growth theory, which enjoyed great popularity in economics during the 50's and the 60's.

The study is organized as follows. Section 18.2 defines the basic model which describes the interactions between the two economic sectors with endogenous accumulations of capital and knowledge. Section 18.3 guarantees the existence of equilibria and provides stability conditions. Section 18.4 studies the effects of a government's infrastructure and network policy upon the system's economic variables. Section 18.5 studies the relationship between creativity and the economic structure.

Section 18.6 analyzes the effects of changes in the savings rate upon the economic system. Section 18.7 concludes the study.

It should be mentioned that by introducing knowledge, infrastructure and networks, we can discuss some important issues and obtain some new insights which cannot be obtained from Uzawa's two-sector growth model. As our model is an extension of the traditional two-sector growth model, the main conclusions from Uzawa's model can be obtained as special cases of our model. The economic system in this study may have either a unique equilibrium like that in Uzawa's model or multiple equilibria, depending upon its characteristics of knowledge utilization and creativity. We show that a government's intervention by constructing infrastructure and networks may either stimulate or depress the national economy. We also prove that an increase in the savings rate may either expand or decrease the national wealth, depending upon the interactions between capital, various forms of infrastructure and knowledge. This conclusion is different to that obtained by the traditional neoclassical growth model.

18.2 THE MODEL

Little laws have bigger laws
From which they're forced to follow,
But a bigger law's bigger guess
That's harder still to swallow.

Anon

It should be noted that the various aspects of economic activities in our model are essentially the same as those in Uzawa's two-sector model, except for the assumption that knowledge and infrastructure explicitly affect the outputs of the two sectors.

We use indexes i and s to denote the industrial sector and the service sector, respectively.

18.2.1 Production of the Industrial Sector

We specify the production function of the industrial sector as follows

$$F_i = Z^m G^x K_i^\alpha N_i^\beta, \quad \alpha, \beta > 0, \alpha + \beta = 1, \ m, x \geq 0, \tag{1}$$

in which Z and G denote knowledge and the infrastructure stock of the society, respectively, and K_i and N_i are the capital stock and the labour force employed by the industrial sector, respectively. Here, Z and G are treated as "public goods" in the sense that the utilization of Z and G does not cost the production sectors anything, though they have to pay tax in order to support the construction and operation of infrastructure.

It should be noted that the concept of knowledge used in this study is not identical to the concept of human capital that is widely used in the literature concerning human capital and economic growth. We may consider that a person who belongs to the society obtains certain objective knowledge. However, one cannot master all of the knowledge created by mankind. In an approximate sense, we may describe his qualifications as a function, $g^*(Z)$, of knowledge Z. For an individual, $g^*(Z)$ is determined by culture, family background, social position, material conditions, education and other life experiences. Similarly, we may define a function $g^*(Z,N)$ as a measurement of country i's human capital. According to this interpretation, we see that our concept of knowledge is identical to the concept of human capital (as far as production is concerned). Hence, the human capital approach should not be considered to be in conflict with our approach to the concept of knowledge. To see this, let us assume that a "qualified" labour force is given by $N_i^* = Z^{m/\beta} N_i$. In the case of x = 0, we can rewrite (1) as

$$F_i = N_i^{*\alpha} K_i^{\beta},$$

which is identical to the neoclassical production function. From the definition of the production function, above, we see that while for fixed knowledge and infrastructure the economic production function is homogenous of degree one with respect to labour and capital, it may have the property of increasing returns to scale when Z is an endogenous variable.

It should be noted that our concept of knowledge may be particularly convenient for certain special problems. It is obvious that knowledge can be embodied both in capital and in humans. In accord with our interpretation of human capital as a function of knowledge, we may describe "effective capital" by f*(Z,K), where K is some basic physical measurement of capital. For instance, we may have the same number of computers in two countries, but the quality of the computers may be different. Hence, the concept of knowledge allows us to introduce differences in "capital quality" into our approach. Indeed, the use of concepts such as human capital, technology and knowledge when studying the impact of knowledge upon economic problems is dependent upon the issues under consideration.

Although we neglect the spatial character of economic reality by aggregation, economic agents are actually connected by different types of infrastructure and networks. It is obvious that economic efficiency is dependent upon the structure of a system's communication and transportation infrastructure. For the simplicity of analysis, we assume that economic efficiency is positively related to the infrastructure stocks of the society, in the way described by (1).

18.2.2 Production of the Service Sector

We specify the production function of the service sector as follows

$$F_s = Z^q G^y K_s^a N_s^b, \quad a, b > 0, a + b = 1, q, y \geq 0, \tag{2}$$

where K_s and N_s are the capital stock and the labour force employed by the service sector, respectively.

As products of the service sector are simultaneously consumed as they are produced, we have

$$C_s = F_s, \tag{3}$$

in which C_s is the consumption of services. This constraint simply states that services cannot be saved.

18.2.3 Markets of Capital and Labour

In this study, prices are measured in terms of the industrial good. Let the price of the industrial good be unity during the study period and denote p(t) the ratio of the price of the service product to that of the industrial good at time t. Let r(t) and w(t) be the interest rate and the wage rate, respectively. Let perfect competition dominate the markets, then each sector's wage rate is equal to the marginal product of labour and the interest rate of capital is equal to the marginal product of capital, i.e.,

$$w = \beta TF_i/N_i = pbTF_s/N_s, \quad r = \alpha TF_i/K_i = paTF_s/K_s, \tag{4}$$

in which (1-T) is the tax rate levied on the output of the industrial and service sectors.

18.2.4 Savings and Capital Formation

The gross national product in terms of industrial goods, $Y(t)$, is given by

$$Y = F_i + pF_s. \tag{5}$$

Let s $(0 < s < 1)$ denote the population's savings rate out of the current gross national product at time t. In this study, we assume that the savings rate is constant. The national savings is thus given by: sY. Accordingly, capital accumulation is given by

$$dK/dt = sY - \delta K, \tag{6}$$

in which δ is the given depreciation rate of capital. Since the investment is in the form of industrial goods, we require: $sY + m_o L < F_i$ at any point of time. Here, m_o is the survival level of the population required for industrial goods. In what follows, we assume that this inequality is always satisfied.

Denote C_i the industrial goods which are consumed by the population. Since industrial production consists of investment and consumption, one has:

$$C_i = F_i - sY. \tag{7}$$

18.2.5 Choice of Consumption Goods and Services

Let N denote the constant population during the study period. We assume that the utility an individual obtains from consuming industrial goods and services can be expressed in the following form: $U(C_i, C_s) = (C_i / N)^v (C_s / N)^u$, where $u > 0$, $v > 0$. The total consumption budget is defined by: $(1-s)Y/N$. Accordingly, the consumer problem is given by:

$$\max (C_i / N)^v (C_s / N)^u, \quad \text{subject to: } C_i + pC_s = (1-s)Y.$$

The unique optimal solution is given by

$$C_i = (1-s)vY/(u+v), \quad C_s = (1-s)uY/p(u+v). \tag{8}$$

18.2.6 Knowledge Accumulation

The more skills people have
The further novelties multiply.

Lao Tzu

According to Zhang (1991, 1992), we propose the following possible form of knowledge accumulation

$$dZ / dt = \tau_i G^{\sigma} F_i / Z^{\varepsilon} + \tau_s G^{d} F_s / Z^{\pi} - \delta_z Z, \tag{9}$$

in which δ_z is the depreciation rate of knowledge, τ_i, τ_s ($0 \le \tau_i, \tau_s$), σ, d, ε, and $\pi(0 \le \sigma, d, \varepsilon, \pi \le 1)$ are positive parameters. The term, $\tau_i G^{\sigma} F_i / Z^{\varepsilon}$, measures the effect of learning by doing by the industrial sector upon knowledge accumulation. The term, $\tau_i G^{\sigma} F_i$, implies that knowledge accumulation is positively related to the production scales and transportation and communication conditions of the society. The term, $1 / Z^{\varepsilon}$, describes the fact that there are return to scale effects in knowledge accumulation. We can similarly interpret the term, $r_s G^{d} F_s / Z^{\pi}$. It should noted that in knowledge accumulation the scale effect parameters, ε and π, may be either positive or negative.

18.2.7 Infrastructure and Governmental Policy

We now design a way to determine G. As we only add one constraint to the government's budget equation, we have one variable for the government to determine. We assume the following simple rule of governmental intervention

$$G = nK, \quad 0 < n < 1, \tag{10}$$

in which n denotes the policy variables fixed by the government. We assume that n is exogenously given. We are interested in the effects of the policy parameters upon the economic structure. It should be noted that the infrastructure policy specified by Equation (10) is a simplified case. The way in which the government determines the structure of infrastructure for communication and transportation is affected by numerous political, military and economic factors.

In this study, we assume that the tax rate is endogenously given in the sense that the government will only tax the producers to satisfy the budget constraint

$$(1-T)(F_i + pF_s) = rG. \tag{11}$$

18.2.8 Full Employment

Let K denote the total capital stock at time t and K_i and K_s the capital stocks of the industrial and service sectors, respectively. The assumption of full employment of capital implies

$$K_i + K_s = (1-n)K. \tag{12}$$

The assumption of full employment is given by

$$N_i + N_s = N. \tag{13}$$

We have now completed the model. The dynamic model consists of two differential equations for K and Z. There are 16 time-dependent endogenous variables, K_i, K_s, G, K, p, r, w, N_i, N_s, Z, F_i, F_s, T, Y, C_i, C_s. Since the system consists of 16 independent equations, the system may have solutions under certain conditions.

18.3 ECONOMIC DYNAMICS, EQUILIBRIA AND STABILITY

First, this section expresses the economic dynamics of the terms K and Z. Then, we guarantee the existence of equilibria and provide the conditions for stability.

From (4), we have $K_i/K_s = (\alpha b/a\beta)N_i/N_s$. Utilizing $K_i/K_s = \alpha F_i/apF_s$, $(1-s)vY(u+v) = F_i - sY$ and $Y = F_i + pF_s$ yields

$$K_s/K_i = au(1-s)/\alpha(us+v). \tag{14}$$

From (14) and $K_i/K_s = (\alpha b/a\beta)N_i/N_s$, we have

$$N_s/N_i = bu(1-s)/\beta(us+v). \tag{15}$$

The ratios of the labour force and the capital stocks employed in the service and the industrial sectors are determined by the technology and taste structure parameters, b, β, u and v, and the savings rate respectively. As $K_i + K_s = (1-n)K$ and $N_i + N_s = N$ we have

$$N_i = \beta_1 N, \quad N_s = b_1 N, \quad K_i = \alpha_1 K, \quad K_s = a_1 K, \tag{16}$$

where $\beta_1 = \beta(su+v)/\{\beta(su+v) + bu(1-s)\}$, $b_1 = bu(1-s)/\{\beta(su+v) + bu(1-s)\}$, $\alpha_1 = \alpha(su+v)(1-n)/\{au(1-s) + \alpha(su+v)\}$, $a_1 = au(1-s)(1-n)/\{au(1-s) + \alpha(su+v)\}$.

Utilizing $G = nK$, $r = \alpha TF_i/K_i = paTF_s/K_s$, (14) and $(1-T)(F_i + pF_s) = rG$, we have

$$1/T = \alpha n(us+v)/\alpha_1(u+v) + 1. \tag{17}$$

As (1–T) is the tax rate, the tax rate is independent of capital and knowledge, though it is determined by the policy parameter n. As n is increased, (1–T) is increased. That is, if the government tends to increase the ratio of infrastructure stock to the capital stock of the society, then the government has to increase the tax rate.

Substituting (10) and (16) into (1), (2) and $Y = (u+v)F_i/(su+v)$, we can express F_i, F_s and Y as functions of K and Z as follows

$$F_i(K,Z) = \alpha_1^{\alpha} n^x (\beta_1 N)^{\beta} K^{\alpha+x} Z^m,$$

$$F_s(K,Z) = a_1^a n^y (b_1 N)^b K^{a+y} Z^q,$$

$$Y(K,Z) = (u+v)\alpha_1^{\alpha} n^x (\beta_1 N)^{\beta} K^{\alpha+x} Z^m / (su+v). \tag{18}$$

Substituting (18) into (6) and (9), we can describe economic dynamics in terms of K and Z as follows

$$dK/dt = m_1 f(K,Z) - \delta K,$$

$$dZ/dt = m_2 f_i(K,Z) + m_3 f_s(K,Z) - \delta_z Z, \tag{19}$$

in which $f(K,Z) = K^{\alpha+x}Z^m$, $f_i(K,Z) = K^{\alpha+x+\sigma}Z^{m-\varepsilon}$, $f_s(K,Z) = K^{a+y+d}Z^{q-\pi}$, and

$$m_1 = s(u+v)\alpha_1^{\alpha} n^{x} (\beta_1 N)^{\beta} / (su+v),$$

$$m_2 = \tau_i n^{\sigma} \alpha_1^{\alpha} (\beta_1 N)^{\beta},$$

$$m_3 = \tau_s n^{d} a_1^{a} n^{y} (b_1 N)^{b}. \tag{20}$$

An equilibrium is defined by a positive solution of

$$m_1 K^{\alpha+x} Z^{m} = \delta K, \; m_2 K^{\alpha+x+\sigma} Z^{m-\varepsilon} + m_3 K^{a+y+d} Z^{q-\pi} = \delta_z Z. \tag{21}$$

From the first equation, we have

$$K = (m_1 / \delta)^{X} Z^{mX}, \tag{22}$$

in which $X = 1/(\beta - x)$. As α and x are the marginal production rates of capital and infrastructure in the industrial product, the requirement of $\alpha + x < 1$ means that with other conditions fixed, industrial production has the property of decreasing returns to scale with respect to capital and infrastructure stocks. We see that it is reasonable to require $\alpha + x < 1$. As $\beta - x = 1 - (\alpha + x)$, $\beta - x > 0$. In the remainder of the study, we assume that $\beta - x > 0$.

Substituting (22) into the second equation in (21) yields

$$H(Z) \equiv \Phi_1(Z) + \Phi_2(Z) - \delta_2 = 0, \tag{23}$$

in which

$$\Phi_1 = m_2 (m_1 / \delta)^{(\alpha+x+\sigma)X} Z^{x_1}, \quad \Phi_2 = m_3 (m_1 / \delta)^{(a+y+d)X} Z^{x_2},$$

where $x_1 = (\alpha+x+\sigma)mX + m - \varepsilon - 1$, $x_2 = (a+y+d)mX + q - \pi - 1$.

An equilibrium of the system is given by a positive Z such that $H(Z) = 0$. When $x_1 > 0$ and $x_2 > 0$, $H(Z) = 0$ has a unique positive solution as $H' > 0$ for any positive Z, $H(0) < 0$ and $H(\infty) > 0$. Similarly, if $x_1 < 0$ and $x_2 < 0$, $H(Z) = 0$ has a unique positive solution. If $x_1 = 0$ and $x_2 = 0$, then the equation has infinite solutions. It is easy to check that if either $x_1 = 0$ or $x_2 = 0$, then the system has a unique positive solution under certain conditions.

We now prove that if $x_1 > 0$ and $x_2 < 0$ (or $x_1 < 0$ or $x_2 > 0$), then the system has either two solutions or no solution. It is sufficient for us to examine one case, for instance, of $x_1 > 0$ and $x_2 < 0$. As $H(O) > 0$, $H(\infty) > 0$, we see that $H(Z) = 0$ cannot have a unique solution. That is, $H(Z) = 0$ has either multiple solutions or no solution. On the other hand, as $H'(Z) = 0$ has a unique positive solution, we conclude that $H(Z) = 0$ has two solutions if H(Z) has solutions. The necessary and sufficient condition for the existence of two solutions is that there exists a positive value Z_1, of Z such that $H(Z_1) < 0$ and $H'(Z_1) = 0$.

The Jacobian at an equilibrium is given by:

$$J = \begin{pmatrix} -\delta(x-\beta) & \delta m K/Z \\ n_1/K & -n_2/Z \end{pmatrix}, \tag{24}$$

in which $n_1 = (\alpha+x+\sigma)f_i + (a+y+d)f_s > 0$ and $n_2 = f_i(1+\varepsilon-m) + f_s(1+\pi-q)$.

The two eigenvalues, ϕ_i, are determined by

$$\phi^2 + \{\delta(x-\beta) + n_2/Z\}\phi + \delta(x-\beta) + n_2/Z - n_1\delta m/Z = 0, \text{ or}$$

$$\phi_{1,2} = -\{\delta(x-\beta)+n_2/Z\}/2 \pm [\{\delta(x-\beta)-n_2/Z\}^2/2 + n_1\delta m/Z]^{1/2}. \tag{25}$$

From (25) and definitions of n_1, n_2, x_1 and x_2, we see that if the term Φ

$$\Phi \equiv - x_1\Phi_1/Z - x_2\Phi_2/Z, \tag{26}$$

is negative, then the equilibrium is unstable; if it is positive, then the equilibrium is stable. Summarizing the above analytical results, we prove the following proposition.

Proposition 1. If $x_1 < 0$ and $x_2 < 0$, then the system has a unique positive equilibrium. Moreover, the dynamic system is stable.

If $x_1 > 0$ and $x_2 > 0$, then the system has a unique positive equilibrium. Moreover, the system is unstable.

If $x_1 > 0$ and $x_2 < 0$, ($x_1 < 0$ and $x_2 > 0$), then the system has either two positive equilibria or no equilibrium. When the system has two equilibria, the one with low values of K and Z is stable; the other one is unstable.

We only interpret the stability condition $x_1 < 0$ and $x_2 < 0$, i.e.

$$(\alpha+x+\sigma)m/(\beta-x) + m < \varepsilon + 1,$$

$$(a+y+d)m/(\beta-x) + q < \pi + 1. \tag{27}$$

The parameters $\alpha(\beta = 1 - \alpha)$ and a are marginal production rates of capital in the industrial sector and service sector, respectively. The parameters x and y are marginal production rates of infrastructure in the industrial sector and service sector, respectively. The parameters m and q are marginal production rates of knowledge in the industrial sector and service sector, respectively. The parameters σ and d are "marginal creativity rates" of infrastructure related to industrial and service activities, respectively. The parameters ε and π describe return to scale effects in knowledge accumulation in the industrial sector and the service sector, respectively. Accordingly, we can interpret the stability condition (27) as follows. The system is stable when decreasing scale effects in knowledge accumulation are relatively strong (i.e., ε and π are small) and the marginal production rates of knowledge and infrastructure in the two sectors are relatively small (i.e., m, q, x and y are small).

In the remainder of the study, we carry out a comparative static analysis with some parameters under presumed stability. It should be noted that when the system has two equilibria, our conclusions are valid for the equilibrium with low values of K and Z.

18.4 A GOVERNMENT'S INFRASTRUCTURE POLICY

This section examines effects of changes in a government's infrastructure policy n upon the system.

Taking derivatives of (22) and (23) with respect to n yields

$$(1/XK)dK/dn = (m/Z)dZ/dn + x/n - \alpha/(1-n),$$

$$\Phi dZ/dn = \{(\sigma\beta+x)x/n - \alpha(1+\sigma)/(1-n)\}X\Phi_1 + \{d/n + y/n - a/(1-n) + x(a+y+d)X/n - \alpha(a+y+d)X/(1-n)\}\Phi_2, \tag{28}$$

in which Φ is a positive constant defined by (26).

The impact upon knowledge accumulation is given by complicated interactions between various forces (i.e., different marginal values). For simplicity, we assume that the term Φ_2 is extremely small. We can thus omit the term associated with Φ_2 in (28). This condition is satisfied, for instance, when τ_s is extremely small. The sign of dZ/dn is the same as that of the term $\{(\sigma\beta+x)x/n - \alpha(1+\sigma)/(1-n)\}$. We may conclude that when the ratio n of infrastructure is very low, an increase in n tends to increase the knowledge stock of the society in the long term. When the marginal production rate x of infrastructure in the industrial sector is large, an increase in n also tends to increase Z.

From $(1/XK)dK/dn = (m/Z)dZ/dn + x/n - \alpha/(1-n)$, we can see that it is only when n is appropriately small that an increase in n may cause the capital stock of the society to expand in the long term. We can thus conclude that it is only when the ratio n of infrastructure is very low that it can be economically effective for the government to increase tax for the construction of infrastructure. When the ratio becomes relatively high, it is very difficult to assess the effects of changes in n upon the system.

The remainder of this section requires that n is very small, i.e., dK/dn > 0 and dZ/dn > 0. The effects upon F_a,, F_i, Y, p, r and w are given by

$$\begin{aligned}
(1/F_i)dF_i/dn &= \{(\alpha+x)/K\}dK/dn + (m/Z)dZ/dn + x/n - \alpha/(1-n),\\
(1/F_s)dF_s/dn &= \{(a+y)/K\}dK/dn + (q/Z)dZ/dn + y/n - a/(1-n),\\
(1/Y)dY/dn &= (1/F_i)dF_i/dn,\\
(1/w)dw/dn &= (1/F_i)dF_i/dn + (1/T)dT/dn,\\
(1/r)dr/dn &= (1/T)dT/dn < 0,\\
(1/p)dp/dn &= (1/F_i)dF_i/dn - (1/F_s)dF_s/dn,
\end{aligned} \tag{29}$$

in which $dT/dn = T^2(1 - 1/T)/(1-n) < 0$.

We see that when n is very small, the production of the industrial sector and the service sector and the total output tend to be increased as the government increases the infrastructure stock. The wage rate is increased only when the positive impact upon the output of the industrial sector is stronger than the negative impact of the increased tax rate, i.e., $(1/F_i)dF_i/dn > -(1/T)dT/dn$. The price of services is increased only when the increase of the industrial product is larger than that of the service product due to expanded infrastructures.

18.5 CREATIVITY AND THE ECONOMIC STRUCTURE

It is no paradox to say that in our most theoretical moods we may be nearest to our most practical applications.

A.N. Whitehead

This section examines the impact of changes in the creativity of economic activities upon the economic structure of a society. As changes in the creativity parameters, τ_i and τ_s, in knowledge accumulation have similar effects upon the system, it is sufficient for us to examine the creativity τ_i of the industrial sector.

Taking derivatives of (22) and (23) with respect to τ_i yields

$$\Phi dZ/d\tau_i = \Phi_i/\tau_i > 0, \quad dK/d\tau_i = (mXK/Z)dZ/d\tau_i > 0. \tag{30}$$

As the creativity of the industrial sector is increased, the capital and knowledge stocks of the society always tend to increase. We can check directly that as the creativity of the industrial sector is increased, the output of the industrial sector and the service sector, total output, and the consumption level of industrial goods and services tend to be increased, i.e., $dF_i/d\tau_i > 0$, $dF_s/d\tau_i > 0$, $dY/d\tau_i > 0$, $dC_i/d\tau_i > 0$ and $dC_s/d\tau_i > 0$. It is easy to check $dw/d\tau_i > 0$, $dr/d\tau_i = 0$. Moreover, examine

$$(Z/p)dp/d\tau_i = m\{(b-y)X - q/m\}dZ/d\tau_i. \tag{31}$$

As $b - y = 1 - (a+y)$ in which a and y are the marginal production rates of capital and infrastructure in the service sector, respectively, it is reasonable to require $(a+b) < 1$, i.e., $b - y > 0$. From (31) we conclude that if the marginal rate of knowledge in the service sector is low in comparison to the marginal rate of knowledge in the industrial sector (i.e., $q/m > (b-y)X$), then an increase in τ_i will increase the cost of services.

18.6 THE COMPLEXITY OF EFFECTS OF THE SAVINGS RATE

There is a time for everything.

Ecclesiastes

This section provides some analytical results to illustrate the complexity of the impact of the savings rate upon economic development. It is worthwhile to emphasize that different economic theories, such as Keynesian and neoclassical economics, have various points of view about the effects of the savings rate upon economic development. Moreover, from empirical studies one can hardly obtain a definite conclusion about the link between the savings rate and economic development (see, e.g., Chenery and Srinivasan, 1988, 1989).

Taking the derivatives of (22) and (23) with respect to s yields

$$\Phi dZ/ds = \{m_2^* + (\alpha+x+\sigma)X m_1^*\}\Phi_1 + \{m_3^* + (a+y+d)X m_1^*\}\Phi_2,$$
$$dK/ds = (mX/Z)dZ/ds + X m_1^*,$$

in which

$$m_1^* = v/(su+v) + \alpha\alpha_1^* + \beta_1^*, \quad m_2^* = \alpha\alpha_1^* + \beta_1^*,$$
$$m_3^* = aa_1^* + bb_1^*,$$
$$\alpha_1^* = u/(su+v) + u(\alpha-a)/\{au(1-s) + \alpha(su+v)\},$$
$$\beta_1^* = u/(su+v) + u(b-\beta)/\{\beta(su+v) + bu(1-s)\}$$
$$a_1^* = u(u-\alpha)/\{au(1-s) + \alpha(su+v)\} - 1,$$
$$b_1^* = u(b-\beta)/\{\beta(su+v) + bu(1-s) - 1, \tag{33}$$

in which * expresses the rate of change of a variable with respect to s, for instance, $\alpha_1^* = (1/\alpha_1)d\alpha_1/ds$.

From (32) and (33) we see that it is very difficult to interpret economic conditions for the signs of dK/ds and dZ/ds explicitly. The savings rate affects the economic variables in complicated paths of interactions. This may provide insight into the reason why it is difficult to reach definite conclusions about the effects of the savings rate upon economic development. Although it is easy to analytically express the effects of changes in s upon F_i, F_S, Y, w, r and p, we omit any further analysis as the expressions are too complicated.

18.7 CONCLUDING REMARKS

This study suggested an economic growth model to analyze economic structural change with endogenous knowledge and infrastructure. Our model is an extension of Uzawa's two-sector growth model in the sense that if we fix infrastructure and knowledge, our model is identical to Uzawa's two-sector model. We examined the effects of governmental policy upon infrastructure and the creativity of economic activities upon the economic variables of a system. We also showed the complexity of the impact of changes in the savings rate upon the economic variables of a system.

It should be noted that economic systems with increasing returns to scale have recently generated great interest in the literature about economic development and trade. This study models two important factors - communications and transportation infrastructures and knowledge - for economies with returns to scale within the framework of the neoclassical growth theory.

REFERENCES

Andersson, Å.E., C. Anderstig and B. Hårsman, 1990, "Knowledge and Communications Infrastructure and Regional Economic Change", *Regional Science and Urban Economics* 20:359-376.

Andersson, Å.E., T.R. Lakshmanan and W.B. Zhang, 1992, "Technology and Development - Lessons from Success Stories", *Journal of Scientific & Industrial Research* 5: 157-171.

Batten, D.F., K. Kobayashi and Å.E. Andersson, 1989, "Knowledge, Nodes and Networks: An Analytical Perspective" in *Knowledge and Industrial Organization* by Å. E. Andersson, D. F. Batten and C. Karlsson, (eds.), Springer-Verlag, Berlin.

Beckmann, M.J. and T. Puu, 1985, *Spatial Economics: Potential, Density, and Flow*, North-Holland Publishing Company, Amsterdam.

Chenery, H. and T.N. Srinivasan, (eds.), 1988, *Handbook of Development Economics*, Vol.I, North-Holland, Amsterdam.

Chenery, H. and T.N. Srinivasan, (eds.), 1989, *Handbook of Development Economics*, Vol.II, North-Holland, Amsterdam.

Johansson, B. and C. Karlsson, (eds.), 1990, Innovation, Industrial Knowledge and Trade, CERUM, University of Umeå.

Lakshmanan, T.R., 1989, "Infrastructure and Economic Transformation" in *Advances in Spatial Theory and Dynamics* by Å.E. Andersson et al., (eds.), North-Holland, Amsterdam.

Snickars, F., 1987, "The Transportation Sector in the Communications Society: Some Analytical Observations" in *Transportation Planning in a Changing World* by P. Nijkamp and S. Reichmann, (eds.), Cower, Aldershot, Hampshire.

Uzawa, H., 1961, "On a Two-Sector Model of Economic Growth I", *Review of Economic Studies* XXIX:40-47.

Uzawa, H., 1962, "On a Two-Sector Model of Economic Growth II", *Review of Economic Studies* XXX:105-118.

Westin, L., 1990, "Vintage Models of Spatial Structural Change", *Umeå Economic Studies* No. 227, University of Umeå.

Zhang, W.B., 1990, "Brain Drain and Economic Cycles with International Migration", *Journal of Development Economics* 32:191-203.

Zhang, W.B., 1991, "Regional Dynamics with Creativity and Knowledge Diffusion", *The Annals of Regional Science* 25:179-191.

Zhang, W.B., 1992, "A Development Model of Developing Economies with Capital and Knowledge Accumulation", *Journal of Economics* 55:43-63.

AUTHORS' INDEX

SUBJECT INDEX

L. **Lundqvist**, **L. O. Persson** (Eds.)

Visions and Strategies in European Integration

A North European Perspective

1993. IX, 290 pp. 34 figs. 24 tabs. ISBN 3-540-56615-5

The 90s have witnessed thorough political and economic changes in northern Europe. In this volume, scholars in regional science and related disciplines (geography, economics, environmental and political sciences, planning) have brought together important material on the current processes that are reshaping northern Europe. Visions and strategies on local, national and supranational levels are penetrated in depth. A "mosaic" vision of the regional development pattern emerges highlighting the importance of cooperative and competitive strategies affecting the local conditions of European regions.

R. Thord (Ed.)

The Future of Transportation and Communication

Visions and Perspectives from Europe, Japan and the U.S.A.

1993. VIII, 265 pp. 44 figs. 19 tabs. ISBN 3-540-56377-6

Transportation and communication are indispensible to society, they are the elements which bind all economic systems together. Without networks and communication all social and economic life would be reduced to isolated phenomena.
The focus of this book is to analyze and structure the future development of transportation and communication and to highlight their significance for society, economy and for the people and their environment. It gives the reader visions and perspectives from Europe, Japan and the USA.

T. R. Lakshmanan, P. Nijkamp (Eds.)

Structure and Change in the Space Economy

Festschrift in Honor of Martin J. Beckmann

1993. VIII, 314 pp. 41 figs. 18 tabs. ISBN 3-540-56490-X

This book brings together a collection of recent advances in the field of spatial systems analysis from the perspective of a solid and rigorous research methodology. The topics discussed have a wide coverage, ranging from land use and infrastructure issues to spatial chaos theory. Especially the evolution of spatial economic systems is given due attention in this book, with a strong emphasis on behavioural foundations originating from economic theory.

FSC
www.fsc.org
MIX
Papier | Fördert
gute Waldnutzung
FSC® C083411

Kolibri 360 GmbH
Ferdinand-Jühlke-Straße 7
99095 Erfurt, Deutschland
produktsicherheit@kolibri360.de